I'd rather wake up in the middle of nowhere than in any city on earth.

Steve McQueen

Oceanology: An Introduction

Dale E. Ingmanson
William J. Wallace
California State University, San Diego

Wadsworth Publishing Company, Inc.
Belmont, California

Designer: Steve Renick
Editor: Mary Arbogast
Biological drawings and maps: Darwen & Vally Hennings
Charts and graphs: Carleton Brown

ISBN–0–534–00130–0
L. C. Cat. Card No. 72–91087
Printed in the United States of America
4 5 6 7 8 9 10—77 76 75

Acknowledgments

The quotations on the cover, end papers, and chapter openings are from the following works.

Matthew Arnold: An excerpt from "Dover Beach." **Lewis Carroll:** An excerpt from "The Walrus and the Carpenter." **Joseph Conrad:** An excerpt from "The Lagoon." **Stephen Crane:** An excerpt from "The Open Boat." **Richard Henry Dana:** An excerpt from *Two Years Before the Mast.* **Robert Hayden:** An excerpt from "The Diver" by Robert Hayden. From *Selected Poems.* Copyright © 1966 by Robert Hayden. Reprinted by permission of October House Inc. **James Joyce:** An excerpt from *A Portrait of the Artist as a Young Man.* Reprinted by permission of The Viking Press, Inc., Jonathan Cape Ltd. on behalf of the Executors of the James Joyce Estate, and The Society of Authors as the literary representative of the Estate of James Joyce. **Galway Kinnell:** An excerpt from "Spindrift" from *Flower Herding on Mount Monadnock.* Copyright © 1964 by Galway Kinnell. Reprinted by permission of Houghton Mifflin Company. **Anne Morrow Lindbergh:** Excerpts from *Gift from the Sea* by Anne Morrow Lindbergh. Copyright © 1955 by Anne Morrow Lindbergh. Reprinted by permission of Pantheon Books/ A Division of Random House, Inc., and Chatto and Windus Ltd. **Thane McCulloh:** An excerpt from *U.S.G.S. Professional Paper 679,* U.S. Government Printing Office, 1969. **Steve McQueen:** A quotation from *On the Loose* by Terry and Renny Russell, Sierra Club, 1967. **Herman Melville:** Excerpts from *Moby Dick.* **Edward L. Meyerson:** "La Jolla" and "Crossing the Great Salt Lake" from *Chameleon* by Edward L. Meyerson. Copyright by Branden Press, 1970. **Theodore Roethke:** An excerpt from "The Long Waters." Copyright © 1962 by Beatrice Roethke as Administratrix of the Estate of Theodore Roethke. From *Collected Poems of Theodore Roethke.* Reprinted by permission of Doubleday & Company, Inc. **Victor B. Scheffer:** An excerpt from *The Year of the Seal.* Copyright © 1970 by Victor B. Scheffer. Reprinted by permission of Charles Scribner's Sons. **Henry David Thoreau:** An excerpt from *Cape Cod.* **Walt Whitman:** "The World below the Brine."

The biological illustrations in this book are based on the following sources.

Ralph Buchsbaum, *Animals without Backbones,* 2nd ed., University of Chicago Press, 1948. Ralph Buchsbaum and Lorus J. Milne, *The Lower Animals,* Doubleday, 1960. Libbie H. Hyman, *Hyman Series in Invertebrate Biology,* vols. 1–4, McGraw-Hill. Sol F. Light, *Intertidal Invertebrates of the Central California Coast,* Ralph I. Smith *et al.,* eds., University of California Press, 1954. Edward F. Ricketts and Jack Calvin, *Between Pacific Tides,* 3rd ed., Joel W. Hedgpeth, ed., Stanford University Press, 1952. Grzimeks Tierleben, *Enzyklopädie des Tierreiches,* vols. 1, 3, 4, 5, Kindler Verlag AG, 1968–1970. *Audubon* magazine, January 1972.

Preface

The great American satirist Ambrose Bierce wrote: "Ocean, a body of water occupying about two thirds of a world made for man—who has no gills." In spite of the irony that so little of the planet is his natural habitat, man has tended to view the earth as his possession, and this view, along with man's negligence, has been largely responsible for the ecological crisis we face today. One reason for writing OCEANOLOGY was to interweave the environmental problems with the facts and theories of the various oceanology disciplines. These problems are presented not only in the three chapters covering the marine environment, the contaminated ocean, and man and technology, but are also treated throughout the book, wherever the problems bear on the topic under discussion. For example, the marine geology chapter describes extensively how man has affected different types of shoreline. And the chapter on water motion discusses DDT buildup and its movement in the ocean.

Many of our students indicate a strong interest in learning about the diversity of marine organisms and how physical factors influence these organisms. This provided us with another reason for writing the book. To present a balanced and integrated treatment of the ocean's physical and biological factors, we discuss, for example, in Chapter 4 how upwelling combined with the nutrient-rich Humboldt Current supports one of the world's most abundant populations of marine life on the west coast of South America. Again, Chapters 3 and 7 describe the importance of temperature ranges to organisms and the effects of thermal pollution. We are grateful to Marvalee Wake of the University of California, Berkeley, for writing the marine diversity chapter. The chapter strives to give the student a feel for marine diversity, without overwhelming him in a mass of detail for each phylum. To help achieve this goal, the chapter contains many photographs and line drawings, including three full-page drawings of marine organisms commonly found in deep water, tidepool, and coral reef ecosystems.

The book has an emphasis on visual materials. There are over 400 line drawings and photographs, more than 300 of which are original. Illustrations are essential to an oceanology text because some students have never experienced the ocean in person, while many others have seen it only from the shoreline or from the deck of a boat, usually in only a few locales. If properly executed, illustrations can be good substitutes for direct observation. As another visual feature, a detailed chapter outline appears at each chapter opening to help the student recognize and keep in mind overall organization and interrelationship of topics. This outline should provide the student with a quick overview, serving as a kind of road map when he reads the chapter.

To give the book another dimension, we display at each chapter opening an original photograph along with quotations from the works of Joyce, Melville, Lindbergh, Conrad, Whitman, and others. We hope these quotations and photographs will convey the variety of man's experience with the ocean in a stimulating way that few textbooks achieve.

Among the many who generously contributed time and ideas in reviewing the manuscript in detail, we especially thank: Wolfgang H. Berger, Scripps Institute of Oceanography; Herbert Curl, Oregon State University; Tom S. Garrison, Orange Coast College; Cadet Hand, University of California, Berkeley, and Bodega Marine Laboratory; Rivian S. Lande, Long Beach City College; J. Robert Moore, University of Wisconsin; Richard Phleger, California State University, San Diego, and Scripps Institute of Oceanography; and Robert Riffenburgh, Naval Underwater Research and Development Center, San Diego, and California State University, San Diego. We are indebted to Steve Renick of Wadsworth for the many excellent photographs he took for the book, and to Mary Arbogast for her professional editing of the manuscript. Our deepest thanks go to the many students whose comments helped us write a book suited to their needs.

Contents

To our parents

D. E. I.
W. J. W.

The Ocean in Perspective

1

From the promontory
There is no terminal
Across the sea.

Only a canvas of sky
Water and haze;
And returning waves

Reaching for the cliffs,
Like waterfalls, churning
Their course to shore, foaming

Then limply
Drenching mounds of foam
On the respite of beach

Before sliding back
Into an ageless
Endless odyssey.

Edward L. Meyerson

He was alone. He was
unheeded, happy, and near to
the wild heart of life.
He was alone and young
and willful and wild-hearted,
alone amidst a waste of wild air
and brackish waters
and the seaharvest of shells
and tangle and veiled
grey sunlight.

James Joyce

I. Introduction

A. Relative sizes of celestial bodies

B. Earth and ocean characteristics

II. Scientific Branches of Oceanology

A. Geology

B. Physics

C. Chemistry

D. Biology

III. H.M.S. Challenger *Expedition*

IV. Etymology of Oceanology

V. Origins

A. Earth

B. Solar system

C. Atmosphere

D. Ocean

VI. Relation of Atmosphere and Ocean to the Origin of Life

Human beings standing at the edge of the ocean have been impressed and awed by its power and vastness. The ocean is so incredibly large that an individual feels small before it (Fig. 1–1).

How big is the ocean? It covers 70.8 percent of the earth's surface and contains about 1.4×10^9 (or 1.4 billion) km^3 of water. These figures are impressive but hard to imagine. To relate the oceans and the earth realistically from man's viewpoint it might be valuable to begin by thinking about the universe as a whole.

No one, on this planet anyway, knows how big the universe is. Some estimates suggest that it might be infinite. From Earth, we see scattered through this dark infinity occasional points of light. Clusters of these points are separated by large distances containing virtually nothing. These points of light are, of course, stars and the clusters are galaxies.

Galaxies are massive in size, often more than 100,000 light years across. (A light year is the distance light travels in one year at the rate of about 186,000 mi/sec.) Millions of these galaxies have been photographed. The closest galaxy to Earth is Andromeda, 1,800,000 light years away. If the universe is infinite, an infinite number of galaxies may exist therein.

Galaxies are separated into types, depending on their viewed shape: spiral, elliptical, barred, and irregular. One galaxy, called the Milky Way by earth astronomers, looks like a whirling Roman candle from the top and a wagon wheel from the side. The Milky Way, which is about 100,000 light years in diameter, is the relatively dense band of stars that can be seen across the sky on a clear night. This band is simply the result of viewing space through the Milky Way.

Our sun, one of the stars in the Milky Way, is not particularly large relative to other stars, probably medium-sized. The sun is classified as a type G star. This is based on its surface temperature of about 6,000°C. Its composition is mainly hydrogen and helium, with appreciable amounts of carbon and nitrogen. The nearest star to the sun in our galaxy is about 35,000 billion mi away. The vast majority of the energy that heats the earth comes from the sun.

While viewing the tiny points of light in the sky, man noticed over 2,000 years ago that several of them moved quite rapidly, but most appeared to be in fixed positions relative to each other. In the last 400 years men have known that these rapidly moving points are planets. Most of the planets can be seen with the naked eye, especially Mars, Jupiter, Venus, Saturn, and Mercury. Revolving around the sun in the same direction and in nearly the same plane are nine planets, a number of asteroids, and a few comets. The asteroids vary in size but are smaller than the planets. The largest asteroid is Ceres, with a diameter of about 865 km.

The third planet from the sun is Earth, moving around the sun at an average speed of almost 120,000 km/hr. The earth is 165,000,000 km from the sun at its closest point, and 170,000,000 km at the farthest. Although recent information shows the earth to be slightly oblong or tangerine-shaped, it looks spherical from a distance (see Fig. 4–38). The polar radius is 6,357 km and the equatorial radius is 6,378 km. This difference of 21 km is so slight that from space it cannot be seen by eye.

From the moon, the earth looks like a patchwork of white, blue, and green. Its color is partially due to the uniqueness of this planet. Earth is the water planet of our solar system. Other planets such as Mars and Jupiter may have some water but only in its solid form, ice. Of the nine, only on Earth is liquid water thought to exist. There is so much water on our planet that if it had first been viewed from space, it surely would have been called Water rather than Earth.

The compound water is the most abundant molecule in liquid state on the earth and, as one might expect, most of the water (97 percent) is in the ocean; another 2 percent is in rivers and lakes, 0.9 percent in snow and glaciers, and less than 0.1 percent in the atmosphere. Water covers 60.7 percent of the northern hemisphere and 80.9 percent of the southern hemi-

1–1 A surfer at Makaha, Hawaii. (Photograph by Art Brewer. Courtesy of Surfer Magazine.*)*

sphere. The oceans cover 254 million km^2 with a volume of approximately 1.4 billion km^3 and a mass of 1,600 million billion tons.

The average depth of the world ocean is about 4 km. Even though this is a tremendous amount of water, it represents 1/790 of the total volume of the earth. More aptly phrased, this water cover is only a thin film on the surface of the planet, somewhat analogous to the skin of an apple. In fact, the film of water left on a basketball after being dunked in water easily represents both the ocean and the atmosphere (Fig. 1–2). In this thin film of water and atmosphere exists all life known to man.

To aid in the study of the oceans, which is now a necessity for man's survival, man has arbitrarily divided this study into geological, physical, chemical, and biological oceanography. Geological oceanography studies the configurations of the ocean basins, the origin of the basins, the nature of the rocks and minerals found in the basins, and the geological processes in the ocean. Physical oceanography deals with the motion of the water, the temperature distribution in the ocean basins, the transmission of light, sound, and other types of energy through the water, and air–sea interactions. Chemical oceanography is the study of all the chemical reactions that take place and the distribution of elements in ocean water. Biological oceanography investigates the types of life that are found in the sea and their interrelationships. These same divisions exist in all scientific studies of the physical world. This total scientific study of the ocean is most often called oceanography. The arbitrary divisions exist mainly because graduate universities are divided that way. Oceanography is an interdisciplinary field because, for example, geological processes may affect the physical, chemical, and biological characteristics of the ocean. This interaction is the rule rather than the exception and requires research oceanographers to be knowledgeable in many areas of science.

A few words about the *Challenger* expedition will give insight into the science of oceanography. The steamship H.M.S. *Challenger* (Fig. 1–3), under the direction of Sir Wyville Thompson, made the first large oceanographic expedition, between December 1872 and May 1876. In this time, she covered 68,890 nautical mi.[1] She was the first steamship to reach the Antarctic ice

---

[1] A nautical mile equals one minute or 1/21,600 of a great circle of the earth; 6,080.20 ft; 1,853.248 m.

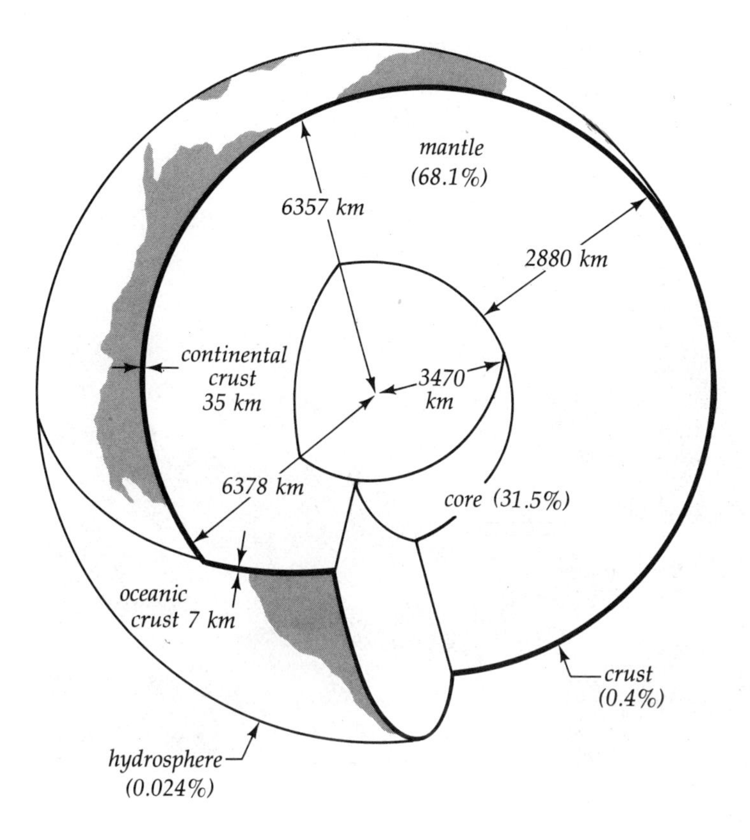

1–2 *Cross section of the earth.*

barrier and the first to cross the Antarctic Circle. During this part of the voyage, however, she was under sail. The *Challenger*'s voyage remains to this day probably the most important single expedition ever undertaken in the study of the sea.

Based on the wealth of information the expedition gathered (50 large volumes in quarto were published), J. Y. Buchanan, the chemist for the *Challenger* expedition, said,

> *The history of the* Challenger *expedition is well known to all students of oceanography, which, as a special science, dates its birth from that expedition. It must be remembered that when the* Challenger *expedition was planned and fitted out, the science of oceanography did not exist. . . . The work of the expedition proper began when the ship sailed from Teneriffe, and the first official station of the expedition was made to the westward of that island on 15th February, 1873. It was not only the first official station of the expedition, but it was the most remarkable. Everything that came up in the dredge was new; the relation between the result of the preliminary sounding and that of the following dredging was new; and, further, from the picturesque point of view, it was the most striking haul of the dredge or trawl which was made during the whole voyage.*
>
> *Consequently, it may be taken that the Science of Oceanography was born at Sea, in Lat. 25° 45′ N, Long. 20° 14′ W, on 15th February, 1873. . . .*
>
> The Birth-day of Oceanography. *When the* Challenger *sailed from Portsmouth in December, 1872, there was no word in the Dictionary for the Department of Geography in which she was to work, and when she returned to Portsmouth in May, 1876, there was a heavy amount of work at the credit of the account of this department, and it had to have a name. It received the name Oceanography. It follows that the science of Oceanography owes its birth to the* Challenger *expedition.* (Buchanan, 1919)

These comments help show the value of the *Challenger* expedition to ocean study. Like any science, however, the roots of oceanography go much deeper and can hardly be based on one event, even such an important one.

The word *oceanography* was introduced and discussed by Sir John Murray in his book *The Oceans:*

> *The term Thalassography has been used, largely in the United States, to express the science which treats of the ocean. The term Oceanography is, however, likely to prevail. The Greeks appear to have used the word Thalassa almost exclusively for the Mediterranean, whereas the almost mythical "oceanus" of the ancients corresponds to the ocean basins of the modern geographer. In recent times I believe the word Oceanography was introduced by myself about 1880, but I find from Murray's English Dictionary that the word "oceanographie" was used in French in 1584, but did not then survive.* (Murray, 1910)

The German word *Ozeanographie,* now largely replaced by *Meereskunde,* is a few years older than the English version. The suffix *-graphy* pertains to processes of drawing, describing, or reporting—as in *biography* and *geography.* The suffix *-ology* pertains to any

1–3 *H.M.S.* Challenger *at St. Paul's Rocks.*

science or branch of knowledge. Surely the study of the sea has progressed beyond the purely descriptive. As a consequence, *oceanology* is a far more meaningful term than the older but less definitive word *oceanography*.

The term *hydrography* has sometimes been wrongly used as a synonym for *oceanography*. From the above, one can see that oceanography, or oceanology from this point on, is concerned with the application of the physical and biological sciences to the study of the sea. Hydrography, on the other hand, deals primarily with charting coastlines, bottom topography, currents, and tides for practical uses in ocean navigation. *Limnology* is the name given to the study of similar processes in lakes and streams.

Before one talks about the age and origin of the ocean, it is probably better to discuss the age and origin of the earth itself. Presently, the planet Earth is considered by scientists to be about 4.7 billion years old. Two thousand years ago, it was thought to be no more than a few thousand years old. In other words, in the minds of men, the planet has aged 4.7 billion years in the last 2,000 years. As man learns more about his planet, it gets older.

Descriptive facts about the solar system have been compiled by many scientists during the past 2,000 years. Ptolemy (fl. A.D. 130), Galileo (1564–1642), Tycho Brahe (1546–1601), Johann Kepler (1571–1630), and Isaac Newton (1642–1727) have made some of the most significant contributions to our knowledge of the solar system. In spite of all that is known, a major question remains unanswered, "What is the origin of the solar system?"

During the past thirty years many famous scientists, including G. P. Kuiper, F. L. Whipple, and H. C. Urey, have wrestled with this question. Most widely accepted now is the idea that the solar system was preceded by a cold (approaching absolute zero) dust-gas cloud. The cloud may have been the accumulated residue from the explosion of a supernova. The cold residue then began to coalesce through gravitational attraction. Discrete planets formed as eddies in the rotational movement of the cloud separated from the rest of the mass. This is an updated version of the Kant–Laplace theory.

H. C. Urey has recently stated that his ideas are in a state of flux as a result of the analysis of lunar samples.

His attitude reflects the flexibility necessary when viewing new sources of information. Because the origin of the solar system remains unknown, scientists are anxious to accumulate data from the moon, meteorites, and planets. Developing a theory of the solar system's origin remains one of the most challenging parts of science.

Analysis of the light spectra of the planets in our solar system reveals an atmosphere surrounding each. These atmospheres vary, however, in density and composition. Although hydrogen, nitrogen, carbon, and oxygen are present on each, current knowledge suggests that only Earth has water in abundance.

Where can atmospheres come from? One possibility is that as celestial bodies acquire more mass through molecular collisions, their gravitational fields increase, allowing them to capture gases moving through space. A second possibility is that gases present when the planet was formed were held by gravitational attraction. In time, some of the less dense gases would escape into space. A third idea suggests that gases are emitted by planets during volcanic eruptions. If the gravitational field is great enough, the gases become part of the atmosphere; otherwise, they escape. Of course, any combination of these possibilities can exist.

One piece of evidence supporting this third possibility is the lack of neon gas in the earth's present atmosphere. A neon atom and a water molecule have the same mass. From current knowledge of elements that can be and were generated from natural radioactive decay on the planet, there should be more neon in the atmosphere than there is. Evidently, in the early chaotic stages of the earth's development, these neon atoms simply drifted off into space. Because the water molecule has the same mass as the neon atom, it is reasonable to suppose that any water existing on the earth's surface before volcanic activity began also floated off into space.

Our present atmosphere contains predominantly nitrogen gas (78 percent), oxygen gas (21 percent), and argon gas (1 percent). On the other planets, methane (CH_4), ammonia (NH_3), hydrogen (H), and carbon dioxide (CO_2) are the most common. The formation of all planetary atmospheres might have been similar, but clearly unique factors must have affected the atmosphere of the earth.

Volcanic gases have been analyzed. About 85 to 95 percent of the gases emitted from volcanoes is water vapor. By radioactive decay dating of volcanic rocks it is known that volcanic activity has occurred on Earth for more than 3 billion years. The extent or rate of this activity is unknown. If, as presently believed, the earth's crust is constantly being recirculated deep in the interior, the extent and rate during the earth's history might be impossible to know. Volcanic activity is, however, an undeniable source of water vapor, which when cooled in our atmosphere forms water that falls to the surface due to gravity.

The current addition of water to the earth's surface from volcanoes is 0.1 km^3/year. The average rate of this addition in the past must have been several times the present rate in order for the oceans to have been formed in less than 4 billion years by condensation and precipitation alone. This idea also presumes that the total amount of water on the earth's crust is constantly increasing. Unfortunately, the periodic continental ice sheets of the past million years have obliterated evidence of changes in the overall amount of water.

The average amount of water in siliceous meteorites is about 0.5 percent. If, as geologists believe, the earth's mantle is similar to siliceous meteorites, it would contain ample water to have released through volcanism the amount that is now found in the ocean. This belief is based primarily on the similar speeds of sound waves through both.

Another known source of new water introduced to the earth's surface is the chemical reaction of hydrogen gas and oxygen gas:

$$2H_2 + O_2 \rightarrow 2H_2O$$

This reaction occurs only when large amounts of energy are applied to the hydrogen and oxygen gases to trigger the reaction. If most of the water on the earth's surface resulted from some high energy source related to the formation of the earth itself, this process could have been significant. Because most present evidence supports the condensation hypothesis of the earth's formation, a volcanic origin of water seems most likely.

Because nitrogen is now present on the sun and all the planets, it has probably been present since the formation of the solar system at least. Some of the nitrogen in our atmosphere may be a residue of the primeval dust cloud. Significant amounts of nitrogen, however, are produced by the decay of organic material and are emitted by volcanoes. There is now no conclusive evidence supporting any one theory of origin in preference to the others.

Scientists currently believe that most of the oxygen gas in our atmosphere was produced by plant photosynthesis. Oxygen combines readily with many elements at the temperatures and pressures prevalent at or near the earth's surface. If iron is exposed to the atmosphere, iron oxide forms. This reaction removes oxygen from the atmosphere. Other elements such as aluminum also oxidize readily when exposed to the atmosphere. Because these elements are abundant on the earth's surface, oxygen is and has been rapidly removed from air. The reducing atmosphere of the early earth, containing ammonia, methane, and water vapor, would have spurred the combination of oxygen with other elements. This evidence would suggest that little oxygen gas was present in the early atmosphere or oceans of the earth. Thus, the oxygen gas presently found in the air was probably formed in rather recent times, mainly by plants. When photosynthetic plants evolved, the oxygen gas added to the atmosphere began to exceed the oxidation rate and oxygen began to accumulate in the oceans and migrate into the atmosphere. It continued to increase in proportion to plant growth and survival. Eventually animal life evolved, probably about 1 billion years ago. The animals used the oxygen gas in respiration, thus establishing a limit to oxygen accumulation. It can be said that life or life processes changed the earth's atmosphere and oceans and was significantly responsible for the present amount of oxygen gas in both.

A number of scientists have experimented with passing an electrical spark through a reducing atmosphere. In these experiments, simple amino acids were synthesized. Protein molecules, consisting of complex chains of amino acids, have also been synthesized in the laboratory. Recently, scientists from Stanford University synthesized a DNA molecule, which is essential for life as we know it. All these experiments use compounds believed to have been present in the earth's atmosphere or ocean 2 to 4 billion years ago.

Did the first life originate in a lake, a river, an ocean, or a geyser-like hot spring? All the necessary elements were probably present in each of these. Because the ocean is largest and has the most stable conditions over long periods of time (millions of years), the first life probably formed in the ocean. However, geyser-like hot springs emanating from the earth would probably have contained large amounts of methane, ammonia, and phosphate in addition to hot water. No one knows the complete answer to the origin of life, but most experts presently suspect the ocean was a major partner.

The significant elements in the development of life —carbon, nitrogen, oxygen, hydrogen, and phosphorus —are present on all the planets. As far as is now known life is present only on earth. The important fact is that water, liquid H_2O, is present in large amounts only on the earth. Rivers, lakes, and oceans do not exist on other planets.

The oldest known fossils are found in sediments that were deposited under water. The calcium carbonate protective covering of marine algae has been found in sediments 600 million years old. Phytane and pristane, carbon compounds that are known decay products of the chlorophyll in photosynthetic plants, have been found in sediments 3 billion years old.

The origin and dating of the ocean are currently the subject of spirited discussions and much experimentation and exploration by scientists. The preceding summary is at best a simplified view of the topic, meant mainly to indicate the significance of water in events on and near the earth's surface for more than 3 billion years. No attempt has been made to imply that the ocean, the atmosphere, and life evolved just this way. It should be remembered that these ideas are theories. A theory is simply a statement of current scientific thought.

Summary

The oceans are small and insignificant relative to the universe. Although they dominate the external appearance of the earth, the oceans are small relative to its total mass and volume. They are extremely important, however, to an understanding of the processes that occur within the area 20 mi above and below the surface of the earth. If the ocean is viewed as a complex chemical solution of finite though slightly variable volume occupying an ever-changing, irregular basin that contains interrelated living organisms, it is possible to begin developing a reasonable perspective of oceanology.

Further Reading

Buchanan, J. Y., *Accounts Rendered of Work Done and Things Seen.* Cambridge: Cambridge University Press, 1919.

Bullard, Sir Edward, "The Origin of the Oceans." *Scientific American,* September 1969, pp. 66–75.

Cotter, C. H., *The Physical Geography of the Oceans.* New York: Elsevier, 1965.

Dubach, H. W., and R. W. Taber, *Questions about the Oceans.* Washington D.C.: National Oceanographic Data Center, Publication G-13, USNOO, 1968.

Duxbury, A. C., *The Earth and Its Oceans.* Reading, Mass.: Addison-Wesley, 1971.

Engel, Leonard, *The Sea.* New York: Time-Life Books, 1961.

Fairbridge, R. W., ed., *Encyclopedia of Oceanography.* New York: Van Nostrand-Reinhold, 1966.

Gross, M. G., *Oceanography.* Columbus, Ohio: Charles E. Merrill, 1967.

King, C. A. M., *An Introduction to Oceanography.* New York: McGraw-Hill, 1965.

Murray, Sir John, *The Oceans.* New York: Holt, 1910.

Revelle, Roger, "The Ocean." *Scientific American,* September 1969, pp. 54–65.

Ross, D. A., *Introduction to Oceanography.* New York: Appleton-Century-Crofts, 1970.

Spar, J., *Earth, Sea and Air: A Survey of the Geophysical Sciences.* Reading, Mass.: Addison-Wesley, 1965.

Strahler, A. N., *The Earth Sciences.* New York: Harper & Row, 1963.

Turekian, K. K., *Oceans.* Englewood Cliffs, N.J.: Prentice-Hall, 1968.

Weyl, P. K., *Oceanography: An Introduction to the Marine Environment.* New York: John Wiley & Sons, 1970.

Wooster, W. S., "The Ocean and Man." *Scientific American,* September 1969, pp. 121–130.

Marine Geology

2

On this tree thrown up
From the sea, its tangle of roots
Letting the wind go through, I sit
Looking down the beach: old
Horseshoe crabs, broken skates,
Sand dollars, sea horses, as though
Only primeval creatures get destroyed,
At chunks of sea-mud still quivering,
At the light as it glints off the water
And the billion facets of the sand,
At the soft, mystical shine the wind
Blows over the dunes as they creep.

Galway Kinnell

The ocean rolls on, untouched by words.
It rolls to the turning of the earth,
and the heat and pull of the sun,
and the drag of the moon,
and the influences of all the
solid and gaseous matter in the
universe. The deep and dark blue ocean
penetrates in time the smallest crannies
of the caves of earth and sucks
out slowly the chemical ions of
the old earth-rock, wafts them away
in salty streams, and drops them
in gray-purple oozes on its bed, where
they lie softly stirring for a thousand
years until other layers of
ooze press them down, and they subside
into soft rock—and in a hundred
million years into stone.

Victor B. Shaeffer

I. Continents and Ocean Basins

- A. *Topography*
- B. *Composition*
- C. *Density*
- D. *Sediment thickness*
- E. *Location*
- F. *Definition*

II. Shorelines

- A. *Beach terminology*
- B. *Shoreline processes*
- C. *Case studies*
  1. *Cape Cod*
  2. *Mount Desert Island*
  3. *Everglades National Park*
  4. *Louisiana Gulf Coast*
  5. *southern California*
- D. *Effects of man*

III. The Continental Shelf

- A. *Location*
- B. *Processes and resulting features*
- C. *Origins*
  1. *New England—glacial*
  2. *Mid-Atlantic shelf—geosyncline*
  3. *Florida Bank—carbonate*
  4. *Mississippi River delta*
  5. *Texas shelf*
  6. *southern California shelf—faulting*

IV. Submarine Canyons

- A. *Definition*
- B. *Formation*
- C. *Types*
  1. *Scripps–La Jolla Canyon*
  2. *Hudson Canyon*
  3. *San Lucas Canyon*
  4. *Tongue of the Ocean*

V. Obtaining Ocean Bottom Data

- A. *Depth recording*
- B. *Dredging*
- C. *Cores*
- D. *Seismic profiling*
- E. *Heat conduction*
- F. *Magnetism*
- G. *Free vehicles*

VI. Abyssal Plains and Seamount Provinces

- A. *Topography*
- B. *Minerals*
- C. *Sediment*
- D. *Submerged mountains*
- E. *Fracture zones*
- F. *Coral atolls*

VII. Island Arcs and Trenches

- A. *Distribution*
- B. *Composition*
- C. *Trenches*
- D. *Gravity anomalies*
- E. *Structure*
- F. *Origin*
- G. *Heat flow*
- H. *Case studies*
  1. *West Indies*
  2. *Aleutian Islands*
  3. *Peru–Chile Trench*

VIII. Oceanic Ridges

- A. *Volcanism*
- B. *Earthquakes*
- C. *Composition*
- D. *Rift valley*
- E. *Sediment*
- F. *Transform faults*
- G. *Magnetic bands*
- H. *Gravity anomalies*
- I. *Age*
- J. *Heat flow*
- K. *Other ridges and rises*

IX. Global Plate Tectonics

- A. *Hypothesis*
- B. *Similar theories*
- C. *Supporting facts*
- D. *Summary: significance for geologists*

An automobile trip through the Bighorn Mountains of north central Wyoming provides opportunity for spectacular views of valleys, plains, mountain peaks, forests, and wildlife. One of the most impressive views is from U.S. Route 14 looking back toward Steamboat Point (Fig. 2–1). What does this have to do with marine geology? Steamboat Point is capped by a thick layer of dolomitic limestone called the Bighorn Formation, which contains abundant fossils. The fossils include calcareous algae and corals. This indicates that the great cliff is composed of rock material that was deposited millions of years ago under marine conditions. It is now more than 2,500 m above sea level and 1,500 km from the nearest ocean. Even more impressive is the fact that the great Himalaya Mountains contain marine fossils.

The broad significance of marine geology is often overlooked, but it quickly comes into focus with the realization that more than 50 percent of the rocks presently exposed on continents were once deposited below sea level under marine conditions. This branch of geology, however, usually refers to the study of present submarine topographic features and the rocks and sediments that are presently found beneath the ocean.

Continents and Ocean Basins

Topography. In geography, the continents and ocean basins have traditionally been classified separately. Relative elevation compared to sea level is used as the main criterion. With the possible exception of parts of Greenland and Antarctica, the elevations of most prominent locations on islands and continents have been measured in great detail. The average elevation of land for the world is about 0.4 km above present sea level. The highest elevation on land is the peak of Mount Everest in the Himalaya Mountains (8,800 meters above sea level) and the lowest is at the Dead Sea in Israel (−400 meters). The shallowest areas of the oceans are the continental shelves and the deepest known location is the Challenger Deep in the Mariana Trench, 11 km below sea level (Fig. 2–2).

The difference between ocean basins and continents extends beyond comparative topography. Significant differences in rock composition and density, sediment thickness, and related properties have been discovered. The only significant similarity seems to be an even average heat flow emanating from both the ocean basins and continents. However, unusual heat flow characteristics in some parts of the ocean basins are significant in theories of geological processes.

Composition. Two general descriptive terms for rock types, simatic and sialic, can be used to differentiate the composition of the earth's crust under the oceans and under the continents. Simatic rocks contain abundant amounts of the elements iron, magnesium, calcium, silicon, and oxygen. Sialic rocks contain mostly aluminum, potassium, silicon, and oxygen. Both sialic and simatic rocks contain similar amounts of sodium. The ocean basin crusts are composed largely of simatic rocks, but the continental crusts are less homogeneous, having a top layer of sialic rocks mainly and an underlayer of intermediate and simatic rocks. The zones between the crusts of the continents and ocean basins, like the middle layer under the continents, contain appreciable sodium, aluminum, potassium, iron, magnesium, silicon, and oxygen. For this reason they are often referred to as transition zones with intermediate composition. These intermediate rock types are called andesitic rocks.

Mapping the distribution of the rock types has not been simple because sampling technology for the ocean bottom has progressed slowly. In the 1870s, the *Challenger* expedition collected the first representative samples from each ocean basin except the Indian and the Arctic oceans. The samples were collected by tedious dredging operations. For almost 100 years, the main samples have been collected by dredges or small

2–1 Steamboat Point in the Bighorn Mountains west of Sheridan, Wyoming. The cliff is composed of Bighorn dolomite of Ordovician age. (Courtesy of the United States Forest Service.)

"grab" samplers using spring-set sediment traps. The recently built *Glomar Challenger* can drill more than 1,000 m into the ocean bottom to recover rock or sediment cores. Another important and less expensive way to discover rock or sediment composition is to send sound waves from a device towed behind a ship. The speed of the sound waves moving through the rocks is recorded and compared in a laboratory to the speed of sound waves through known materials.

Density. The continental and ocean basin crusts differ in density (density equals mass/volume). The density of the ocean basin crust is greater than that of the continental crust. On the average, the density of the continental crust is about 2.6 g/cm^3, while the density of the ocean basin crust is about 2.9 g/cm^3. This difference occurs because the oceanic crust has more iron and magnesium than the continental crust. The numbers seem to be similar; however, two liquids with different densities will stratify when immersed even if the density difference is only 0.1 g/cm^3. Also, if two objects having equal volume but different masses are placed an equal distance from the center of a fulcrum, the object with the greater mass moves down while the other moves up. Thus, the relative elevations of continents and ocean basins may be partly related to density differences.

Sediment Thickness. The thickness of sediment on the upper part of the earth's crust is highly variable. Generally, sediments in the ocean basin are from 50 to 1,000 m thick. On the continents, sediments usually vary from 500 to 3,000 m in thickness. Regions on continents and in ocean basins where crystalline rock has formed by the cooling of molten material may have no sediment cover. The thickest sediments, however, are found in mountain ranges like the Rocky Mountain cordillera and in coastal basins like the lower Mississippi River basin, where sediment up to 10,000 m thick has been measured. Drilling and comparison of sound wave speeds are again the principal source of this information.

Location. The location of the various continents and ocean basins is important (see map on page 326). The continents are North America, South America, Asia, Europe, Africa, Antarctica, and Australia. The ocean basins are the Pacific, Atlantic, Indian, and Arctic, each with many subdivisions. In addition there are many smaller seas including the Gulf of Mexico, Caribbean Sea, Mediterranean Sea, Hudson Bay, Sea of Okhotsk, Black Sea, Caspian Sea, Red Sea, and the seas covering the continental shelves along the fringe of the Pacific Ocean from Alaska to Australia. The question "Why are ocean basins and continents located where they are?" is not simple to answer. A computer analysis of the probability of finding a point in an ocean basin antipodal to a point on a continent has indicated that the continents and ocean basins have only one chance in fourteen of being randomly located. In other words, it is probable that some physical process has determined their relative locations.

Definition. An ocean basin is defined as a vast depression filled with a continuous body of salt water. A continent is defined as a vast continuous area of land. The vague part of this definition is in the word *vast.* Why is Australia a continent and Greenland an island? Why is the Caribbean a sea and the Arctic an ocean? This naming is arbitrary and has evolved in daily usage over the past 300 years. The term *sea* now refers either to a mass of water with physical and chemical characteristics that distinguish it from a nearby ocean or to a large salty landlocked body of water. Thus, the Caribbean Sea is distinguished from the Atlantic Ocean, and the Black Sea is landlocked.

Before the fifteenth century, "the seven seas" was used by the Arab world for the known bodies of water —Mediterranean Sea, Red Sea, West African Sea, East African Sea, China Sea, Indian Ocean, and Persian Gulf. Rudyard Kipling popularized the term in the western world by using it as the title of a book of poetry. According to the International Hydrographic Bureau, there are six oceans—Arctic, North Atlantic, South Atlantic, North Pacific, South Pacific, and Indian—and 54 seas. The "Antarctic Ocean" is not recognized by the IHB as a separate ocean.

From a geological point of view, the ocean basins do not comprise a single unit, as they do from a geographic point of view. Geologically, the following features are distinctly identifiable: the shoreline, the continental shelf, the continental slope, mid-ocean ridges, island arcs, trenches, abyssal plains, broad basins, and fracture zones.

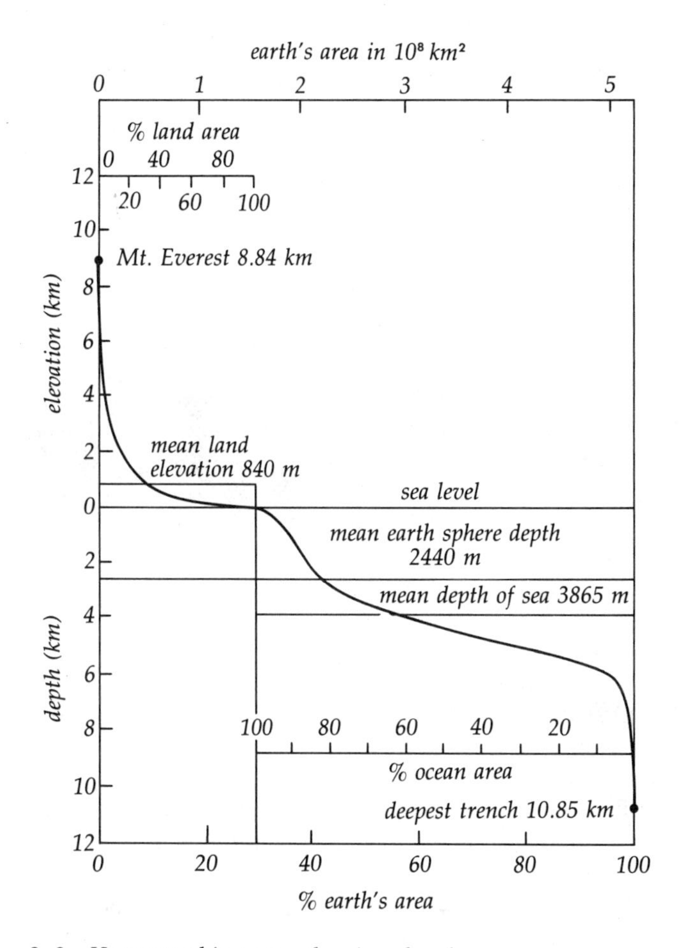

2–2 *Hypsographic curve, showing abundance distribution of elevations and depths.*

Shorelines

The shoreline is usually defined as the general region where sea level intersects land. In some areas, like the Bay of Fundy and the northern Gulf of California, the shoreline is an exceptionally broad area having a width of several miles because of large tidal fluctuation. In other areas, the shoreline width may be very narrow, as on some oceanic islands, where sheer cliffs enter the sea.

Beach Terminology. Shorelines have been studied intensively by marine geologists. Some reasons for intensive study have been the erosion of land (southern California), the creation of land (Netherlands), and the need for improved harbors (Newport Beach, California). Shorelines will be emphasized in this book because they are the part of the ocean that the reader will most likely encounter.

Some widely accepted nomenclature (Fig. 2–3) is:

I. Backshore (above high tide mark)
   A. Sea cliffs (large cliffs of exposed rock)
   B. Dunes (unstable piles of sand)
   C. Berm (flat upper part of beach)
II. Foreshore (area exposed at low tide)
   A. Berm crest (edge of flat upper part of beach)
   B. Beach face (sloping section of beach below high tide mark and washed over by waves)
   C. Beach scarp (vertical slope produced by wave erosion)
   D. Low tide terrace (broad flat area exposed at low tide)

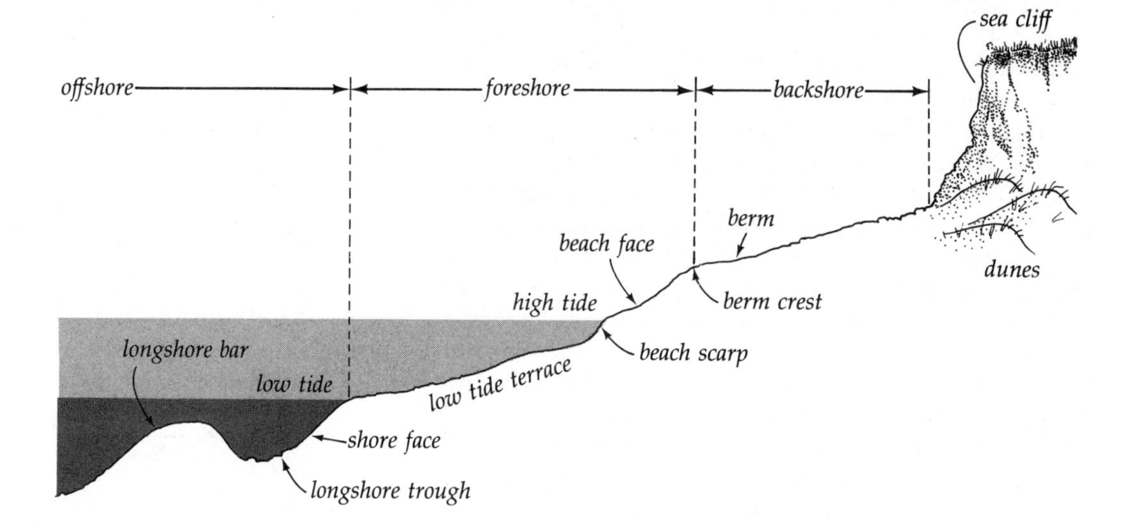

2–3 Beach profile. (From Submarine Geology, *Second Edition, by Francis P. Shepard, Harper & Row, 1963.)*

III. Offshore (from low tide mark to wave-breaking zone)
   A. Shore face (slope below low tide mark)
   B. Longshore trough (depression parallel to beach between low tide mark and wave-breaking zone)
   C. Longshore bar (sand ridge parallel to beach, on which waves break)

Many more terms have been used in shoreline research. The preceding are basic and necessary for further discussion of shoreline features and processes.

Shoreline Processes. Shoreline features, topography, and dynamic processes are closely related. Although some authors have designed intricate classifications for shorelines, it is useful to discuss the processes simply and then to describe the features associated with various combinations of these processes.

Basically, four factors interact to form all shorelines: (1) dissipation of energy, (2) weathering and erosion of bedrock, (3) movement of the earth's crust or parts of it, and (4) short- or long-term fluctuation of sea level. Besides these natural processes, of course, man plays an increasingly significant role.

Energy. All types of energy are constantly affecting shorelines. Waves are an obvious source of energy. They periodically pound or lap against all land areas. The amount of energy they expend is highly variable (Fig. 2–4). Every year storms originating in the North Pacific Ocean generate 20 m waves that roar into Waimea Bay on the north coast of Oahu Island in Hawaii. These waves are capable of rolling 10 ton lava boulders as though they were marbles. On a normal day at Waimea Bay and Sunset Beach, the waves are 1 to 1½ m high. At the other extreme, waves slapping against Seahorse Key just south of the mouth of the Suwannee River on the west coast of Florida rarely get higher than 1 m even during hurricanes. Their usual size is about 25 cm. Even these small waves move unconsolidated sediment.

Another fairly obvious type of energy somewhat related to waves is currents. Wind-driven waves sweep from north to south along the California coast helping to initiate a strong longshore movement of water. Strong longshore currents are common in many parts of the world. Copacabana Beach at Rio de Janeiro and Nauset Beach on Cape Cod are two excellent examples. Currents at these localities move fast enough (2 knots) at times to carry sand and silt in suspension, to erode shorelines, and to roll boulders and pebbles. Monomoy Point, Cape Cod, is comprised mostly of

(a)

(b)

2–4 (a) Five-meter wave smashing the pier off Imperial Beach, California. (Photo by William J. Wallace.) (b) Low energy conditions at Seahorse Key, Florida. (Photo by Dale E. Ingmanson.)

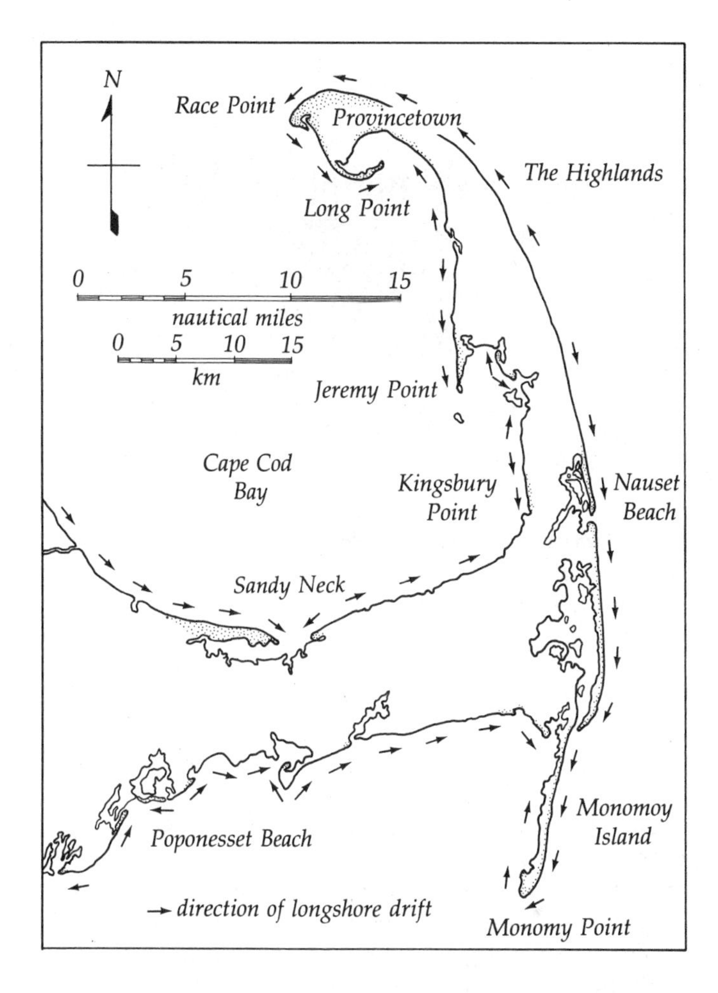

2–5 Cape Cod, showing the direction of longshore currents.

sand eroded from the cliffs near Nauset Beach and transported southward by the longshore current (Fig. 2–5).

Wind is a third source of energy affecting shorelines. Air movement is one of the major processes propelling sand from the foreshore to the backshore. An excellent example of this process and the dunes produced can be seen at Cape Hatteras, North Carolina (Fig. 2–6).

The relative amounts of wave, current, and wind energy will, in many cases, have a profound effect on the shoreline. Of course, other less obvious sources of energy may be significant in particular localities. For instance, the extent of rain could be important in some places.

Weathering and erosion of bedrock. The composition of bedrock or other rock material exposed at the coast will affect the topographic features. From this point of view, the resistance or susceptibility of the rocks to erosion and weathering is most important. For example, granite, a crystalline rock, is much more resistant to erosion than sandstone. Granite is composed mostly of quartz (SiO_2) and feldspar (mostly

2–6 Dunes at Cape Hatteras, North Carolina. (Courtesy of the United States Department of the Interior, National Park Service Photo.)

(*a*)

(*b*)

2–7 (a) Sea cliff and boulder beach. Note the alternating beds of shale and sandstone, which vary in their resistance to erosion. Point Loma, San Diego, California. (b) Granite, weathering and eroding slowly. It is still sharp after thousands of years of wave pounding. Dyce's Head, Castine, Maine. (Photos by William J. Wallace.)

$KAlSi_3O_8$), both of which are less likely to dissolve than olivine ($FeMgSi_3O_8$) and pyroxene ($Ca(Mg,Fe)Si_2O_6$), common in basalt. Where granite or other resistant rock outcrops at the shoreline, rocky, rugged topography is expected (Fig. 2–7). The coast of Maine is composed mostly of granite. Where unconsolidated sand outcrops at a shoaling shoreline, smooth, shifting beaches are expected because the sand is susceptible to erosion (Fig. 2–8).

Moving the land. Less obvious than the type of rock region and the type of energy is the stability of the land region. In other words, to what extent has the shoreline been affected by land movements? Because it has relatively nonresistant rocks, strong currents, and high waves, California's shoreline should be broad and flat. Instead, high cliffs are common. This is caused by uplift of the land. For at least 20 million years, vertical land movements have produced cliffs and horizontal movements have produced points (Fig. 2–9). In other areas, like Lake Pontchartrain near New Orleans, Louisiana, and in the Netherlands, the land is sinking and the battle against ocean water encroachment is continuous.

Sea-level fluctuation. The fourth major factor is sea-level fluctuation. Short-term sea-level changes are caused mostly by the periodic tides at typical shorelines. Along most of the U.S. coast these tides vary less than 2 m daily. In some areas, tidal variation may be as much as 9 m, which daily exposes whole harbors (Fig. 2–10) or two-mile sand flats (Fig. 2–11). These tidal fluctuations may cut channels and

2–8 Crane's Beach, Ipswich, Massachusetts. A typical east coast wide sandy beach. (Photo by William J. Wallace.)

2–10 Low tide at Windsor, Nova Scotia. The bay fills entirely at high tide. (Photo by Dale E. Ingmanson.)

2–9 Point Reyes National Seashore, California. (Courtesy of the United States Department of the Interior, National Park Service Photo.)

transport tons of sediment (Fig. 2–12). In other areas such as the west coast of Florida, fluctuations may be less than 1 m.

Occasional sea-level changes, caused by catastrophic events such as tsunamis and hurricanes, can significantly alter coasts. *Tsunami* is a Japanese word designating a seismic wave system formed in the ocean by any large-scale disturbance of the free surface as might occur after an earthquake. A tsunami at least 30 m high stripped trees and other life to a height of several hundred feet above normal sea level in a remote Alaskan bay. Tsunamis have been the principal natural cause of shoreline changes in Japanese bays during recorded history.

Hurricanes have caused major changes on the shorelines of the eastern United States and Gulf of Mexico. Besides strong winds and high waves, hurricanes also create extreme low pressure areas, which allow sea level locally to rise from 1 to 7 m above normal. In the East Pakistan hurricane of 1970, the effects were devastating. The shoreline changes resulting from such a rise in sea level are appreciable, especially in areas with nonresistant rocks and low topography.

Long-term changes in sea level have been extremely important in producing shoreline features. Many localities show evidence that during the past million years sea level has been at least 100 m below its present level and at least 30 m above its present level. These shifts are associated with continental glaciation, interglacial episodes, and tectonic earth movements. During the glacial periods, water was stored as ice on the continents and over the seas, causing sea level to lower. During the interglacials, sea level rose. The rises carved platforms in sea cliffs and deposited shell fossils of

2–11 Tidal flats at East Brewster, Cape Cod, Massachusetts. Note the diving towers in the foreground, which are high and dry at low tide. (Photo by Dale E. Ingmanson.)

2–12 Converging sand spits and tidal channels at Martha's Vineyard, Massachusetts. (Photo by William J. Wallace.)

marine organisms that are still found on the terraces. The lowering of sea level allowed land fossils such as plant remains, mammoth teeth, and Indian shell middens to be deposited on areas now covered with ocean water.

During the past 10,000 years, since the most recent advance of continental glaciation, sea level has been

2–13 Sand dunes at Crane's Beach, Ipswich, Massachusetts. Sand deposited by glaciers is now reworked by wind. (Photo by William J. Wallace.)

following a broad trend of rising with small-scale fluctuations. It is rising because of the continued melting of polar sea ice and the ice sheet on Greenland and expansion from the increasing average temperature of ocean water. This thermal expansion is analogous to the expansion of iron or mercury when heated.

Case Studies. In analyzing a particular shoreline, each of these major processes must be considered and the clues sought carefully. Some case studies of shorelines bordering the United States are useful in seeing how these processes interact to produce observable features and what kind of clues may be present.

Cape Cod. During the past million years, continental glaciers have periodically covered appreciable parts of the North American continent. The most recent continental ice sheet extended southeastward to what is now Cape Cod dumping sand, gravel, silt, and clay where major melting occurred. This unconsolidated sand, gravel, silt, and clay provides the surface sediment of Cape Cod (Fig. 2–13).

One point to consider in analyzing the shoreline is the nature of the outcropping rock. In this case, the unconsolidated rocks are nonresistant. Thus, the shoreline is impermanent and susceptible to erosion. Wide

2–14 Heavy surf at Coast Guard Beach, Eastham, Massachusetts, on Cape Cod National Seashore. (Courtesy of the United States Department of the Interior, National Park Service Photo.)

beaches, extensive longshore movement of sand and silt, offshore bars, tidal channels, and other types of transient features would be expected.

A second point to consider is the amount of energy expended on the shoreline. Because the eastern side of Cape Cod is exposed to the Atlantic Ocean, continuous moderate wave action can be expected. In addition, the New England area including Cape Cod has long been known for its "nor'easters" (Fig. 2–14). These storms occur in the fall and winter when low pressure systems form off Cape Hatteras and move northeastward. They frequently are accompanied by a rise in sea level of up to 1 m because of the low air pressure, winds up to 40 knots, rain, snow, and 3 to 7 m waves on shorelines facing the Atlantic Ocean. Obviously, the combination of high energy conditions and unconsolidated sediments will accentuate the impermanence of the shoreline.

The third point is the stability of the land. No major earthquakes or earth movements have occurred in the Cape Cod region during the last 300 years. The area may have arched less than 1 m as an elastic response to the retreat of the glaciers and subsequent removal of heavy ice, but little evidence is available to prove this. Cape Cod can be called a stable land area.

The fourth point in analyzing a shoreline is sea-level fluctuation. The frequent rise in sea level of about 1 m accompanying low pressure storms has been previously mentioned. The normal tidal range on the north side of Cape Cod in Cape Cod Bay is about 5 m. Tidal currents, then, can be expected to be strong enough to cut tidal channels in offshore sand shoals and bars.

Do the expectations from this general analysis hold true? At Nauset Beach, facing the Atlantic Ocean, is a cliff of unconsolidated sediments between 30 and 70 m high. It has been extensively undercut by waves and currents during fall and winter storms. A few kilometers to the north, the cliff has been eroded almost 70 m since 1911. The longshore currents during these storms move sediment in the foreshore area south forming Cape Monomoy. The hook near Provincetown is continuously being extended. In the summer, when low energy conditions prevail, wide, sandy beaches quickly develop. The answer to the question at the beginning of the paragraph is "yes" (Fig. 2–15).

Mount Desert Island. Mount Desert Island, most of which is now Acadia National Park, is located off the coast of Maine. In a general analysis of the shoreline at Mount Desert Island, the energy, land stability, and sea-level fluctuation are similar to the Cape Cod conditions. The major difference is the resistance of the outcropping granite bedrock. Steep slopes at the shoreline have remained relatively unchanged for hundreds of years. The island has a number of long, narrow harbors bordered by steep slopes that are fjords. Fjords

2–15 Sandy hook and baymouth barrier beach at Martha's Vineyard, Massachusetts. (*Photo by William J. Wallace.*)

2–16 Wave-cut cave in sedimentary rock. (*Photo by William J. Wallace.*)

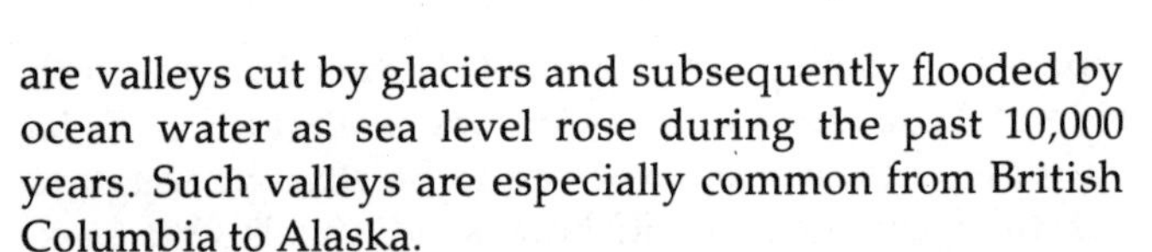

are valleys cut by glaciers and subsequently flooded by ocean water as sea level rose during the past 10,000 years. Such valleys are especially common from British Columbia to Alaska.

At some locations along the shoreline the waves generated by winter storms have eroded even the resistant cliffs. Wave-cut caves are produced in the foreshore area (Fig. 2–16). These are found especially where fractures from ancient stresses exist. Careful examination during the winter shows water seeping into the fracture. At low tide, the water freezes and expands, prying apart the granitic rock along the fracture lines. (This process is called frost wedging.) At subsequent high tides, waves then loosen and eventually extract large blocks of the rock. These boulders roll into the foreshore and offshore area providing an ideal environment for the popular but endangered Maine lobster. Most of the shoreline features other than frost-wedged features on Mount Desert Island, then, are actually remnants of glacial erosion that have been largely unaffected by normal shoreline processes because of the extreme resistance of the bedrock (Figure 2–17).

Everglades National Park. Everglades National Park, Florida, is in many ways a unique part of the U.S. natural heritage. This park contains a fantastic assemblage of terrestrial wildlife including alligators, cougars, egrets, ibises, rosette spoonbills, and many more. Few people, however, are aware of its unusual and remote shoreline (Fig. 2–18).

The southern Florida coastline is exposed to low energy conditions during most of the year. The warm, milky waters of Florida Bay are protected from large waves by the Florida Keys, an exposed fossil reef, and the shallow bottom of the bay, generally less than 10 m deep. The presence of the Florida Keys also diverts the Florida Current to the deep channel between the keys and Cuba, reducing the effect of currents on the shore. The highest energy dissipated on the coast of the Everglades occurs during severe thunderstorms and hurricanes when wind, rain, and high tides batter the shoreline. Local thunderstorms often produce winds that exceed 50 knots and dump more than 15 cm of rain in a three-hour period. The occasional hurricanes not only create high winds and abundant rain but also cause a rise in sea level of nearly a meter.

Everglades National Park is underlain by a hard, relatively nonporous limestone called the Tamiami Limestone. Its surface has been dissolved by acidic rainwater and then recemented as the rainwater evaporated or flowed away. This has produced a "caliche" soil, really a cement-like pavement over which water flows. The broad topography of the area is gently sloping from north to south. Rain runoff trickles slowly from the north, originating in the region of the Great Cypress Swamp and south of Lake Okeechobee. In the

2–17 Mount Desert Island, Acadia National Park, Maine. (Courtesy of the United States Department of the Interior, National Park Service Photo.)

past, the runoff was not funneled into a river but flowed across country in a "river" more than 180 km wide and only a few centimeters to a meter deep, depending on the amount of rain. Presently, the natural runoff is funneled into flood control canals built by the U.S. Army Corps of Engineers. The altered ecological conditions have endangered animal and plant life in the park.

The low topography and resistant bedrock produce a highly irregular shoreline with practically no beaches. The irregular surface of the limestone comprising the shoreline has created thousands of islands on which mangroves grow. Scattered among these islands are oyster bars and Indian shell middens, now forming islands themselves in some cases.

Florida is one of the most stable land areas in the United States. Because of this, it has been an ideal place to study the changes in sea level that have occurred during the past million years or more. Thus, any effects of faulting or earthquakes found on shorelines elsewhere are nonexistent on the Everglades coast.

Tidal fluctuation, however, has been and still is significant in the Everglades. Long-term fluctuation, caused by alternating continental glaciation and interglacials, has repeatedly inundated and exposed all of the Everglades and even Florida Bay. A 10 m rise in sea level, which has occurred several times during the past million years, would completely flood Everglades National Park. Because of the gently sloping topography, even short-term rises of about 1 m associated with hurricanes can flood most of the coastal islands and cause an invasion of the salty water several miles inland.

The ecological balance of this coastal area is especially delicate. Many proposals have been made to preserve and protect the area including federal acquisition of Great Cypress Swamp.

Louisiana Gulf Coast. The shoreline of Louisiana is dominated by the Mississippi River delta complex and is exposed to the Gulf of Mexico. Under normal conditions, waves are less than 1 m high and coastal longshore currents are weak. Tidal currents are also weak. The primary sources of high energy are the principal channels of the Mississippi River during flood stages and occasional hurricanes. The sediments outcropping on the shoreline are mostly alternating layers of unconsolidated sand, silt, and mud transported by the Mississippi River. The rocks are nonresistant and susceptible to erosion. The combination of low energy and low resistance prevents large-scale shifting, as

2–18 Mangrove swamp along the coast of Everglades National Park, Florida. (Courtesy of the United States Department of the Interior, National Park Service Photo.)

found at Nauset Beach. Instead, rich soil develops and abundant terrestrial vegetation extends almost to the berm in many places.

The land stability of southern Louisiana is very different from other coastal regions in the United States. The Mississippi River continually deposits sediment at the coast, building the shoreline outward into the Gulf of Mexico. Geologists call this process "accretion." At the same time, however, as the loose sediment settles and becomes compacted by the continual addition of more sediment, New Orleans and its vicinity are sinking, due to the loading of material deposited in the Mississippi River delta. Today, there may be additional causes, including the pumping of oil from deep wells in the area. The fact remains: southern Louisiana is building southward and at the same time slowly sinking.

Sea-level fluctuation, both long- and short-term, and, in this case, flooding, are important in producing the features of the Louisiana coast. The sediment and topographic features are the result of deposition and erosion by the Mississippi River. Long-term sea-level changes caused by glaciation and short-term changes caused mainly by hurricanes then modify the topography by cutting terraces and redistributing large amounts of sediment.

Southern California. The southern California coast between Oceanside and San Diego is appreciably different from the four previous examples, although it shares some similarities with the Atlantic Ocean side of Cape Cod.

Except in the coastal bays and estuaries, the southern California coast is exposed to continuous high energy conditions (Fig. 2–19). Normally, waves are about 1 m high with frequent periods of 2 m high surf. During two or three periods each year, waves from 3 to 6 m or larger, generated by Pacific Ocean storms, travel more than 1,800 km to batter the coast. Even tsunamis are not unknown along this coast. In addition, the strong wind-driven waves sweep from north to south along the coast. As changes occur in the size and direction of swell and surf movement, the amount of sediment in suspension carried by longshore currents, the steepness of the beach, and the movement and location of offshore bars also change (Fig. 2–20). These high energy conditions produce wide beaches all along the coast with rapid beach transport in the offshore surf zone. The highest energy conditions occur during the winter, when wave action removes beach sand from the foreshore to the offshore, leaving a large percentage of boulders behind. The sand is slowly replaced in the foreshore during the summer, when lower energy prevails (Fig. 2–21).

The sediments outcropping along the coast between Oceanside and San Diego range from unconsolidated barrier beaches across the mouths of estuaries to well cemented sandstones forming impressive cliffs. The combination of well and poorly cemented sediment forms precipitous cliffs with undercut arches (Fig. 2–22).

The instability of the southern California area is notorious. Several frequently active faults produce earthquakes almost daily 50 to 180 km east of the coast (the Elsinore and San Andreas faults). At least one fault, which has been active in recorded history although quiet for 200 years, runs through downtown San Diego and intersects the coast at San Diego Bay, Mission Bay, and La Jolla. These earth movements have caused land uplift during the past million years to produce the cliffs on the coast.

Because uplift of the land relative to sea level has so dominated the area, the effects of sea-level fluctuation are difficult if not impossible to determine. Certainly, short-term tidal fluctuation has some effect, especially when high tides in January and February coincide with high energy wave trains.

The significant result of these processes is extensive longshore currents and sediment transport. Millions of tons of sand are transported in the offshore and foreshore areas. What makes this condition unique is the proximity of submarine canyons to the shoreline. They act as funnels, moving sand from the shore to deep water about 21 km out in the San Diego Trough at a depth of more than 800 m.

Effects of Man. So far, the features discussed in this chapter have been natural. Man has also had a tremendous effect on the shorelines of many land areas. Some of these effects have been beneficial; many have been harmful.

Five general activities of man have had great effect, some positive and some negative, on the configuration of shorelines: (1) damming rivers, (2) land recovery, (3) inlet dredging (Fig. 2–23), (4) jetty construction (Fig. 2–24), and (5) development of dune areas.

The effect on shorelines of damming rivers is often overlooked by politicians and engineers who are interested in conserving water and building power plants. Most rivers flow eventually into the ocean, although a few empty into desert basins or inland seas and lakes. Sediment transported by rivers is deposited in a base-level basin, which in most cases is the ocean. Ordinarily, the sediment would be deposited in the nearshore area where waves and currents then distribute it. Longshore currents, for instance, often have a near-capacity load of sediment supplied by rivers. This means that the current is capable of carrying very little additional sediment without increasing the speed or volume of the current. When a sediment-supplying river is dammed, the sediment is deposited on the upstream side of the dam, preventing the normal supply to coastal currents. Coastal currents carrying a sediment load far below their capacity are called "starved." A starved longshore current will readily pick up sand from the foreshore and offshore areas, causing excessive beach and shoreline erosion. The only way to prevent this is to supply the starved current artificially. Much of the Washington, Oregon, and California shoreline has been experiencing increased erosion because of river damming.

Land recovery is a costly endeavor but has been worthwhile in several parts of the world. Hundreds of square kilometers of shallow coastal marshes and bays have been turned into fertile farmland in the Netherlands. This has been accomplished by building dikes,

2–19 Looking north from the National Marine Fisheries Service Building at La Jolla, California. Scripps Canyon originates in the surf zone in the foreground. (Photo by Thomas H. Foote.)

(*a*)

(*b*)

2–20 Torrey Pines State Park at La Jolla, California. (a) Note gravel left on the beach by waves separating finer-grained material. (b) A close-up of the beach foreslope during the winter. (Photos by William J. Wallace.)

dredging and relocating sediment, and pumping out salty water. In spite of this effort, occasional storms have damaged or broken sections of dikes, causing extensive flooding and loss of life. Although land recovery may be an economic necessity in some places, it destroys marshes and bays, which are important habitats for many forms of wildlife. Hopefully, economic needs and ecologic needs are carefully evaluated. In the San Francisco Bay area, the clash between those who wished to fill parts of the bay for economic gain and those who wished to preserve the same parts for wildlife sanctuaries was especially bitter. Presently, a moratorium on land fill in the bay exists. In New England, beginning in the 1700s with Boston's Back Bay, coastal marshes have been filled at such a rapid rate that they are now protected environments in a number of communities like Guilford, Connecticut.

In the past 40 years, inlet dredging has become increasingly popular. Pleasure boating especially has prompted many communities to seek out the U.S. Army Corps of Engineers to ease access to desirable open water. As with land recovery, economic and ecologic needs must be carefully balanced. The U.S. Navy has its largest Pacific port facility in San Diego, California, and the city itself has encouraged increased commercial shipping. Dredging the channel in San Diego Bay has been a worthwhile enterprise because no known wildlife habitats were affected. At Matanzas Inlet, Florida, between Daytona Beach and St. Augustine, the channel was also dredged, presumably to increase the value of land in Crescent Beach and at Marineland. The effect was incredible. Tidal currents rushing from the new

(a)

(b)

2–21 (a) "Winter" beach at Scripps Institution of Oceanography; (b) a "summer" beach. During a "summer" beach, all but the tops of the highest rocks are covered with sand. (Photos by William J. Wallace.)

channel diverted the strong longshore current and caused sediment to be deposited on the north shore of the inlet. When the starved longshore current swept by the south shore, erosion rapidly consumed not only the shorefront property but also a highway. There is now a new road. The beach has been extended inland and is often used by surf fishermen and swimmers. No preventive action was taken.

2–22 Sunset Cliffs at San Diego, California, in 1951. (Courtesy of the Historical Collection, Title Insurance and Trust Co., San Diego.)

Jetties or groins are commonly built to prevent longshore currents from transporting sand or to cause deposition from transporting currents. This works well for the property immediately down current from the jetty but accentuates the problem for properties further down current. The traditional behavior of shoreline property owners is to build a whole group of jetties. The former longshore current now becomes a zig-zag current and the coast loses its linear form. Shoreline property owners should understand that where the rock is nonresistant the shore will inevitably erode.

The fifth cultural effect on shorelines is "developing" dune areas. As evidenced by their existence, dunes are transient unless stabilized by vegetation, but houses or other structures are by nature permanent. The two are obviously incompatible. In addition, any stability dunes have is provided by usually sparse vegetation, which can easily be destroyed even by dune buggies. Often they are a part of the backshore that is occasionally flooded by storms. As a result of development, the unstable dune areas deteriorate rapidly and the shoreline migrates landward.

Most of man's changes to the shoreline have been careless and damaging. Generally, the impermanence

(*a*)

(*b*)

2–23 (*a*) *Pumping station on the north side of Boynton Inlet, Florida.* (*b*) *Sand and water slurry draining into the south side of Boynton Inlet, Florida. (Photos courtesy of D. W. Lovejoy.)*

of the shoreline and its delicate marine–nonmarine transition zone suggest that it should be left largely undeveloped and available for recreation and study purposes. There are, of course, exceptions like the Netherlands and a number of city ports where development and alteration by man is worthwhile and not overwhelmingly harmful to a valuable recreational resource (Fig. 2–25).

The Continental Shelf

The continental shelf is the part of the continents that is presently submerged beneath ocean water.

Location. Each continent is partly submerged and its shelf is usually located on the boundary between a continent and a major ocean, although sometimes it can be landlocked, as in the case of Hudson Bay, Canada. These shallow regions form 18 percent of the earth's total land (Fig. 2–26) and 7.5 percent of the total area of the ocean. Only about 10 to 15 percent of this area has been studied. The continental shelf is up to 1,500 km wide with an average width of between 50 and 100 km.

Often the outer edge of the continental shelf is found at a depth of about 200 m (Fig. 2–27). This is close to the average depth at which an appreciable increase in the angle of slope of the bottom occurs. The increase would be from about 1 m drop per 1,000 m on the shelf to 1 m per 100 m. A number of exceptions to this exist, however. Glacier-eroded shelves generally extend to a depth of nearly 500 m before the slope increases, while many coral shelves extend only about 20 m.

Formation. Eight major factors have produced the majority of features found to date on continental shelves around the world: (1) glaciation, especially continental, (2) sea-level fluctuation, often related to glaciation, (3) energy from waves and currents, (4) sedimentation from a variety of sources, (5) faulting, (6) volcanism, (7) reef building by organisms, and (8) global plate tectonics.

Glaciation and sea-level fluctuation. During the Pleistocene epoch, extensive glaciation periodically

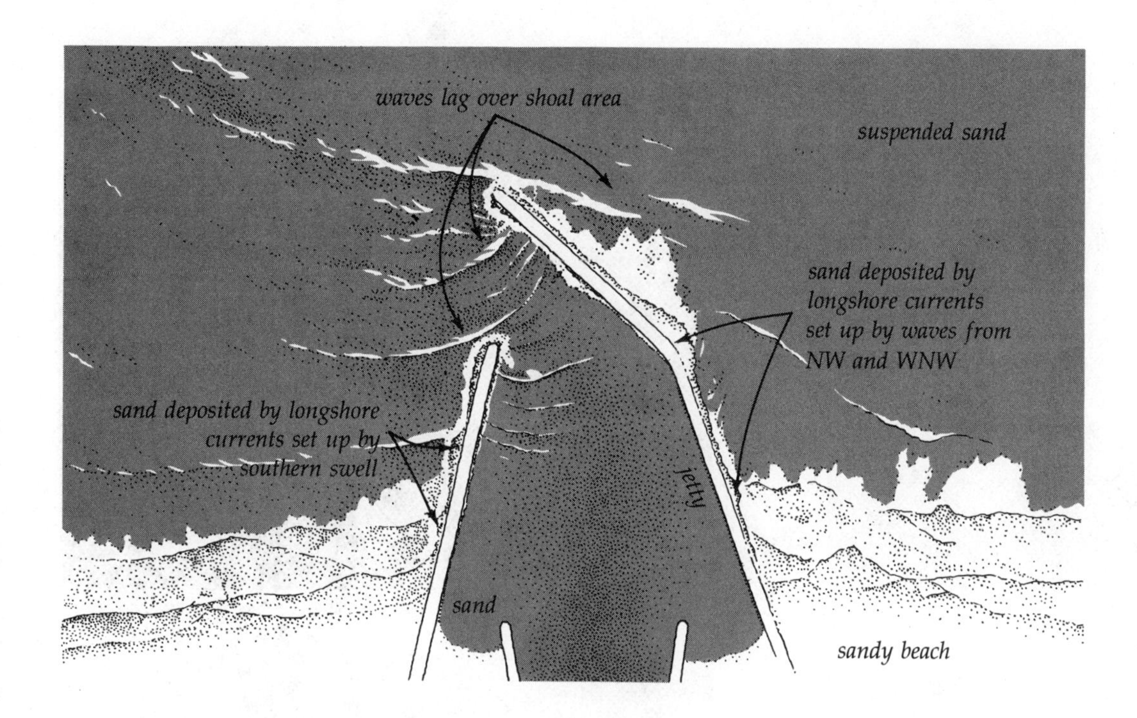

2–24 Sediment and wave patterns around a man-made jetty system.

occurred in the northern hemisphere. Present evidence indicates that glaciers were continuously present in Antarctica and Greenland during this time. North America, Europe, and Asia were subjected to periods of partial blanketing by glaciers thick enough to be called ice sheets. These periods were separated by relatively mild interglacial periods. Many local and short-term advances and retreats of the ice sheet have been described, yielding complex classifications of glacial history. Because climate patterns vary from region to region, glaciation also varied (Table 2–1).

Table 2–1 Generalized Pleistocene Epoch Glacial Chronology in North America

Period	Years ago
Holocene epoch	10,000 to present
Wisconsian glaciation	10,000 to 100,000
Sangamonian interglacial	100,000 to 250,000
Illinoian glaciation	250,000 to 400,000
Yarmouthian interglacial	400,000 to 600,000
Kansan glaciation	600,000 to 750,000
Aftonian interglacial	750,000 to 850,000
Nebraskan glaciation	850,000 to 1,000,000

Theories concerning the initiation of the glacial periods are difficult to test. Initial conditions probably included a lack of sea ice covering the Arctic Ocean and surface water temperatures several centigrade degrees warmer than at present. Conditions in the atmosphere must have favored the precipitation of snow, which then fell on the continents fringing the Arctic Ocean. The climate, however, must have been cool enough to prevent melting of all the snow that fell during each year. As tens of thousands of years passed, the snow accumulated, became compacted, and covered the northern parts of North America, Asia, and Europe. As the ice accumulated, stream and river runoff of melt water did not replenish the ocean. Slowly, as the thickness of snow and ice increased on the continents, the average level of the ocean decreased.

The effects of the alternating glacial and interglacial periods on the earth's surface have been appreciable. The effects on the ocean water, the atmosphere, the organisms living in the oceans, and the continents have been especially significant.

Present evidence suggests that sea level was probably 180 m below its current level. This obviously would have exposed large parts of what is now the continental shelf. One extraordinary evidence of this is the

2–25 (a) San Diego Bay at San Diego, California, in 1969. The two barrier islands in the lower left are land fill. The North Island Naval Air Station in the center was formerly a marine marsh. (Courtesy of the City of San Diego.) (b) Mission Bay entrance (top) and Sunset Cliffs (bottom) at San Diego, California, in 1968. (Courtesy of the City of San Diego.) (c) Aerial view of Mission Bay at San Diego, California, in 1948. Note the configuration of the bay entrance in comparison to b. (Courtesy of the Historical Collection, Title Insurance and Trust Co., San Diego.) (d) The multi-million dollar Mission Bay Aquatic Park in San Diego encompasses 4,600 acres of land and water, providing facilities for fishing, boating, waterskiing, sailing, swimming, picnicking, and every other type of water sport. Accommodations in Mission Bay range from luxury hotels to bay-side trailer parks and picnic grounds. (Courtesy of the San Diego Chamber of Commerce.)

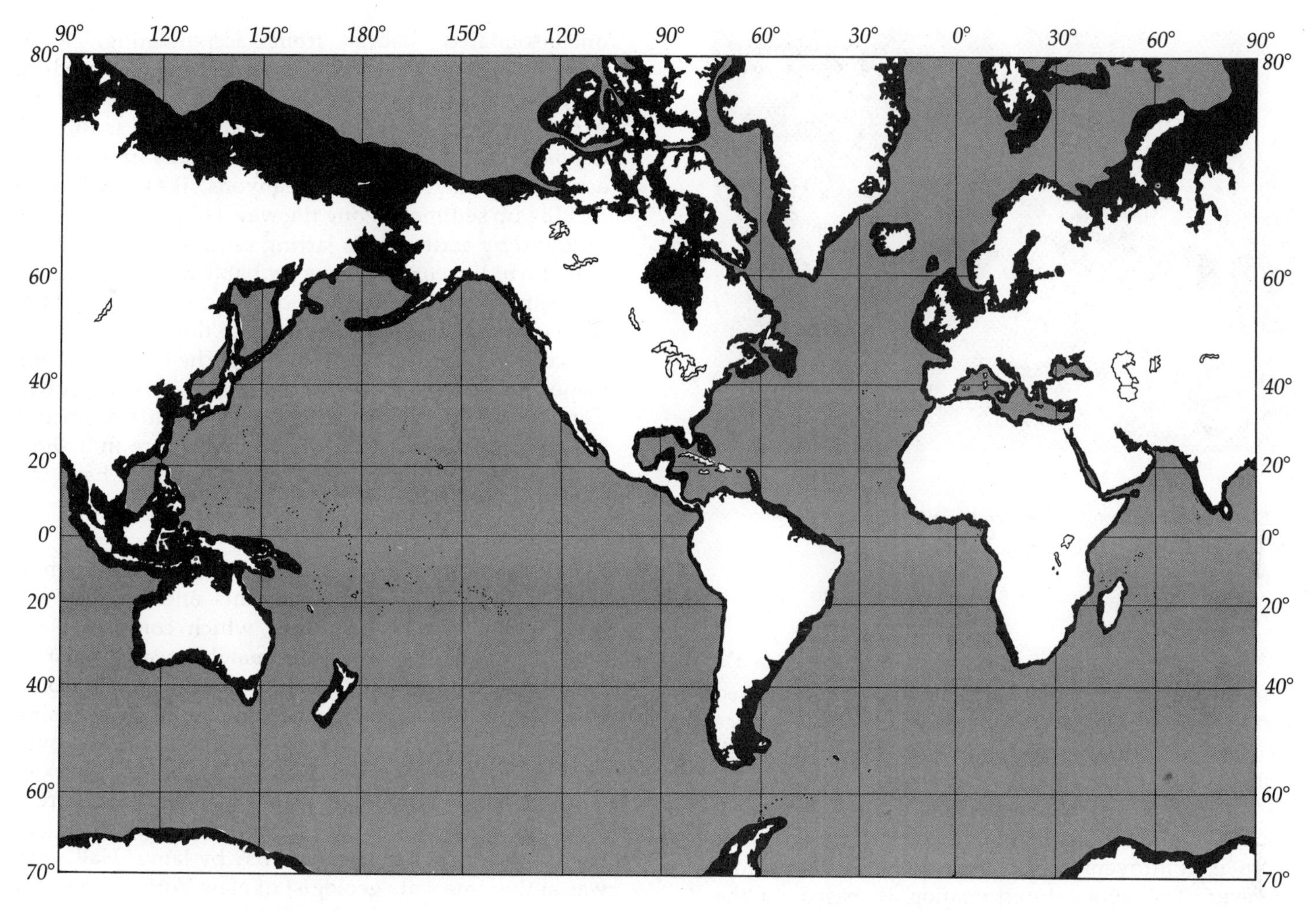

2–26 *The continental shelves.*

frequent discovery of fossil mammoth teeth in dredgings from the continental shelf off New England. In addition, by coring the bottom sediment geologists have discovered submerged ancient peat bogs containing fossil plants and pollen grains. The plants are composed largely of carbon. Carbon has a number of forms, called isotopes, that occur in a known ratio in live plants. One isotope (carbon 14) decays at a known rate to a stable element, nitrogen, because it is radioactive. As a result, fairly precise dates of the times when the plants were living can be obtained. This establishes how long ago the glacial periods were. Because distinct faunas are associated with warm and cold ocean water and because the ratio of oxygen isotopes in shells varies with water temperature, the effect of glaciation on ocean water temperature has also been deduced. During the Pleistocene, the temperature fluctuation of the surface waters was about 4°C for the Pacific and about 7°C for the tropical Atlantic.

During interglacial periods sea level is relatively high. The most recent interglacial period commenced about 10,000 years ago and continues to date. Much of the evidence for higher levels of the oceans is based on the presence of Pleistocene marine fossils on what is now land. By definition, if about 95 percent of the species found in the fossil assemblage still have living

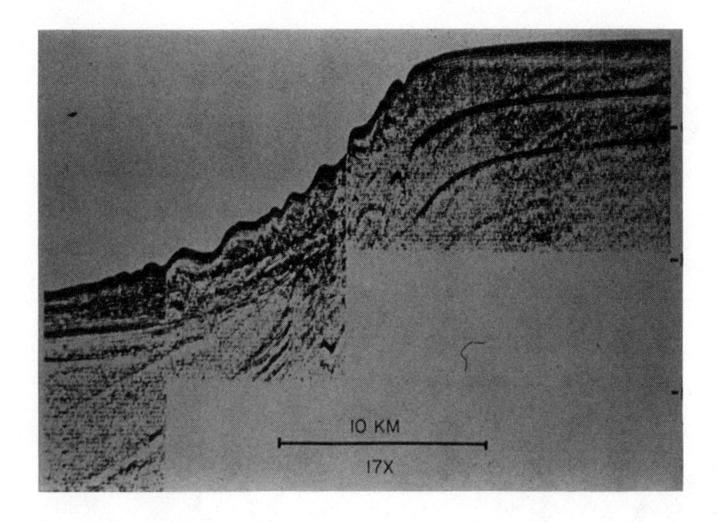

2–27 *Subbottom profile of the continental slope in the Gulf of California. (Courtesy United States Navy.)*

counterparts in the ocean, the fossils were deposited during the past 1,000,000 years. Based on such evidence sea level has been much higher than its present level, probably just before the Nebraskan glaciation. During the succeeding interglacial periods, sea level probably reached maxima of about 10 m above the present level. There is controversy about the extent of fluctuation, however. The above interpretation is based on the probable Pleistocene history of Florida.

During periods of glaciation and sea-level fluctuation, parts of the continental shelf were covered and eroded by advancing glaciers, some areas received sediments from the glaciers, others were eroded by waves as sea level changed, parts received alternating terrestrial sediments (with land fossils) and marine sediments (with shell fossils), and still others had rivers flowing over them.

Energy. The continental shelf is affected by various sources of natural energy. As sea level changed, wave action and even wind produced features that are now submerged. The present shelf is affected by longshore, river, and turbidity currents, tides, tsunamis, surface and internal waves, and storm surges (rises in sea level that accompany low pressure storms). In some areas, fast-moving tidal currents funnel between land masses, sweeping the shelf and preventing unconsolidated debris from accumulating. River currents may occasionally be strong enough to produce features on the shelf. The Congo River of west central Africa has carved channels into the loose sediments deposited on the shelf. Turbidity currents have moved across the shelf, funneled by canyons, like an avalanche picking up sediment along the way. Usually caused, it is thought, by earthquakes jarring sediment into suspension, turbidity currents have probably transported large amounts of shelf sediments out into the ocean basin.

Tides and tsunamis both affect the whole column of ocean water from top to bottom. The extent of their respective effects on the continental shelf is unknown. Certainly tsunamis can shift large amounts of unconsolidated sediments long distances inshore in a short period of time. Storm surges are thought to have the same type of effect.

Sedimentation. Because the continental shelf is adjacent to the continents and because sea level is the base level toward which continents are being eroded, the shelves ultimately receive most of the sediments eroded from the continents. In some cases, these sediments may accumulate in large basins called geosynclines by geologists. These may be over 1800 km long, 360 km wide, and 10 km thick. The existence of these impressive basins may not be apparent from the shelf bottom topography. In fact, their existence was first described in 1859 by James Hall, who was at the time state geologist of New York. He noticed the linearity of the feature, the fact that marine sediments were present and could be traced from Alabama to New York, and the great thicknesses of the sediments. Now that geophysicists have techniques such as seismic profiling by which to map the subbottom of the shelves, geosynclines have been recognized in various stages of formation on a number of continental borderlands.

All the major river deltas of the world have a shelf portion: Mississippi, Amazon, Orinoco, Nile, Congo, Indus, Ganges, Irawaddy, and numerous others. Some, like the Mississippi River delta, are located in a geosyncline. Herodotus, who lived about 450 B.C., wrote:

The nature of the land of Egypt is such that when a ship is approaching it and is yet one day's sail from the shore, if a man try sounding, he will bring up mud even at a depth of 11 fathoms. (Quoted in Emery, 1969)

The "birdfoot" pattern of deltas such as the Nile, Amazon, or Mississippi may extend many kilometers from the present shoreline. In addition to their configuration, these sediments have a common origin from rocks exposed on the land. Because of the preponderance of quartz and feldspar in continental rocks, the deltas also contain these minerals in abundant supply. In deltas that contain sediments eroded from mountains rich in gold, like the Columbia River delta in Washington, concentrated pockets of gold nuggets are found. Many other oil and mineral resources located in the sediments on the continental shelf are especially concentrated in deltas.

Carbonate deposits are not common on the continental shelf bordering the United States. They are found bordering southern Florida and in many other parts of the world, especially adjoining tropical volcanic islands, atolls, and northeast Australia. Carbonate deposits must have been much more common in the past than now because great thicknesses are found underlying Florida, in the Rocky Mountain chain, in the Alps, and in the Himalayas. In Florida, the carbonate deposits below the surface of the land and the shelf are more than 3,000 m thick.

Carbonate deposits contain compounds composed of carbon and oxygen (CO_3^{-2}, called the carbonate ion) combined usually with calcium (Ca^{+2}) to form calcite ($CaCO_3$), with calcium and magnesium (Mg^{+2}) to form dolomite ($CaMg(CO_3)_2$), and solid solution mixtures of the two. Most of these carbonate deposits are the broken remains of the shells of marine animals and plants. Because a large number of organisms with shells or hard parts composed of calcium carbonate or calcium magnesium carbonate live on the continental shelf, in shallow water over the shelf, or near oceanic islands, carbonate sediments are found in the same locations.

On some continental shelves, former terraced shorelines formed during glacial periods are now submerged (see Fig. 2–32). Some of these terraces contain shell concentrations accumulated by wave action when they were beaches. Others contain phosphate (PO_4^{-3}) enrichment. This may be partly fossil bird guano, which accumulated where birds roosted when the terrace was a shoreline. Potentially commercially valuable phosphate deposits are known on such terraces on the continental shelf off southern California. Evidence indicates that these nodules are continuing to form, thus confirming a precipitation origin involving complex interactions between water and sediment.

Stability and faulting. Because continental shelves are the surface expression of the transition zone between continents and ocean basins, some instability might be expected. Looking at a map of earthquake zones in the earth's crust, continental shelves are indeed areas of extensive instability (Fig. 2–28). Because of the earthquakes, rocks on the shelf fracture and slide or are pushed along the fractures. This type of movement is called faulting. Every continental shelf explored to date has had some faulting in the past. Many shelves and continental borderlands, like those along the west coast of the United States, still have active faults whose movements produce frequent earthquakes. During the great Alaskan earthquake of Easter 1964, segments of Kodiak Island fell 2 m or more, in one case moving a supermarket down to the foreshore where it was partially submerged at each high tide. On Capri, located on the continental shelf off Naples, Italy, a large fault system slices through the middle of the island.

Volcanism. Volcanoes are major features of some continental margins. Many are still active. Especially active is the "Ring of Fire" bordering the Pacific Ocean. Active volcanoes are also common along the European continental margin in the Mediterranean Sea. These volcanoes will be discussed in detail in the section of this chapter on island arcs.

Origins. Six examples of continental shelf origins have been recognized: (1) tectonic dams, (2) reef dams, (3) diapir dams, (4) no dams, (5) ice eroded, and (6) wave eroded. The latter two are usually superimposed on one of the others. A shelf with no dams is in fact a shelf whose sediments have spilled over the dam, burying it with a broad drapery of sand, silt, and clay derived from the continent. Tectonic dams are formed by the faulting and folding of rocks on the outer part of the continental shelf. Reef dams are built by the accumulation of shells or hard parts of marine organisms like coral or algae. Diapir dams are caused by the intrusion of low-density salt domes squeezed upward out of deep sediments (Fig. 2–29).

To illustrate some characteristics and origins of continental shelves in more detail, it may be useful to examine some shelves bordering the United States.

New England. The New England coast stretches from Connecticut to Maine. Its shoreline

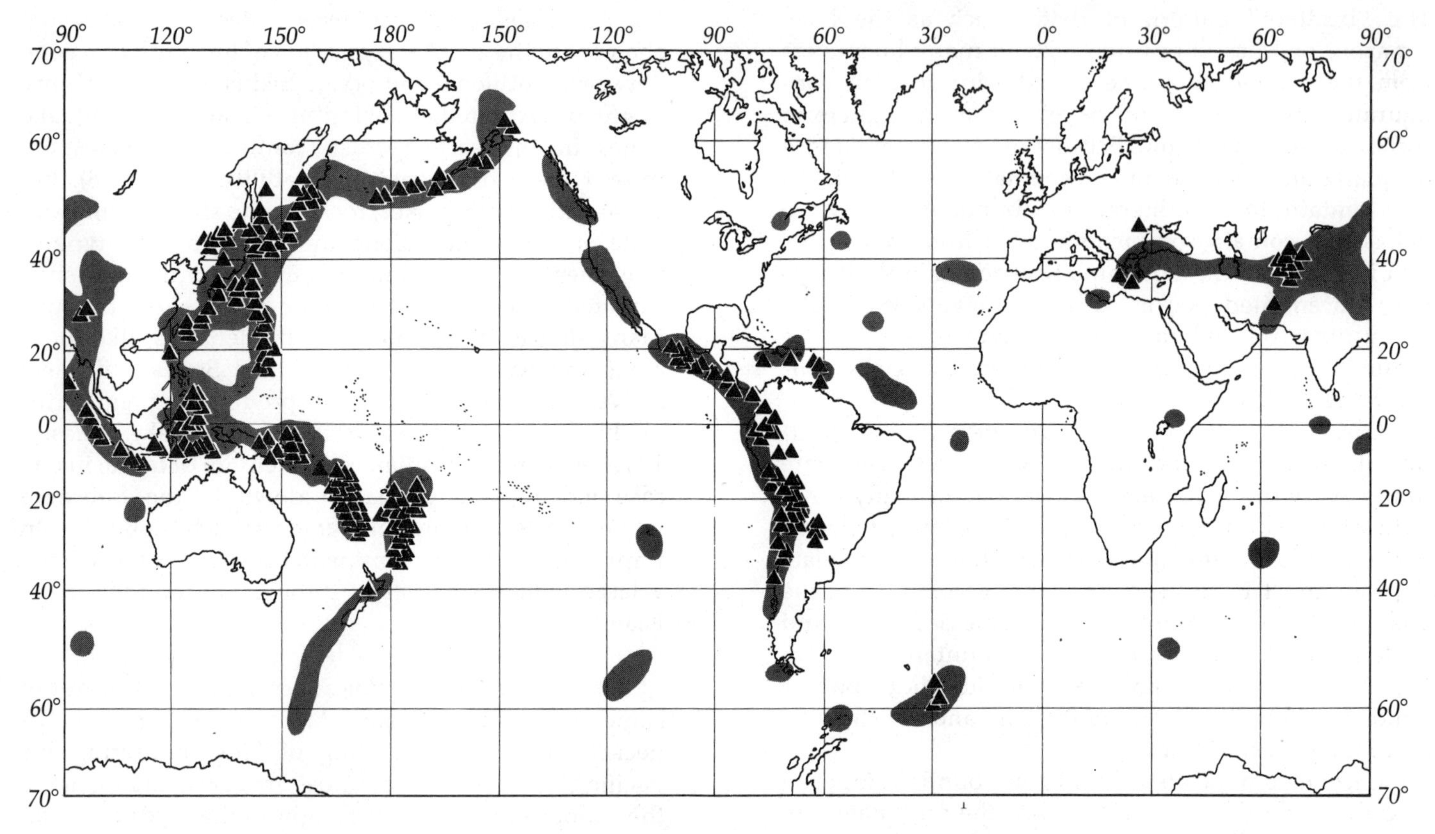

2–28 *World-wide distribution of shallow and deep (below 70 km) earthquakes (1904–1952). The triangles show the locations of deep earthquakes.*

consists of irregular crystalline rocks with occasional strips of loose sand deposited by glaciers. The continental shelf extends seaward a maximum of 140 km east of the present shoreline. Thus, the area of New England would be nearly doubled were its continental shelf to be exposed.

Long Island Sound lies between the Connecticut shoreline and Long Island. It is from 5 to 35 km wide and is connected on both ends with the Atlantic Ocean. Its deepest point, between Long Island and New London, Connecticut, is only a little more than 100 m deep. In the central part of the sound, the depth is less than 50 m. A continuous depth record from Bridgeport, Connecticut, to Port Jefferson, Long Island, would show no spot deeper than 40 m. For the most part, the bottom of Long Island Sound is relatively smooth, consisting of mud with some local areas of sand and gravel. These sediments form a thin veneer over ancient crystalline rocks, mostly granites similar to those seen along the present Connecticut shoreline. In some parts of the sound, Tertiary sediments lie between the crystalline rocks and the glacial deposits. Most sediment now found on the bottom of the sound was transported to the vicinity by glaciers during the past million years. Since then, rivers, currents, and waves have reworked and redistributed them to their present location.

At the times when ice sheets advanced over part of the North American continent, a mass of ice perhaps 100 to 300 m thick covered all of Connecticut and Long Island Sound and terminated where Long Island is now. As the ice melted there, the sand, silt, and gravel suspended in the ice were dumped. This glacial dump of unconsolidated sediment located at the forward edge of a glacier is called a terminal moraine. Long Island is a remnant terminal moraine. During the glacial periods, what is now the continental shelf southeast of Long Island was a broad, gently sloping land area whose shoreline was about 75 km southeast of Long Island's present Atlantic shore. The land area was covered with grass, trees, and streams from melted glacial ice.

Scattered remnants of Cretaceous and Tertiary sediments have been found on Long Island, Cape Cod, and Martha's Vineyard.

Cape Cod is similar in structure and origin to Long Island, but is is a complex of several moraines that formed when three lobes of the ice sheet converged (Fig. 2–30). The continental shelf directly east of Cape Cod is 140 km wide. In some places 75 km east of the present shoreline, the shelf is only 10 m below sea level. This shallow shelf is called George's Bank. The mixing of cold water from the Labrador Current moving south along the New England coast and warm water from the Gulf Stream moving northeast parallel to the U.S. east coast produces nutrient-rich water conducive to the growth of many marine organisms. As a result, crab, lobster, flounder, sole, and haddock are abundant on the bottom and tuna, bass, and herring swim the water column above the shelf, producing one of the richest fishing grounds in the world.

North of Cape Cod is a deep basin called Wilkinson's Basin. It formed long before the glacial periods, at a time when stresses were producing the now deeply buried tectonic dams found further south. The shelf surrounding this basin has been scoured by ice movements across the rocks, producing numerous valleys and some strips of sand and gravel. A map of the submarine contours of this portion of the shelf looks identical to a map of the contours of the present land area in northeast Massachusetts, New Hampshire, and Maine.

The continental shelf off New England has been important to the United States and other countries as a source of marine organisms for food. The area is now also becoming a center for research in aquaculture. A number of attempts are being made, especially by the U.S. Fisheries Commission, to raise lobsters and oysters under controlled conditions in laboratories and in protected nearshore bays. A number of research institutions in the New England area are involved in aquaculture programs.

Mid-Atlantic shelf. No glaciation has affected the continental shelf south of New Jersey. From New Jersey to North Carolina, the shelf slopes gently with no apparent dams. Recently, deep seismic profiles have revealed ancient tectonic dams. Sediments, apparently eroded off the eastern slope of the Appalachian Mountains, have buried and overlapped the dams.

Longshore drift currents have produced numerous longshore barrier islands, which are found from Sandy Hook, New Jersey, to Miami Beach, Florida. In some locations, these currents have built submerged, shifting, sandy shoals. Shoals in the vicinity of Cape Hatteras are scattered across the continental shelf near the shoreline and several kilometers offshore. Mariners have called the shelf immediately off Cape Hatteras the "graveyard of the Atlantic" because many ships have been grounded on the shoals, especially during storms, and pounded to bits by the waves.

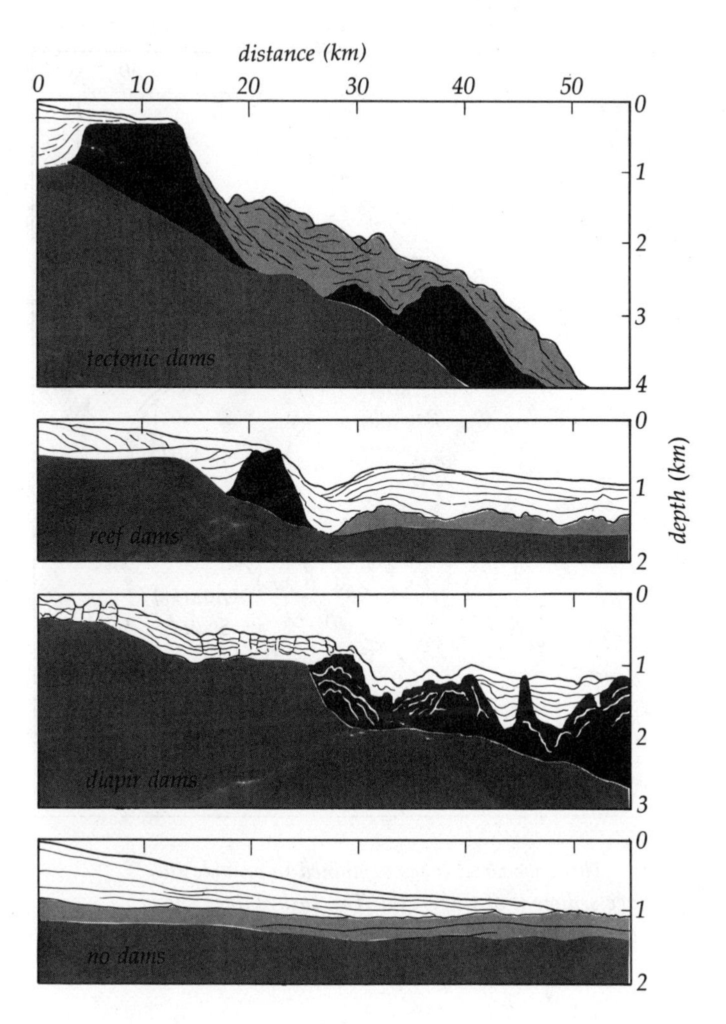

2–29 Origins of continental shelves. (From K. O. Emery, "The Continental Shelves," Scientific American, *September 1969. Copyright © 1969 by Scientific American, Inc. All rights reserved. Reprinted by permission.)*

Florida Bank. All of Florida, including its continental shelf, is a carbonate bank. Clastic (frag-

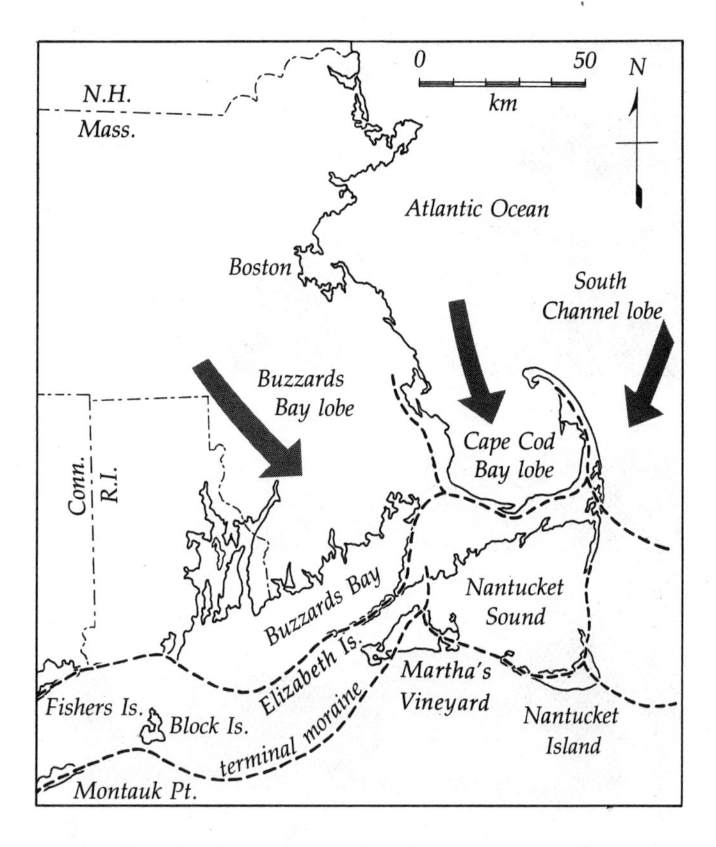

2–30 Three ice sheet lobes combined to provide the surface sediment of Cape Cod. The dashed lines show the two positions of ice standstill.

mented or broken) sediments, however, are known at depth. If sea level were to rise 100 m above its present level, all of peninsular Florida would be submerged. Some noncarbonate sediments form a surface covering over the thick carbonate sediments. Reef dams built of coral and calcareous algal buttresses ring the continental shelf. In fact, reef deposits comprise a large part of the peninsula. Broken shells and dissolved and reprecipitated calcium carbonate fill in between the reefs.

The continental shelf off the east coast of Florida is narrow. Eastward from Miami, the gradient of the shelf is not steep for about 1,000 m. A series of steep scarps with indented terraces formed at lower stands of sea level are found. Eight km offshore, the water is more than 600 m deep. The shelf obviously ends steeply and abruptly. Thirty km further eastward, the depth decreases just as abruptly. This region, the Bahama Bank, is formed basically of the same sediments as Florida. Its shelf also has terraces at about the same depth as on the Florida shelf.

At the Florida Keys on the southern extremity of the state, small patches of coral and coralline algal buttresses grow near the southern edge of the continental shelf. This may be an example of a reef dam in the process of forming. Also on the southern side of the keys, the Florida Current is funneled through the 130 km wide strait between Cuba and the Florida Bank. The combination of few floating planktonic organisms and the lack of sediments from rivers produces extremely clear water ideal for skin and scuba diving. Visibility is often 60 m or more. As a result, Pennecamp State Park, the first submarine state park in the United States, was established at Key Largo. Divers can swim over delightful reefs and banks and nonswimmers can observe the beauty of the undersea world in glass-bottomed boats.

North of the Florida Keys is Florida Bay, a shallow body of water separating the keys from the Everglades. Currents are few in the bay and the shallow waters reach high temperatures, especially in the summer. Many microscopic organisms like foraminifera (protozoans that secrete calcium carbonate shells) and calcareous algae live on the bottom. Because the water is saturated with calcium and carbon dioxide, the shells from the dead organisms accumulate on the bottom to form a fine calcium carbonate mud. When winter winds from the northwest stir up the waters, the tiny shells become suspended in the moving water, giving it a milky color. At such times, the milky bay water moving southward through the inlets to the edge of the Florida Current produces a striking visual contrast of white and blue-green.

The continental shelf west of Florida is 140 km wide and slopes at only about 1 m per 4,500 m offshore. This broad, flat bank was completely exposed during the Pleistocene glacial periods. Mammoths, saber-toothed cats, alligators, turtles, and many other animals lived on the plain. Their bones and teeth are often found by fishermen and divers. This whole bank is also underlain by limestone and fringed by reef dams with possible faults on the westward side. The submarine topographic features of the bank are similar to those presently found on the land adjacent to it. During the lower stands of sea level, the same weathering processes that affected what is now the land also affected the present shelf. Rainwater containing carbonic acid slowly dissolved the limestone, producing sink holes and springs. These dissolved areas give a pock-marked appearance to both the shelf and the land. Fresh water springs still flow into the Gulf of Mexico from far offshore. The extensive phosphate deposits found in central Florida also can be traced out onto the shelf off Fort Myers. One hundred forty km west of the present shoreline, the slope abruptly increases so that less than 10 km further, the water is more than 600 m deep. This is one of the steepest known scarps.

Mississippi River delta. The Mississippi River delta dominates the continental shelf off the coast of Louisiana. No dam has been found to produce the shelf. Presumably it results from the continued deposition of sediment eroded from the entire North American interior bounded by the Rockies, Canada, and the Appalachians and transported to the delta region. The shelf here is prograding or building outward into the Gulf of Mexico. The sediments on the shelf are unconsolidated sand, silt, mud, and gravel, which have accumulated for millions of years.

Texas shelf. Some of the most unusual features on the world's continental shelves are found off the Texas coast. The present nearshore area has barrier islands with shallow warm lagoons. During summer months, high air and water temperatures allow the water to evaporate, increasing the salt concentration above saturation. Salt then precipitates onto the bottom. The sediments offshore also contain extensive salt deposits although they may have had a different origin from the lagoonal deposits. When salt is buried, the weight of overlying sediment is so great that the salt becomes fluid and, because its density is less than the surrounding sediment, the salt migrates upward to form domes. These salt domes force the sediments to fold and push upward to form diapir dams. Sediment has accumulated shoreward from the dams. This continental shelf, then, although similar to others in general topography, has a vastly different origin. Oil and gas often seep through porous sediments against the salt domes and accumulate into commercially important reservoirs because the salt is impermeable. Such domes have been found beneath sediments in the middle of the Gulf of Mexico as well as on the shelf and on land.

Southern California shelf. The continental borderland off the coast of southern California is composed of sedimentary rocks that extend inland and are found either outcropping in cliffs along the coast or beneath the surface, where well drilling has revealed them. About 15 km west of the San Diego shoreline, these sediments form a steep, submerged cliff. Further westward, still another cliff rises to form a ridge that is mostly submerged but occasionally forms islands. Between the cliffs is the San Diego Trough, having a depth of more than 800 m. The cliffs have been produced by faulting and erosion. A series of linear banks with some islands and troughs is characteristic of this entire region (Fig. 2–31). Some of the more recent sediments underlying the area have been transported from mountains to the east. In some of the troughs and basins extensive sediments accumulated during the Pliocene and Pleistocene. These features are part of continuing processes associated with large-scale crustal movements. The present continental shelf features are erosional and tectonic. Deep canyons cut into the shelf and extend from the offshore surf zone to the San Diego Trough. The sheer walls of these canyons would certainly qualify them as national parks if they were exposed to public view. The presence of the canyons is significant for continental margin processes. Sediment carried alongshore is funneled out through the canyons by gravity sliding and periodic, small turbidity currents to the San Diego Trough, where large aprons of nearshore sand, silt, mud, and shells accumulate. This process continuously depletes the beach sand supply, especially in winter when waves and currents increase their sediment load. As previously mentioned, dams on the southern California rivers have cut off the resupply of nearshore beach sand. The longshore currents have become starved and shoreline erosion has increased in the past 20 years. No effective solution to this problem has yet been found. One proposal is to build submarine dams at the head of the canyons to catch the sediment and then pump it back to the beaches. As erosion increases, public pressure will cause further attempts to solve the problem.

Like the Florida eastern shelf, the southern California borderland has submerged wave-cut terraces formed at lower stands of sea level. Some of these terraces may also have been lowered by the faulting associated with crustal movements in the region. Phosphate deposits are known to exist on some of these terraces.

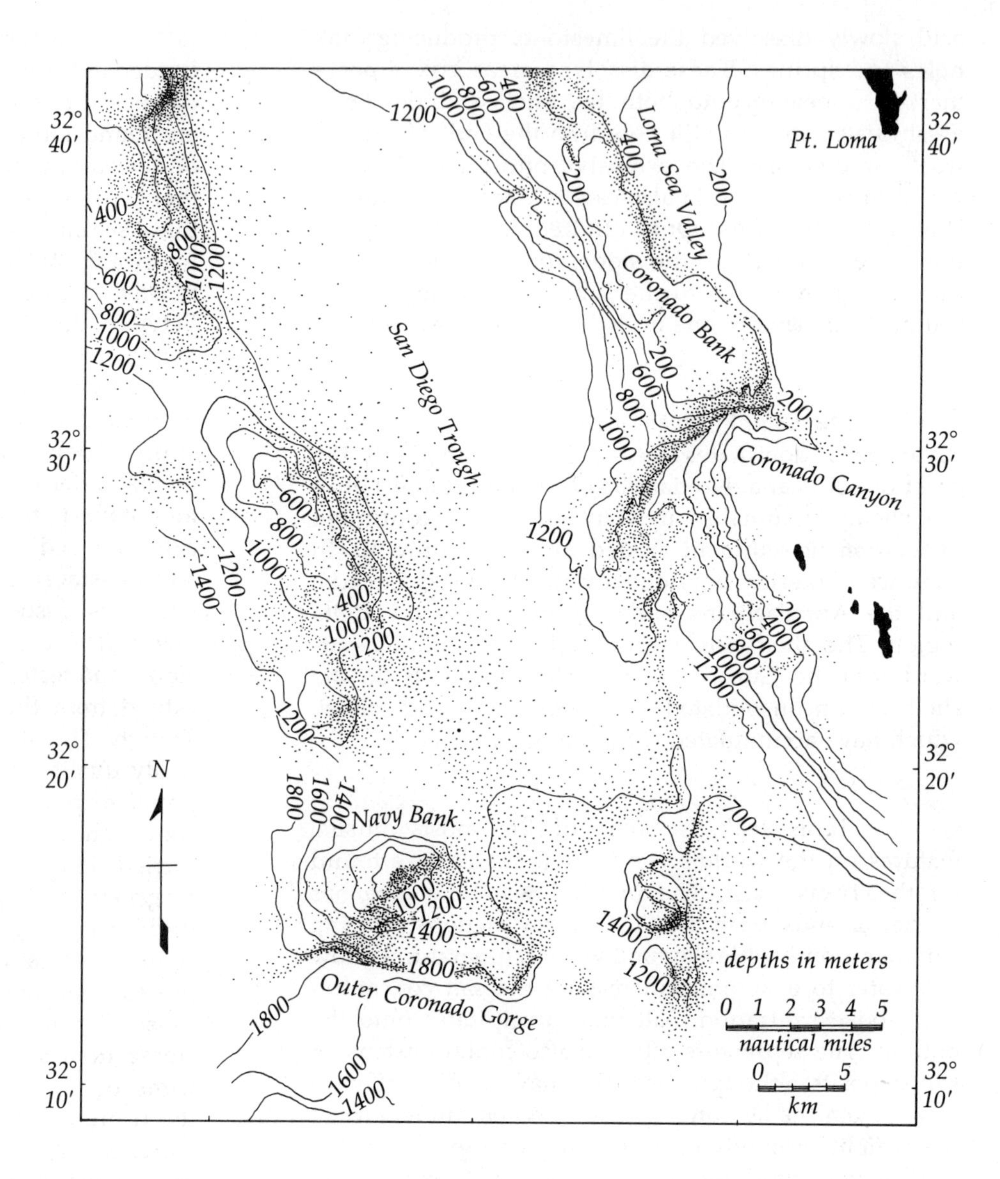

2–31 *Southern California borderland, showing depths in fathoms. (From* Submarine Geology, *Second Edition, by Francis P. Shepard, Harper & Row, 1963.)*

Submarine Canyons

Definition. The continental slope is the region of increased gradient (from a few degrees for the shelf to 7° to 15°) that extends seaward from the continental shelf and down to the ocean floor. These slopes are really the outer parts of the continental shelves and, like the shelves, their features are primarily depositional and tectonic.

Since World War II, the U.S. Navy, the U.S. Coast and Geodetic Survey, and a number of oceanographic educational and research institutions have been investigating the submarine bathymetry and topographic features on the ocean floor. Special attention has been given to the continental shelf and slope because of its military importance, its proximity to land, and its commercial exploitative potential. This extensive exploration has revealed a host of topographic features. The old idea that the sea bottom is flat has been completely disproved. Among the most impres-

2–32 *Divers inspect the wall of a submarine canyon. (Official United States Navy Photograph.)*

sive of these features are the submarine canyons cut into the continental slope (Fig. 2–32). The presence of such canyons off southern California has already been mentioned.

Almost all continental shelves that have been investigated have canyons cut into them. Some may extend many kilometers from shore and can be traced to depths of 1,800 m. Best known are canyons off California, New York, Colombia (South America), southern France, the Mediterranean shore of North Africa, and the Congo.

Formation. In general, the processes forming the canyons are not well understood but are presently being investigated. On land, erosion and faulting are the basic causes of canyons. Marine geologists assume these causes are also important in the formation of submarine canyons. Again as on land, canyons in different regions can probably be expected to have differing developmental histories.

The developmental histories of submarine canyons are thought to involve a number of processes. First, the canyons may have been eroded by rivers or glaciers

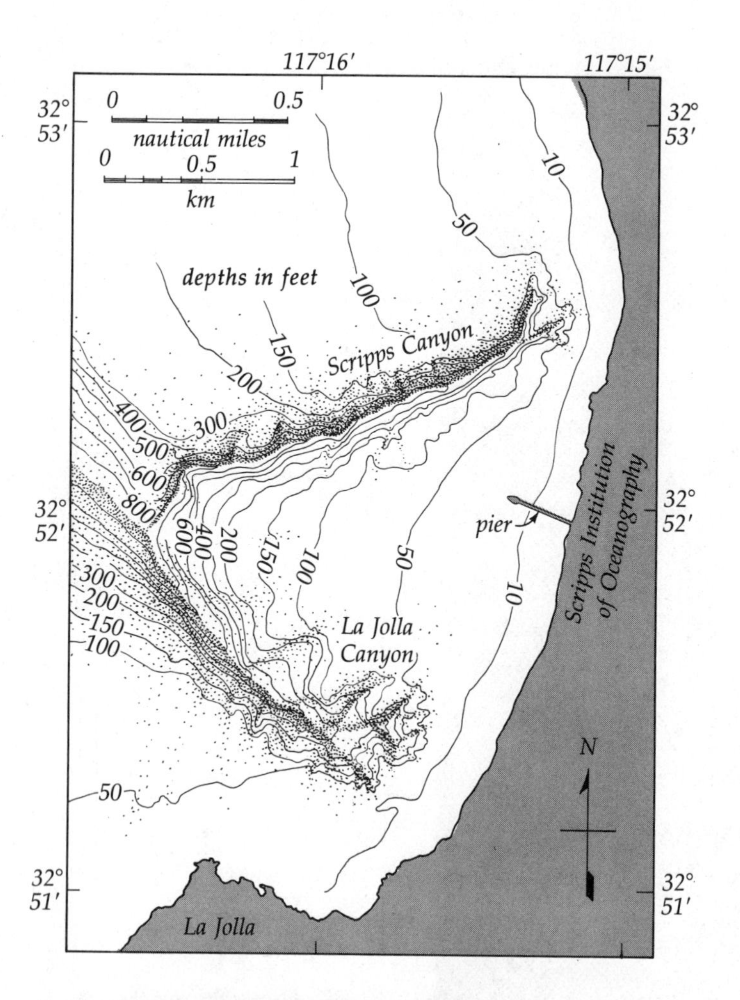

2–33 *Scripps and La Jolla canyons. (From* Submarine Geology, *Second Edition, by Francis P. Shepard, Harper & Row, 1963.)*

when the continental shelf was exposed in the Pleistocene. Sea level is known to have fluctuated appreciably before the Pleistocene, also. An interval of 150 m in the past million years must have had some effect. Second, sediment moving down the canyons from the shoreline may be capable of eroding the canyon walls and bottom by abrasion. Submerged cameras in a few canyons have confirmed that sediment moves and erodes in the canyons. The observed process has been slow and small-scale, hardly enough to imagine that it is responsible for cutting a canyon 500 m deep. One suggestion has been that earthquakes cause large-scale turbidity currents that carry millions of tons of sediment downslope, producing extensive erosion in a short period of time. These great currents have not yet been observed at sea although small-scale currents have been observed in Lake Mead. Thick graded layers at the base of the continental slopes are consistent with such an idea. Third, both of the previous processes may contribute to canyon formation. The canyons, for instance, may originally have been eroded when the shelf was above sea level and then prevented from filling and lightly scoured by moving sediment. Fourth, tectonic processes may have played a role. The continents themselves have been uplifted and downwarped extensively, producing mountains 8,800 m high and valleys 400 m below present sea level. Faults resulting from crustal movements have produced California's Imperial Valley and Wyoming's Grand Tetons. Why could not faulting also raise and lower the continental shelf? Large fault and fracture systems are known to weaken rocks, making them more susceptible to erosion. Thus many canyons form as a direct result of faults or fractures. No satisfactory conclusions can yet be drawn.

Types

Scripps–La Jolla Canyon. One of the best-known canyons is located off La Jolla, California, just north of San Diego. Before the canyon was discovered, the internationally famous Scripps Institution of Oceanography was built on the shore between the two heads of the canyon. The head of the northern canyon, Scripps Canyon, is about 1 km north of the institution and extends almost to the surf zone at Black's Cove. The presence of the cove and canyon contributes to the development of well-formed waves, providing some of the best surfing in the continental United States. At its head, Scripps Canyon is only about 2 m wide. Its walls slope steeply to a depth of 20 to 30 m and are covered with many organisms such as sea anemones, rock oysters, and abalone. Scripps Canyon acts as a drain for the sediment that is continuously moved southward from Oceanside, California, by longshore currents. This sediment accumulates at the head of the canyon and eventually slides down.

The southern tributary, La Jolla Canyon, has its head about 200 m offshore and 1.5 km south of Scripps Institution of Oceanography. The region of the head is characterized by eroded bedrock ledges and mud.

2–34 Sediment fan at the mouth of the San Lucas submarine canyon at the tip of Baja California, Mexico. (Official United States Navy Photograph.)

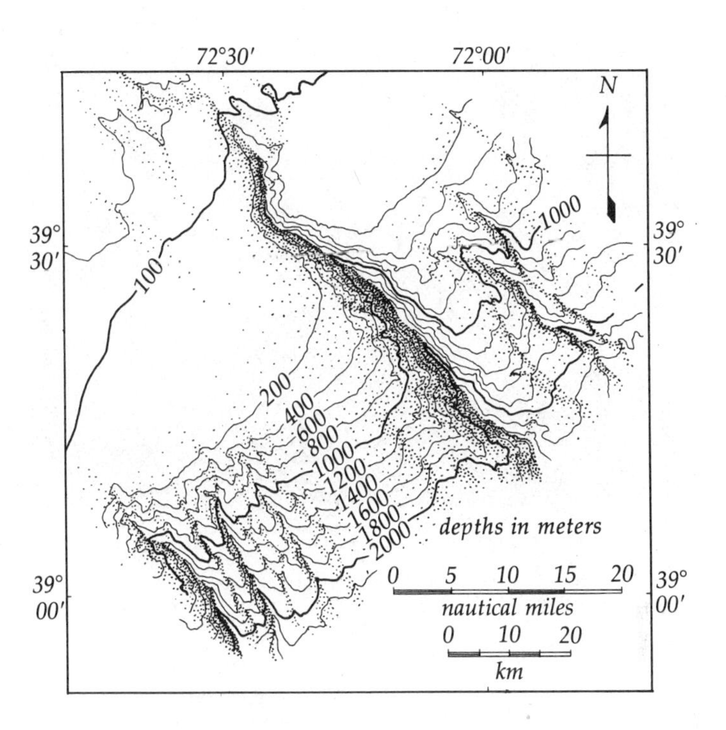

2–35 Hudson Canyon. (From Submarine Geology, *Second Edition, by Francis P. Shepard, Harper & Row, 1963.)*

Farther offshore, it assumes characteristics similar to Scripps Canyon (Fig. 2–33).

About 1,500 m offshore, La Jolla Canyon and Scripps Canyon merge. At this junction, the depth of the bottom of the canyons is approximately 250 m. The canyon meanders across the continental shelf another 12 to 15 km westward, then swings south to empty into the San Diego Trough at a depth of 800 m. A broad fan-shaped delta of sediment spreads from the base of the canyon into the trough (Fig. 2–34). It contains sediments and shells from the nearshore and is reworked by deep sea burrowing organisms.

Hudson Canyon. The Hudson River has drained eastern New York and Vermont for several million years. The valley through which the river flows was formed by a fault and fracture system that extends from the continental shelf south of New York into Canada. During the Pleistocene, periodic advances of ice sheets have further gouged the valley.

A submarine contour map of the continental shelf southeast of New York shows that the Hudson River valley can be traced across the shelf (Fig. 2–35). The valley deepens at a relatively constant rate offshore to a depth of about 100 m. There its slope steepens and widens. At 200 m below sea level, the valley is 20 km from shore; 50 km from shore, its depth reaches 3,000 m and the Atlantic Ocean floor. The valley, then, becomes a spectacular canyon about 20 km from the present shoreline.

San Lucas Canyon. San Lucas Canyon is located off the southern tip of Baja California. It is similar to the Scripps–La Jolla Canyon complex in topographic expression. The unusual feature is that granite walls have been found in deep parts of the canyon. Slowly moving sand may erode sedimentary rock, but it is hard to believe that trickles of sand could erode granite cliffs, even over a long period of time (Fig. 2–36). Based on the characteristics of this

2–36 Sandfall over a scarp in the San Lucas submarine canyon. (Official United States Navy Photograph.)

canyon, the idea that the canyons were cut above sea level and subsequently lowered below sea level gains credibility.

Tongue of the Ocean. The "Tongue of the Ocean," splicing Andros Island in the Bahamas, is one of the most anomalous features in the entire ocean. It differs from most canyons because it does not cut into a continental shelf. At the same time, it is not a trench either because it is not associated with volcanism or earthquakes, does not parallel a continent, and is not arc-shaped. It is, however, deep and spectacular: 30 km wide, 220 km long, and more than 4,000 m deep. All the rocks in the region are limestone, as are the rocks outcropping in the canyon. The most widely accepted theory of its origin is that it is a block fault basin similar to the Imperial Valley in California. How it formed is still a major question.

Obtaining Ocean Bottom Data

When one appreciates how deep the ocean is, he cannot fail to be impressed by the amount of ocean bottom data that man has managed to acquire. Compilation of data is continuing at an accelerated rate. Most of the research is funded by national governments and is carried on by scientists employed by educational institutions, major research institutions, and navies. This approach has been adopted because of the high cost of ships. The data now being compiled include the depth, composition, physical and chemical properties, fossils, and living organisms of the rocks found on the ocean bottom.

Depth Recording. The first accurate physical data obtained from the ocean bottom were depth records. Scientists on H.M.S. *Challenger* obtained depth recordings in the Atlantic, Pacific, Indian, and Antarctic oceans using steam-driven winches designed by the famous physicist Lord Kelvin and one inch circumference hemp rope that would not tangle. Paying out 3,000 m or more of rope and reeling it back in could take as much as twelve hours.

By comparison with the *Challenger*'s method, sonar recordings can now be made by using a depth recorder to make a graph of the bottom. The sonic depth recorder works as follows: (1) a sound or "ping" is sent toward the bottom; (2) the "ping" reflects off the bottom, and (3) a transducer on the ship's hull receives the reflected ping; (4) a timer records the time from transmission to reception; (5) the depth is calculated by multiplying the time by the speed of sound through water (a known quantity): the distance traveled by the ping is twice the depth from the vessel to the bottom. Inaccuracies may result when the ping is reflected off the surfaces of water masses having different densities, schools of fish or other organisms, or what appears to be loosely packed mud and ooze on the bottom. Generally, these inaccuracies produce identifiable patterns on the graph and can be eliminated by the reader. When sound waves penetrate loose muds and reflect off deeper, sub-bottom consolidated rocks, depth recording becomes difficult, but at the same time new information about the sediments underlying the bottom can be obtained. Because sound travels through materials of different densities at predictable speeds based on controlled laboratory studies, the density of the rocks on the bottom can be deduced. Knowing the density, some indication of the composition of the rocks, the degree of cementation, and the degree of compaction can be interpreted.

2–37 *Marine geologists from the University of California's Scripps Institution of Oceanography bring up a light weight shell dredge aboard the research vessel* Spencer F. Baird *during Vermilion Sea expedition to the Gulf of California in 1959. (Courtesy of Scripps Institution of Oceanography, University of California, San Diego.)*

2–38 *Standing in the bucket on the side of a Scripps Institution of Oceanography research vessel, a Scripps scientist starts a Peterson grab sampler on its trip to the ocean floor. The sampler will trip on striking the bottom and then close. (Courtesy of Scripps Institution of Oceanography, University of California, San Diego.)*

Dredging. Even in the days when the Roman Empire controlled the Mediterranean area, nets were dragged by fishermen across shallow sea bottoms to catch crabs and fish. This ancient technique has been adapted to enable scientists to sample organisms and bottom sediments in deep oceanic water. A standard biological dredge consists of a net made of metal chain or nylon mesh in a scoop shape, with the open end attached to a rigid metal frame (Fig. 2–37). A line is attached to the metal frame and held on shipboard. The dredge is lowered and dragged along the bottom. An electric winch then hoists the dredge with its load. This technique is inexpensive and easy to operate. The drawbacks of this method are: small samples of rocks or organisms pass through the mesh; fragile items often break; and precise locations of the samples and their interrelations are impossible to interpret. To overcome these drawbacks, grab samplers have been designed especially for geologic research (Fig. 2–38). Grab samplers have spring-loaded trap doors held open by clamps as they are lowered to the bottom. When the sampler hits bottom, the force of its impact releases the

2–39 Six-foot gravity corer on a research vessel of University of California's Scripps Institution of Oceanography. The corer takes a sample of the bottom sediments for examination by marine geologists. (Courtesy of Scripps Institution of Oceanography, University of California, San Diego.)

clamps. The trap doors snap shut, grabbing a sample of sediment and small or fragile organisms from a single location.

Cores. To interpret the geologic history and origin of the ocean basins, techniques were needed to recover samples of the subbottom for laboratory studies. Because dredges only sample surface mud and clay and sonar pinging produces no samples, coring devices have been designed. One such device is the gravity corer (Fig. 2–39). It is a hollow pipe about 8 cm in diameter with a sharp, core-cutting lower end that is smaller in diameter than the rest of the pipe. The pipe sometimes has fins that enable it to fall vertically. Usually a pilot weight precedes the pipe. When the pilot weight hits bottom, a messenger releases the pipe to fall into the sediment. A loose line attached to the corer is used to haul it up. Such a technique is useful for cores up to a few meters in length. Another device, the piston corer, can recover cores up to 12 m in length. It is structurally similar to the gravity corer but is longer and has a piston, which helps prevent compaction as the corer enters the sediment. A third technique now being developed is platform drilling from a stationary platform or ship using essentially oil drilling apparatus (see Fig. 8–12). Such a ship, the *Glomar Challenger,* was built in 1968 especially for deep drilling and coring. To date, cores more than 1,100 m long have been recovered.

Seismic Profiling. Because deep ocean coring using ships like the *Glomar Challenger* is so recent and still so expensive, other techniques for acquiring information about the rocks under the ocean have been designed. The technology developed for land-based oil exploration has again been adapted. Explosion detonations produce sound waves that travel laterally across a whole ocean basin and vertically through the ocean water and rocks beneath. A seismograph on shipboard receiving the sound waves can collect large amounts of data in a short period of time at little relative expense. Sound waves move through sand at one speed, through limestone at another, and through simatic crystalline rocks at still another speed. Each speed can be determined in a laboratory before field work, and the subbottom rocks can be deduced by comparing known speeds to the recorded speeds. The recently developed air-gun seismic profilers cost less than explosives in the long run, are less dangerous on shipboard, and are less damaging to marine life (Fig. 2–40). However, air-gun profilers have less bottom penetrating ability than explosives.

Heat Conduction. Recent theories regarding ocean basin and earth crust movements have predicted differences in heat radiation in various parts

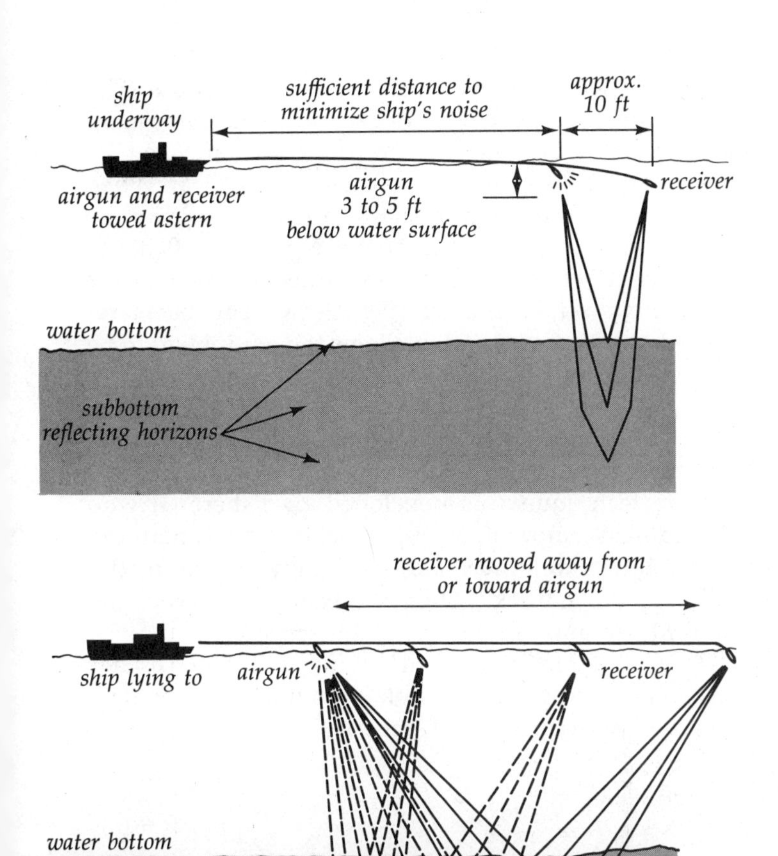

(a)

(b)

2–40 (a) Seismic profiling of subbottom sediments; (b) seismic air gun. (Photo by Roger L. Larson.)

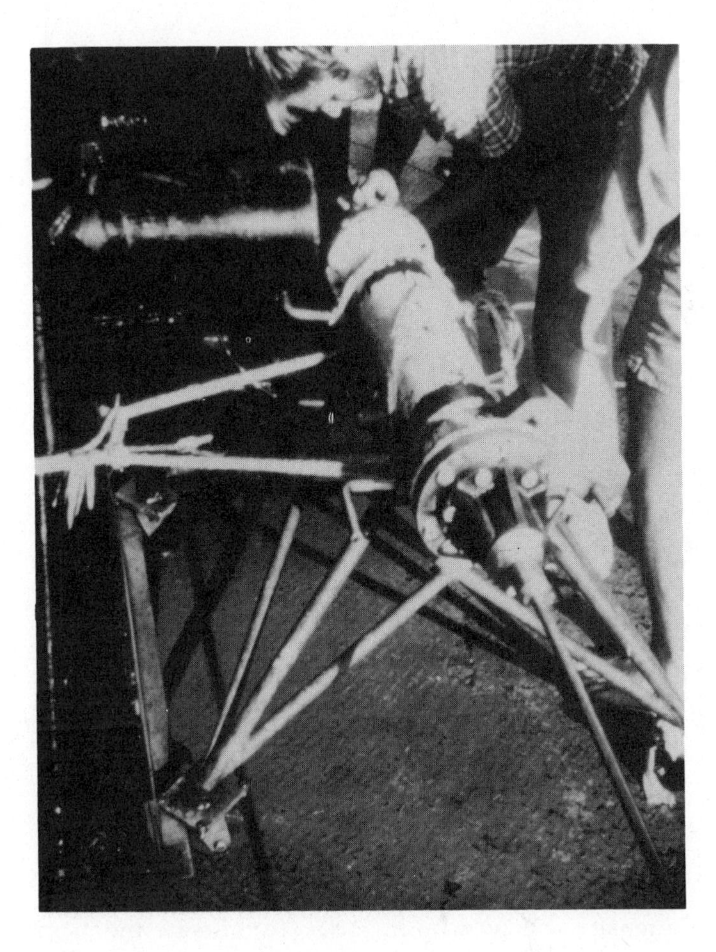

2–41 Heat probe. (Photo by Eric Abramson.)

of the earth (see p. 65). To test these theories, thermistors mounted on probes that can penetrate the ocean bottom have been built to measure the thermal gradient in the sediments (Fig. 2–41). The thermistor operates in the following way. An electric current flows along a metal conductor. The amount of current flowing is measured by a galvanometer and recorded. The resistance of the conductor to the amount of electricity flowing is proportional to the temperature of the conductor. Because the amount of electricity leaving the power source is known, and the resistance offered by the conductor at different temperatures has been previously recorded, the temperature of the thermistor as it lies imbedded in the sediment can be measured. It is assumed that the thermistor will attain a temperature equilibrium with the sediment. Then the temperatures of sediments can also be deduced.

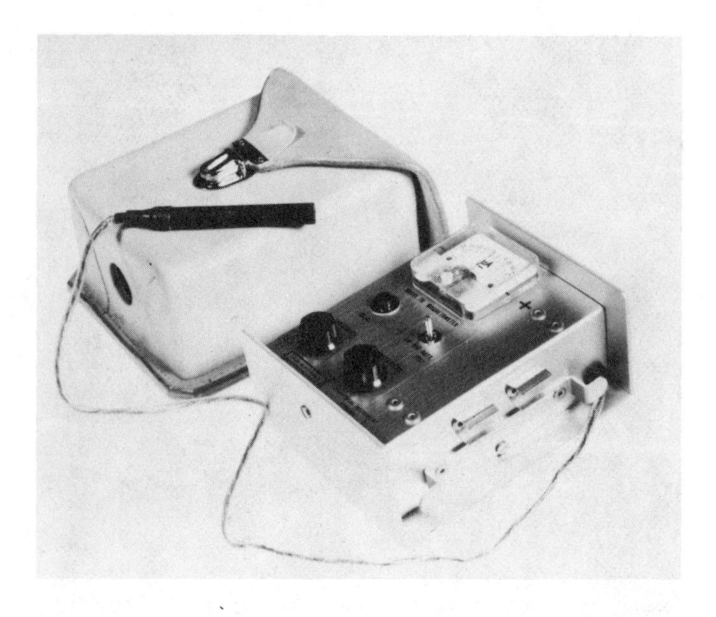

2–42 *Proton-free precession magnetometer. (Courtesy of California Electronics Manufacturing Company, Inc.)*

Magnetism. New theories regarding earth crust movements have also predicted patterns of magnetic intensity and polarity in rocks. To measure the magnetic field in the ocean basin, a proton-free precession magnetometer is towed beneath the sea surface behind a ship (Fig. 2–42). This instrument consists of a copper coil wrapped around a bottle filled with water. A direct current is passed through the coil, causing protons in the liquid to be aligned with the induced field. When the current is shut off, the coil continues to carry a weak current briefly because of the precession of the aligned protons about the earth's magnetic field. This weak current is amplified and measured with a digital counter. When the magnetometer passes over crystalline rocks containing different magnetic properties, the differences can be measured.

Information regarding the magnetic properties of the crystalline rocks can include the relative abundance of magnetic minerals and the direction and magnitude of the magnetic field at the time the minerals cooled below the Curie temperature (575°C).

Free Vehicles. Another recent development has increased the feasibility of many other techniques. Free vehicles have been used to lower cameras (Fig. 2–43), traps, cores, heat probes, and other equipment to the ocean bottom. The equipment is attached by some release mechanism to a weight and a float. When the release mechanism unsnaps or dissolves (as in the case of magnesium releases) the equipment floats to the surface where a radio transmitter attached to the equipment sends a beep (Fig. 2–44). The beep, received by shipboard radio, allows the equipment to be located and recovered. The advantage of free vehicles is that the ship does not have to remain on station, heavy winches do not have to be manned, and no cables are needed to hang onto the equipment. The free-vehicle technique was developed by fishermen who tied capped, empty jugs with attached lines containing baited hooks to a piece of sugar candy, which in turn was tied to a brick. When the candy dissolved, the hooked fish and jug bobbed to the surface.

Visual exploration from submersible vehicles has also become important in obtaining ocean bottom data and will be discussed in Chapter 8.

Abyssal Plains and Seamount Provinces

Abyssal plains are broad, extremely flat areas of the ocean basins or adjacent seas. They are found on the floor of all the oceans but vary in depth from about 4,000 to 5,000 m. Topographically, they are usually bounded by oceanic ridges, sills, or continents. Many submarine canyons funnel out to abyssal plains where large fan-shaped sediment aprons spread onto the plain. Such broad, deep ocean deltas are particularly well-developed on abyssal plains in the Indian Ocean off the Ganges and Indus rivers. The fans probably were produced by sediments transported seaward down the submarine canyons of the rivers by turbidity currents or other types of gravity sliding. Confirming this transport, cores from the fan sediment contain both deep water and shallow water fossils and graded bedding.

Topography. The few topographic variations found on abyssal plains can be categorized as gently sloping sediment, isolated mountains, and earthquake fracture zones.

2–43 *Underwater camera and mounting. (Photo by Roger L. Larson.)*

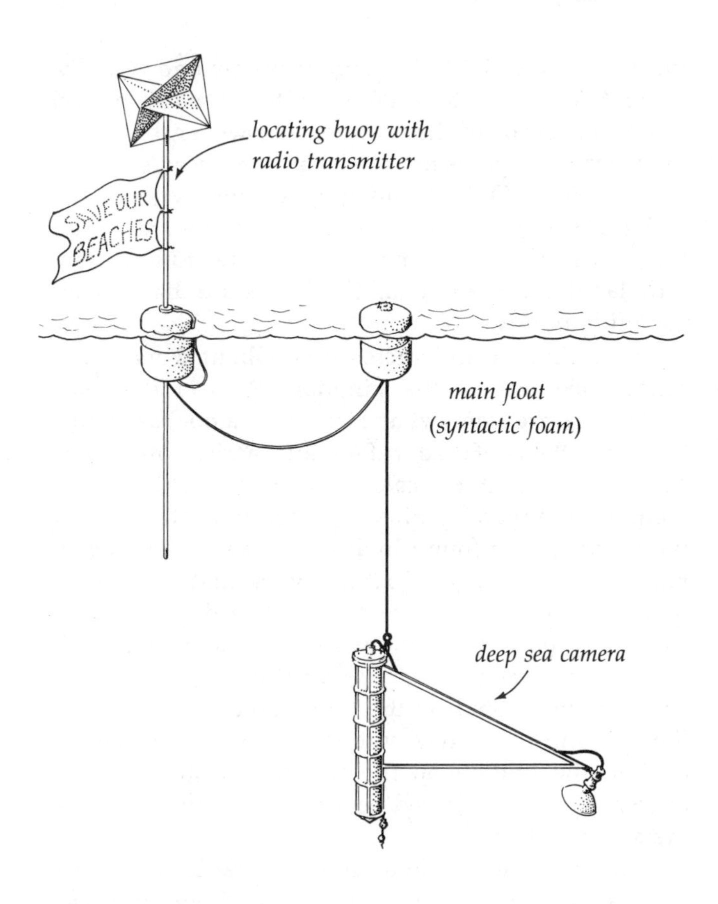

2–44 *Deep sea free vehicle photography system after returning to the surface. (After Scripps Institution of Oceanography, University of California, San Diego.)*

Most of any abyssal plain is gently sloping and covered with relatively unconsolidated sediments. The sediments explored to date have been from 30 to more than 1,000 m thick, presumably underlain by crystalline simatic rocks. The abyssal plain sediments are largely composed of tiny rock particles the size of clay or silt. Sand grains or larger particles are not commonly found except in canyon fans. Some of the small clay particles were carried out into the ocean basin by currents that were strong enough nearshore to pick them up from river mouths, keep them in suspension, and deliver them far out at sea. Some of the particles slid down submarine canyons and were distributed on the plain by bottom currents. Wind probably carried many small particles from desert areas like the Sahara out to sea, where the particles began a slow descent to the ocean floor. Also, some of the particles are volcanic ash that was blasted into the air, where wind transported it to the ocean.

Minerals. The extent to which minerals form on the abyssal plain by precipitation out of seawater or by chemical changes on particles already on the bottom is unknown. In a number of locations on abyssal plains, iron/manganese nodules lie scattered in concentrated beds. Such nodules were first discovered during the *Challenger* expedition. One such area of nodule concentration is southwest of southern California. The composition of the nodules consists mainly of iron oxides and manganese oxides arranged in concentric rings. How these nodules formed is not fully understood. One theory suggests they precipitated out of seawater. This process probably involved the formation of manganese- and iron-rich colloids near the sea surface. As currents carried the colloids to the ocean floor, they settled on solid materials, which acted as a "crystal seed." A number of other theories, however, have also been suggested. These nodules occur in sufficient number to be of interest to mining firms.

Sediment. The sediment on the abyssal plain contains many shells of marine organisms, in some cases comprising more than 50 percent of the sediment measured by weight. Few organisms are known to live at depths of 2,000 to 3,000 m, so few remains of such organisms occur in the sediment. However, billions of microscopic organisms live near the ocean's surface (see Fig. 5–5). Of these organisms, three types of animal shells and two types of plant remains are, by far, the most common. The animals are foraminifera, radiolaria, and pteropoda. The plants are diatoms and coccolithophores.

Foraminifera are microscopic (about 2 mm) one-celled members of the kingdom Protista that have shells and are somewhat similar to amoebas, which have no shells. Foraminifera are marine organisms, however, that secrete calcium carbonate shells or cement tiny mineral grains together into shells called tests. The genera found in abyssal plain sediments are mostly planktonic or floating types that live in the upper 100 m of ocean water and are distributed by surface currents throughout the oceans. Sinking currents, reproduction, and predators cause the globular shells to be vacated by the living organism and allow the shells to fall slowly onto the bottom. Some genera live on the bottom on the continental shelf and are known to be transported down submarine canyons to the abyssal plain.

Closely related to foraminifera are radiolaria, which are also microscopic plankton. The shells of radiolaria are composed of silica (hydrous silicon dioxide). These shells often have long, elaborate, fragile spines radiating from a central chamber. They are distributed on the abyssal plain in the same way as foraminifera.

A third animal shell sometimes found in sediments on the abyssal plain belongs to the pteropod or sea butterfly. Pteropods are members of the phylum Mollusca, class Gastropoda, and are small, swimming snails with wide oceanic distribution. This animal is much larger than the foraminifer, being up to 4 cm in length. It swims by waving a powerful, wing-like "foot." Pteropods reach the ocean bottom in the same way as foraminifera, but are not as widely distributed in the sediments.

Diatoms are microscopic, single-celled, photosynthetic, planktonic plants that have an outer protective coating of silica. They live in the upper 100 m of ocean water where they benefit from sunlight. Although they are almost ubiquitous in both fresh and salt water, the types found in abyssal plains are marine. They, also, have world-wide distribution as living organisms.

A less well-known plant commonly found in deep sea sediments is the coccolithophore. Like the diatom, it is a single-celled, photosynthetic plant with a hard coating. Its shell is composed of calcium carbonate.

The world-wide distribution of foraminifera (carbonate, $CaCO_3$) and radiolaria and diatom (siliceous, SiO_2) oozes, or clay-to-silt sized sediment, reveals some distinct patterns (Fig. 2–45). Carbonate oozes are most common in tropical regions, while siliceous oozes are most common in temperate regions. In addition, carbonates are not abundant in sediments where the ocean bottom is deeper than about 4,000 m. The reason for these patterns is related to the dissolution of carbonates in the cold, deep waters due to increasing pressure and decreasing temperature. If ocean water is saturated with the carbonate ion or is not very cold, calcium carbonate shells will not dissolve. If the water is deficient in carbonate or cold enough, the shells will dissolve. It is assumed, then, that deep water is either deficient in carbonate or is cold enough to cause dissolution because foraminifera shells dissolve and are not found in the sediment.

Submerged Mountains. Submerged mountains, although not dominant features on the abyssal plains, are not uncommon. Some, called abyssal hills or knolls, are quite small, and have not been thoroughly investigated. A number of those rising more than 1 km above the abyssal plain have been cored and dredged. Generally, these mountains fall into two groups, seamounts, which are rounded or irregular on top, and guyots, which are flat on top. All seamounts and guyots thus far investigated are extinct volcanoes. Even though they rise 1 km or more above the plain, their tops may still be considerably below sea level. In spite of this, fossils of shallow water organisms like corals and *Tridacna*, the giant clam (see Fig. 5–16), have been dredged from some of their surfaces. This is strong evidence that they were once located near the surface and have since moved below the surface. In some areas, as south of the Gulf of Alaska, seamounts are so common that the area is called a seamount province. An interesting feature of seamounts is their chain-like pattern, forming a line of seamounts across the ocean bottom. In the Pacific, most of these chains have an approximate NW–SE orientation.

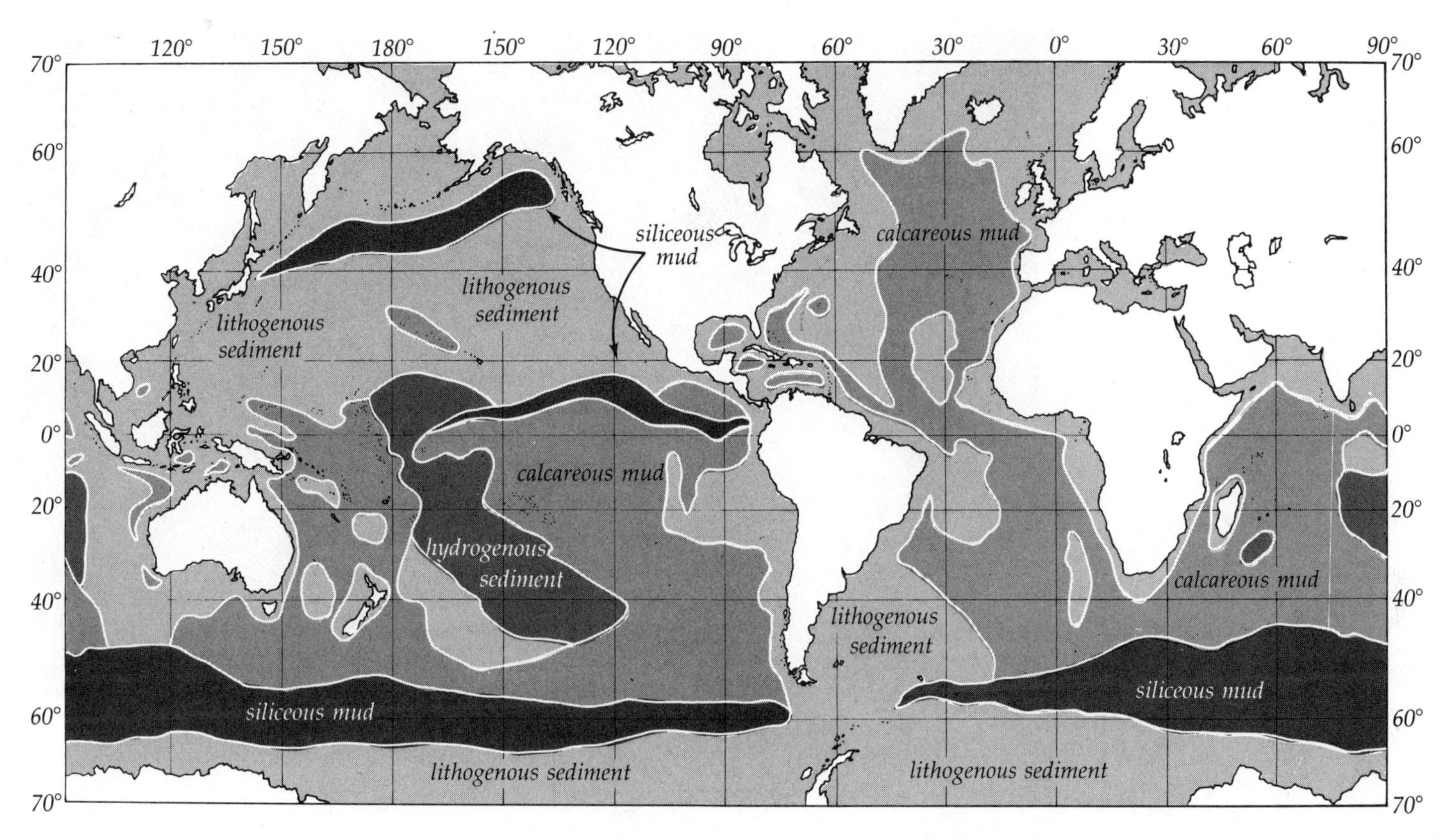

2–45 *Distribution of ocean bottom sediments. (After* Goode's Base Map Series, *Department of Geography, University of Chicago.)*

Fracture Zones. Also located on abyssal plains are large fractures in the rocks on the ocean bottom. On the abyssal plain off southern California, these fractures can be traced for over 500 km at roughly a right angle to the west coast of North America. Some of the fractures also have vertical displacement of up to 200 m between the north and south sides. Their significance and origin is a topic of current research. Present theories relate them to large-scale earth crust movements (see p. 63).

Coral Atolls. In tropical regions, especially in the Pacific Ocean, scattered islands are bordered by living coral reefs. Many of these islands rise from abyssal depths. Many people are fascinated by these islands, mostly for a variety of romantic reasons. Scientists are also fascinated by these islands because their origin has been a point of debate since Captain Cook on the voyage of H.M.S. *Endeavor* (1768–69) first described several between Tahiti and New Zealand.

During the voyage of H.M.S. *Beagle* (1831–36), Charles Darwin made an incredible number of carefully recorded geological and biological observations. He is, of course, best known for his treatise *On the Origin of Species by Means of Natural Selection, or the Preservation of Favoured Races in the Struggle for Life* (1859). One of his other major, but less well-known, contributions was his description of many coral islands in the Pacific and Indian oceans and his analysis of their origin. He later published a book about this entitled *The Structure and Distribution of Coral Reefs* (1842).

In the Pacific, Darwin observed many volcanic islands, some of which were active. He therefore assumed that all islands in the Pacific had a volcanic origin. In traveling westward, he noticed the islands change in character from volcanic, like the Galapagos, to doughnut-shaped coral atolls (Fig. 2–46). He reasoned that the evolution of coral atolls was somehow

2–46 Coral atoll. A lagoon appears in the lower half of the photo. (Photo by Dale E. Ingmanson.)

related to volcanic activity. In his theory he suggested the following sequence of events: (1) a volcanic island (Fig. 2–47) formed; (2) a fringing reef developed on the flank of the island just below sea level (Fig. 2–48); (3) the island began to sink or sea level began to rise and the volcano became inactive; (4) the coral grew upward forming a barrier reef, while broken coral debris and other calcium carbonate shell material accumulated in the lagoon leeward of the reef; (5) eventually the volcano became submerged and completely covered with calcium carbonate debris, and the coral reef continued to grow, producing a coral atoll (Fig. 2–49).

Authors have noticed with surprise, that although atolls are the commonest coral-structures throughout some enormous oceanic tracts, they are entirely absent in other seas, as in the West Indies: we can now at once perceive the cause, for where there has not been subsidence, atolls cannot have been formed; and in the case of the West Indies and parts of the East Indies, these tracts are known to have been rising within the recent period. . . . Taking into consideration the proofs of recent elevation both on the fringed coasts and on some others (for instance, in South America) where there are no reefs, we are led to conclude that the great continents are for the most part rising areas; and from the nature of the coral-reefs, that the central parts of the great oceans are sinking areas. The East Indian archipelago, the most broken land in the world, is in most parts an area of elevation, but surrounded and penetrated, probably in more lines than one, by narrow areas of subsidence.

. . . Throughout the spaces interspersed with atolls, where not a single peak of high land has been left above the level of the sea, the sinking must have been immense in amount. The sinking, moreover, whether continuous, or recurrent with intervals sufficiently long for the corals again to bring up their living edifices to the surface, must necessarily have been extremely slow. This conclusion is probably the most important one which can be deduced from the study of coral formations;—and it is one which is difficult to imagine, how otherwise could ever have been arrived at. Nor can I quite pass over the probability of the former existence of large archipelagoes of lofty islands, where now only rings of coral-rock scarcely break the open expanse of the sea, throwing some light on the distribution of the inhabitants of the other high islands, now

2–47 *Crystallized lava flow at the edge of the sea. (Courtesy of the National Oceanic and Atmospheric Administration.)*

left standing so immensely remote from each other in the midst of the great oceans. The reef-constructing corals have indeed reared and preserved wonderful memorials of the subterranean oscillations of level; we see in each barrier-reef a proof that the land has there subsided, and in each atoll a monument over an island now lost. We may thus, like unto a geologist who had lived his ten thousand years and kept a record of the passing changes, gain some insight into the great system by which the surface of this globe has been broken up, and land and water interchanged.[1]

2–48 *A fringing reef. View from Annaberg ruins of Leinster Bay and Mary Point, U.S. Virgin Islands. (Courtesy of the United States Department of the Interior, National Park Service Photo.)*

Near the end of World War II, Dr. Harry Hess (now deceased), a professor of geology and geophysics at Princeton University commissioned for the duration of the war in the U.S. Navy, began to lobby for permission to drill cores into atolls to test Darwin's theory. Shortly after the war, Hess's suggestion was adopted and, since then, a few atolls have been cored. Each atoll drilled had a volcanic core buried under as much as 800 m of limestone debris. Basically, Darwin was correct. The major question still not completely answered remains, "How did the volcanic core move to a lower position relative to sea level?"

Island Arcs and Trenches

One of the most exciting features found in the ocean basins is the island arc. Island arcs are exciting because a majority of the active volcanoes in the world are located in them. As the name suggests, the overall pattern of the island groups is an arc (Fig. 2–50). Individual islands have a variety of shapes, the most common of which is roughly circular as a result of the

---

[1]Charles Darwin, *The Voyage of the Beagle*. Perhaps the most readily available edition is by Anchor Books (New York, 1962). The book was originally published in 1839.

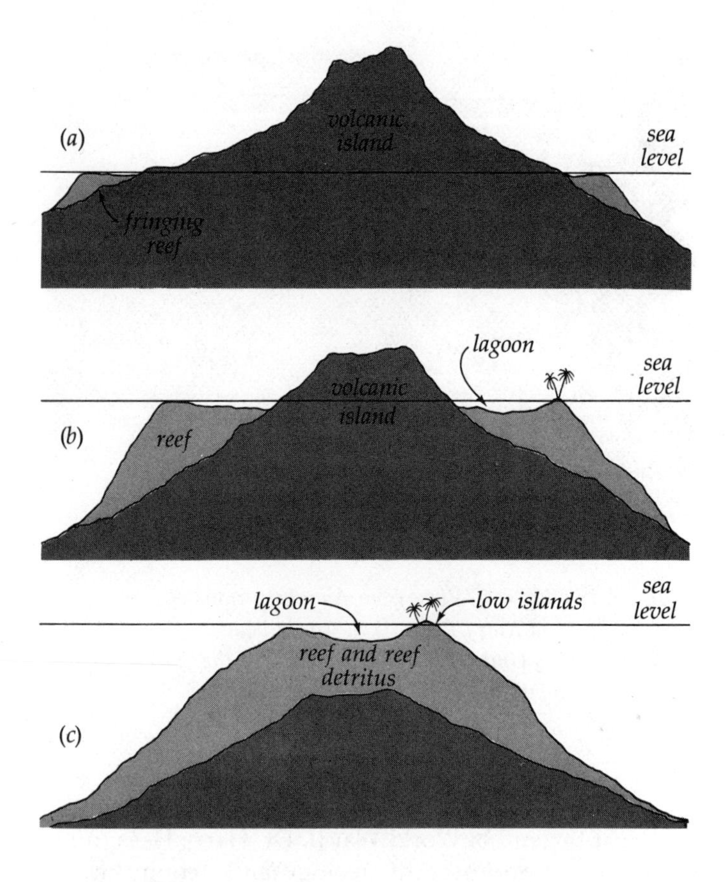

2–49 *Sequence of coral atoll development according to Darwin. (a) Fringing reef; (b) barrier reef; (c) atoll.*

volcanic origin. In most cases, the concave side of the arc faces a continent and the convex side faces an ocean basin. Usually island arcs are located near the outer edges of a continental shelf. Besides volcanic islands, the arcs also generally have deep trenches on the convex or oceanic side. The general pattern from a continent oceanward includes the continent, a relatively shallow marginal sea covering the continental shelf, the volcanic islands, a trench, and the ocean basin.

Distribution. Geographically, most of these island arc–trench systems fringe the Pacific or separate the Pacific from the Indian Ocean. Exceptions include the West Indies, separating the Caribbean Sea from the Atlantic Ocean, and the South Sandwich Islands, at the southern tip of South America. In the Indo-Pacific province, the Aleutian Islands, Kuril Islands, Japan, Ryukyu Islands, Philippine Islands, and the Malay archipelago are a few examples of island arc–trench complexes. This distribution of volcanic islands, many of which are active, is the reason the Pacific Ocean boundary is called the "Ring of Fire." Rarely is there a year without a major volcanic eruption. Eruptions are especially spectacular because they are frequently explosive, producing fiery showers of molten masses, lava flows, clouds of water vapor and other gases, extensive damage to vegetation, and, occasionally, loss of human lives.

Composition. The composition of the molten material, called magma, varies according to the location of the volcanic island. The general composition of the islands ringing the Pacific Ocean is andesitic. The term *andesitic* indicates that the mineral crystals formed when the magma cools are largely the same as those found in the rock andesite, which is dominant in the Andes Mountains of South America. These minerals include several varieties of feldspar, mica, quartz, and hornblende. Most common is a plagioclase feldspar that contains sodium. Quartz is present but not as abundantly as in sialic rocks. Iron and magnesium silicates are also present but not as abundantly as in simatic rocks.

Trenches. The trenches on the convex side of the volcanic islands closely parallel the islands' arcuate form. These trenches are usually named after the islands they border, for instance, the Aleutian Trench. They are rarely more than 130 km wide but may be 1,500 km long and more than 10,000 m deep. They are the deepest parts of the ocean basins. The deepest known spot in all the oceans, called the Challenger Deep, is in the Mariana Trench approximately half way between Guam in the Mariana Islands and Yap in the Caroline Islands of the western Pacific Ocean. This deep area is about 10,850 m below the surface of the ocean. If Mount Everest were cut off at sea level, inverted and inserted into the Mariana Trench, the ocean above it would still be 2,000 m deep. In 1961, the bathyscaphe *Trieste* (see Fig. 8–5), built under the direction of Jacques Piccard, descended to the bottom of the Mariana Trench with Piccard and a U.S. Navy

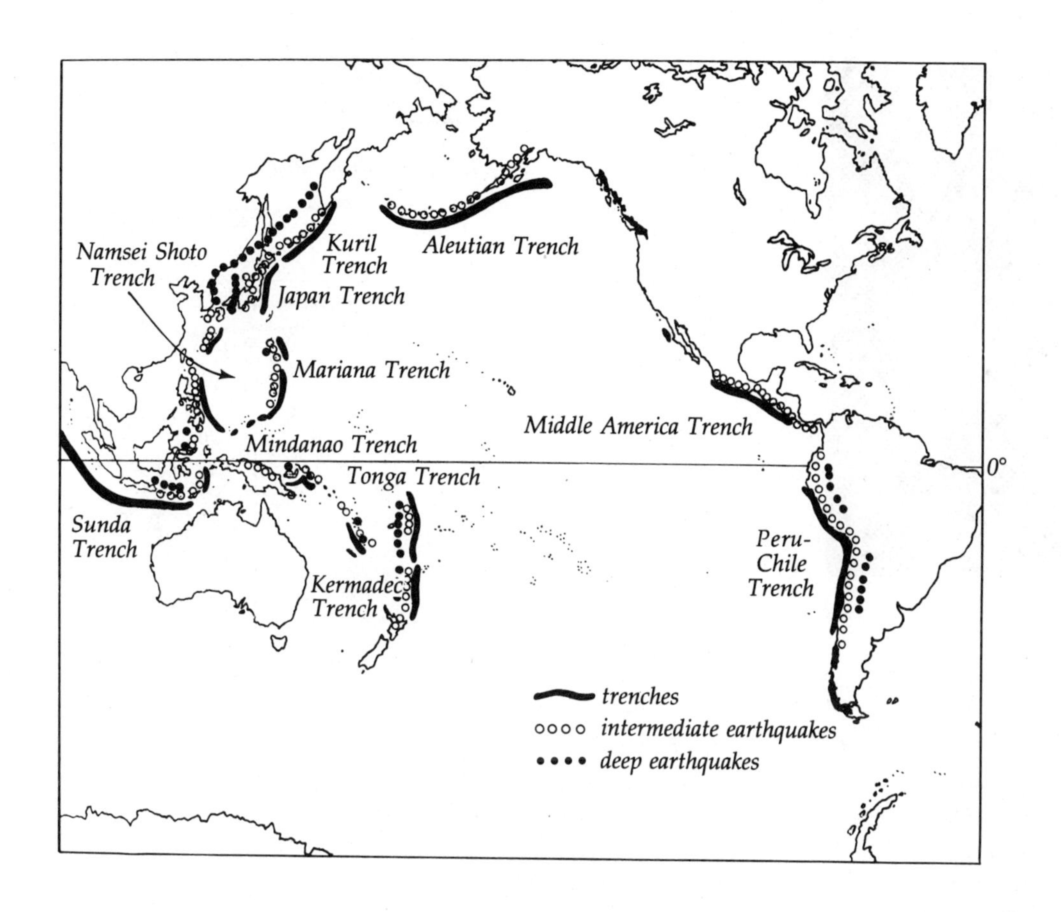

2–50 Island arcs showing earthquake foci. (After B. Gutenberg and C. F. Richter, Seismicity of the Earth and Associated Phenomena.)

officer aboard. They photographed ripple marks and the tracks and trails of organisms, indicating the presence of currents and oxygen even at such great depths.

Gravity Anomalies. To explain movements in the earth's crust, geologists have postulated that the distribution of mass is uneven. Volcanic activity is a type of crustal material movement. To measure the amount of mass per unit area in the crust of the earth, gravimeters have been designed (Fig. 2–51). Essentially, they measure precisely the acceleration due to gravity by measuring the period of a pendulum or the pull of gravity against a spring. Theoretically, adjacent parts of the crust should balance one another and attract the known mass equally if the distance to the center of mass of the earth is equal. The gravimeter checks this balance. Over the trenches, negative gravity anomalies have been measured, indicating that the crust there is deficient in mass. Positive anomalies have been found over the volcanic islands, indicating an overload of mass. The region of island arcs is therefore out of balance (Fig. 2–52).

An appreciable percentage of the earthquakes that occur in the earth's crust each year are located in the vicinity of island arcs. These earthquakes are usually deep-focus types, occurring at depths greater than 70 km below the earth's surface. The crustal imbalance suggested by gravity anomalies may be partly responsible for the earthquakes, as earthquakes are shock waves that emanate from earth movements. Another possibility is that sinking in the ocean basin crust causes both the gravity anomalies and the earthquakes. Because ocean water covers much of the earthquake zones, tsunamis frequently originate in the area (see p. 133).

Structure. A cross section through the island arc system shows the following characteristics: (1) the continent slopes normally to the shoreline; (2) the continental shelf is composed of sediments from 200

2–51 *La Coste-Romberg gimbaled gravimeter. (Photo by Bruce Leyendyk.)*

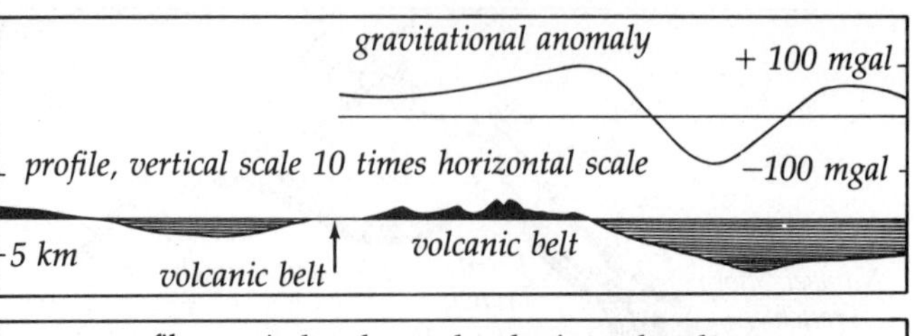

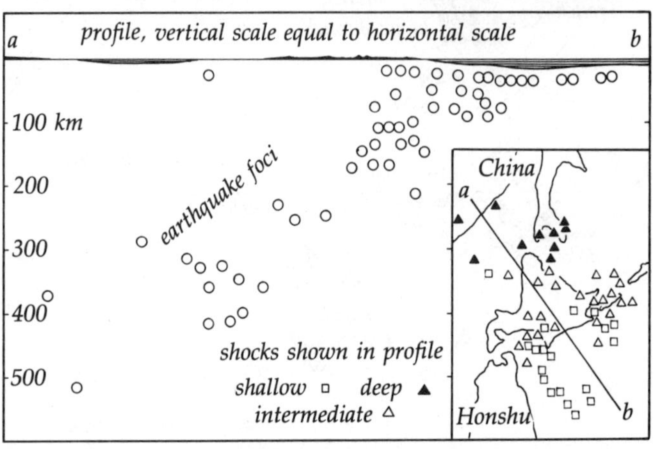

2–52 *Island arc cross section showing earthquake foci and gravity anomalies. (After B. Gutenberg and C. F. Richter,* Seismicity of the Earth and Associated Phenomena.*)*

to 1,000 m thick and has some organic reefs in tropical regions; (3) volcanic islands; (4) trenches; (5) a decrease in the thickness of the crust seaward (see Fig. 2–52).

When the origin points of earthquakes, earthquake foci, are superimposed on the island arc system cross section, a pattern is evident. The shallowest foci are located beneath the trenches and become progressively deeper toward the continent. The region seaward of the trenches is relatively earthquake-free. A graph representing the location of foci in an island arc system would be a curved plane extending from beneath the trench to a region beneath the adjacent continent.

Origin. Because magma rises to the surface forming volcanoes, the source of the magma must be somewhere beneath. It may originate near the intersection of a vertical line descending from the volcano and the zone of earthquake foci (Fig. 2–53). This idea is supported by the measurement of low-velocity seismic waves in the region of the intersection. The low-velocity waves indicate rocks of relatively low density. Liquids, for instance, are usually less dense than solids. Some scientists believe that the rocks may even be "plastic" and capable of flowing. Further confirmation is the lack of earthquake foci in the region of intersection, which indicates that the rocks are responding to crustal stresses by flowing instead of rigid shifting. Just what happens is still the subject of active research and discussion.

Heat Flow. Heat flow emanating from the rocks under the trenches is quite low and is similar to the heat flow found in abyssal plains (0.5 to 1.6 $mcal/cm^2$ sec). In the vicinity of the island arcs (excluding areas of active volcanics), the heat flow is about twice that found in the trenches. Although this comparison is predictable, the general heat flow around the islands is less than might be expected, considering the proximity of magma. Apparently the magma is quite localized at depth and flows to the surface along relatively narrow channels, minimizing the overall temperature effect on surrounding rocks.

Case Studies

West Indies. An exception to the general description of island arcs is the West Indies–Antilles arc. The basic cross-sectional structure is similar to other island arcs, but these islands do not clearly parallel a continental border. Instead, they seem to divide the North and South American continents. In addition, the composition of the rocks found on these islands is sialic as opposed to the andesitic rocks found elsewhere. Recent samples of the crust beneath the Caribbean basin have also been sialic. This is inconsistent with the rest of the ocean basin, which is simatic. The whole region, therefore, may have a unique geological history.

Other factors complicate the generalized discussion at the beginning of this section. Island arcs are much more complex than indicated. A brief description of two other examples hint at this complexity.

Aleutian Islands. The Aleutian Islands are located southwest of Alaska. At the western end, their cross section is similar to the previously described general pattern. At the eastern end, however, a bifurcation produces two parallel island arcs between the mainland and the Aleutian Trench. The arc nearest the mainland is volcanic. The arc nearest the trench is composed of folded sedimentary rocks. Between the two arcs is a minitrench 700 m deep. By what process did this form? One hypothesis suggests that thick sediments accumulated on the continental shelf and buckled when the trench formed, producing the nonvolcanic chain. Current research may shed further light on the problem.

Peru–Chile Trench. Another complex problem is the presence of the Peru–Chile Trench. Its pattern and characteristics match those of other trenches. However, there is no island arc chain between the trench and the South American continent. There are, of course, active volcanoes near the coast and in the Andes Mountains spewing andesitic material. Was the South American continent somehow pushed over the island arc? Such questions intrigue marine geologists and keep them busy.

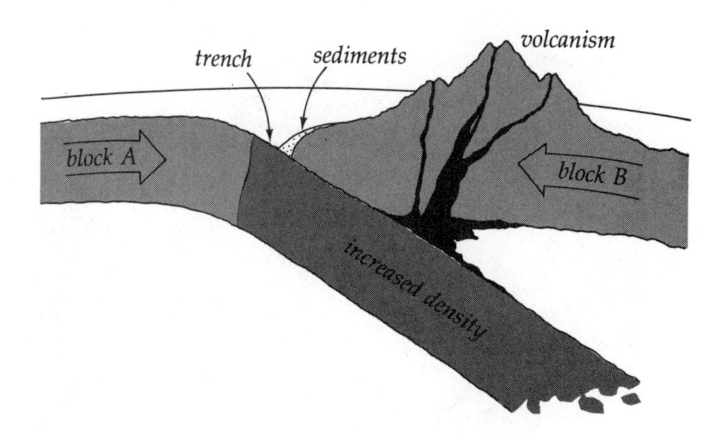

2–53 Cross section of an island arc complex. (After B. Gutenberg and C. F. Richter, Seismicity of the Earth and Associated Phenomena.)

Oceanic Ridges

The mid-ocean ridge system is a major submarine topographic feature whose worldwide extent was first charted by the German oceanographer L. Kober in 1928. Ridges with islands, like Hawaii, have been known by western civilization for more than 200 years. But the discovery of a relatively continuous, linear ridge system extending through the ocean basins (Fig. 2–54) startled marine geologists. The first detailed descriptions of the ridges were published by Maurice Ewing, Bruce Heezen, H. W. Menard, and others between 1950 and 1962. This information was probably the first step in a revolution of geologic thought. Since then, the mid-Atlantic ridge has been found to divide the entire Atlantic Ocean from 55°S to 70°N. In the South Atlantic, the ridge swings east to the Indian Ocean where it becomes the mid-Indian ridge. The mid-Indian ridge disappears near Cape Guardafui at the eastern tip of Somalia in the north, but bifurcates at about 25°S latitude. From there, it can be traced eastward between Australia and Antarctica (where it is called the Pacific–Antarctic ridge) and up through the eastern Pacific Ocean (where it is called the East Pacific rise) to the Gulf of California, where the ridge

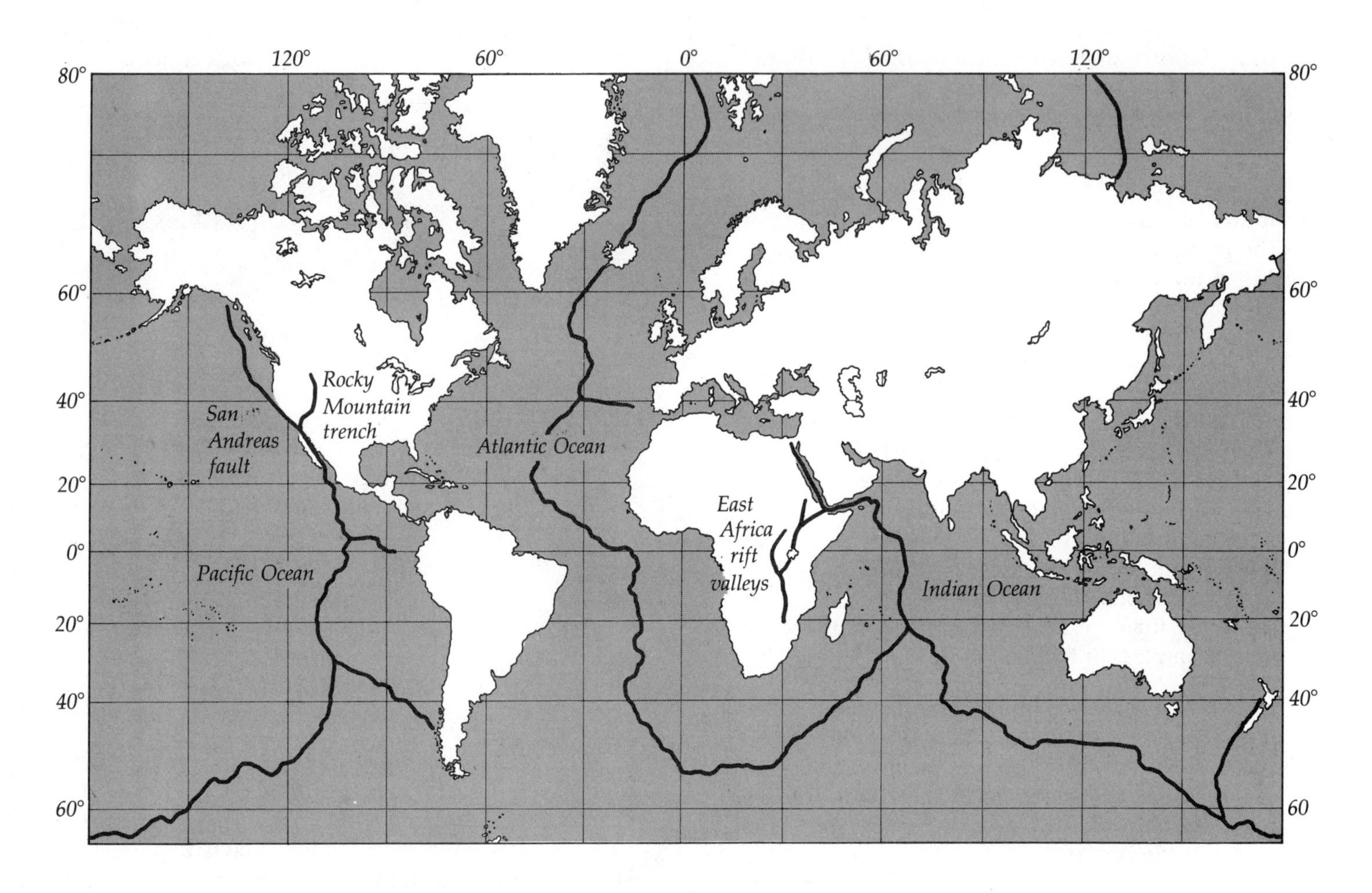

2–54 *Mid-ocean ridges. (Adapted from* Surtsey *by Sigurdur Thorarinsson. Copyright 1964. Copyright © 1966 by Almenna Bokafelagid. Reprinted by permission of The Viking Press, Inc.)*

again disappears, reappearing west of Oregon. Such a ridge system unquestionably represents a major topographic feature of the earth's crust.

Volcanism. The parts of the ridges that have been thoroughly studied all have a volcanic origin. The rocks were formed by submarine volcanoes, by fissure eruptions, or by magma cooling deep in the crust and subsequently moving to the ocean bottom. In some places, parts of the ridges appear above sea level. Iceland and its neighboring islands are part of the mid-Atlantic ridge (Fig. 2–55). Their origin is almost entirely volcanic. Active volcanoes are found on Iceland and several of the adjacent islands. An account of this activity was reported by Sigurdur Thorarinsson:

Iceland is one of the most active volcanic areas in the world; in former centuries, in fact, it was known for little else than its volcanoes and other natural phenomena.

As early as the twelfth century, one of the most famous of Iceland's volcanoes, Hekla, enjoyed a legendary fame on the European continent as the main gateway to Hell, or even Hell itself. Traces of this superstition lingered through the nineteenth century.

In June 1783 the greatest lava eruption in historic times occurred in Iceland, when the sixteen-mile-long Laki fissure opened up south of Vatnajokul, the country's largest glacier. From that tremendous outpouring of lava, a bluish haze caused by volcanic gases spread not only all over Europe but over parts of Africa and Asia as well, producing "red sun" and other atmospheric peculiarities.

In March 1875 some gentlemen walking across Gustaf Adolf Square in Stockholm noticed that their black overcoats had mysteriously turned gray and that the light from the gas lamps in the street was obscured by falling dust. Such dust settled all over southern Norway and central Sweden. The famous Swedish explorer A. E. Nordenskiold, who was the first to circumnavigate the Old World, was extremely interested in the dust, believing at first that it might be of cosmic origin. It was, however, far from heavenly. Microscopic studies revealed it to be volcanic glass, and the packet-boat Diana, *which arrived in Scandinavia from Iceland three weeks later, brought the information that the origin was a volcanic crater in northern Iceland now known as Viti, that is, Hell.*

Although eruptions in Iceland have often caused astonishment in other countries, probably none of them has attracted so much attention in so many places as that which, in November 1963, gave birth to a new island off the south coast of Iceland. This eruption has been in progress ever since, and there is no sign of its ending. The island that appeared on November 15, 1963, continued to grow until the middle of May 1965, at which time it had attained an area of one square mile, or more than half the size of Central Park in New York City. Its height was then 568 feet above sea level and nearly 1,000 feet above the former ocean floor. The Icelandic place-name committee named it Surtsey after the black giant Surtr, the leader of the fire giants who fought against the gods at Ragnarok [Fig. 2–56].

Surtsey was not the only island born during this eruption. When volcanic activity ceased in Surtsey, it broke out in another place in the sea a little ENE of that island, and during the spring and summer of 1965 a new island arose, popularly called Syrtlingur (Little Surtr—it never received an official name).

By the middle of September 1965, Syrtlingur had reached a height of 230 feet above sea level and an area of 35 acres. Yet this sizable island was completely swept away by the breakers during a single week of storms and high seas in late October 1965. On December 28 still another island was born, this time on the opposite side (SW) of Surtsey, half a mile offshore. This island has since continued to disappear and reappear.

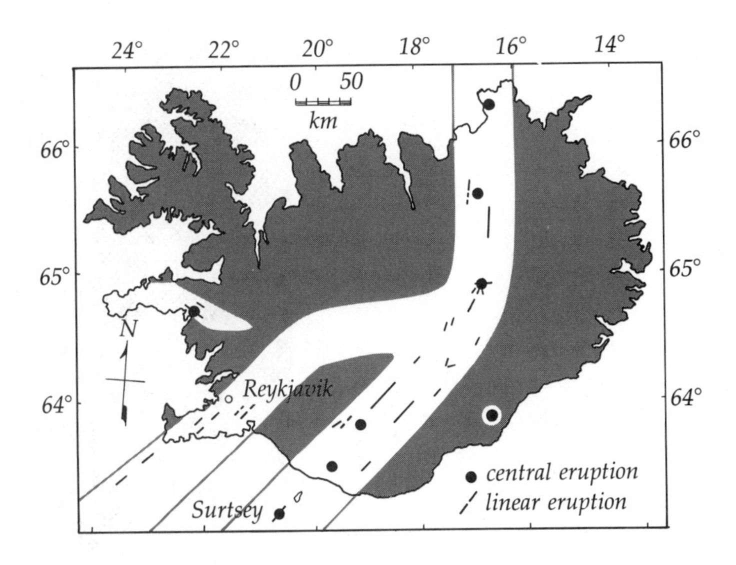

2–55 Mid-Atlantic ridge intersecting Iceland. (Adapted from Surtsey *by Sigurdur Thorarinsson. Copyright 1964. Copyright © 1966 by Almenna Bokafelagid. Reprinted by permission of The Viking Press, Inc.)*

The widespread interest in the Surtsey eruption is due both to the fact that it is a rare phenomenon—a major eruption out of the ocean—and has also presented an ever-changing, magnificent spectacle, sometimes terrifying, sometimes of breathtaking beauty. Moreover, because of good communications by air it has been more accessible to tourists than any other eruption in Iceland. Where thousands of people have seen it from ships, scores of thousands have seen it from the air.

The variety of phenomena presented by this eruption was largely dependent on external factors, especially the easy access of the sea to the magma in the crater vent, which in turn was dependent on wind and weather. For the first few months the eruption was purely explosive. The weather in Iceland in winter is changeable, and the sea continually made inroads into the crater walls on the weather side and water poured into the opening. Although basalt magma of this type is not ordinarily explosive, it becomes very much so when put in contact with water. The explosive activity consisted either of intermittent explosions (the so-called "cock's tail" explosions typical of

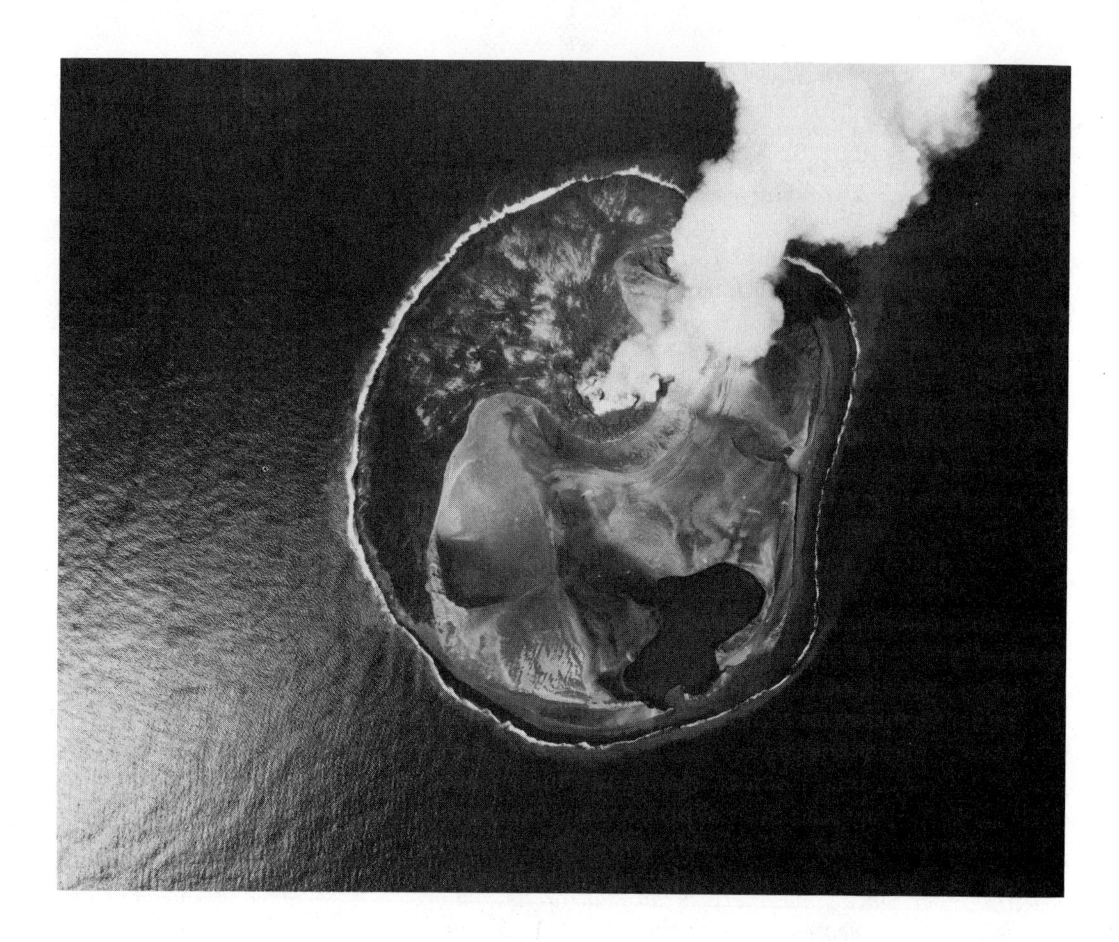

2–56 Surtsey Island. (From Surtsey *by Sigurdur Thorarinsson.*

submarine eruptions) or a continuous uprush of gases and tephra (i.e., bombs, scoriae, and other fragmented material).

The eruption column sometimes attained a height of 30,000 feet, and, towering high above the mountains to the southeast, was often clearly visible from Reykjavik. At this distance of 75 miles it looked quite harmless, but it was menacing enough to the townspeople of the Westman Islands who live on Heimaey Island, the most important Icelandic codfishing center, only fourteen miles away. They have special reason to fear ashfalls, for they have to collect their drinking water from the roofs of their homes. They were afraid that it might become contaminated by fluorine or other poisonous elements in the ash. Fortunately, only a small amount of ash was carried towards Heimaey.

The explosive eruption was often spectacular, especially when there was a continuous uprush in twilight or darkness. The eruption column was then sometimes red-hot to a height of over one mile, and the outer slopes of the crater walls were completely covered by glowing bombs that rolled down until they were extinguished by the white surf on the shore. When the column gradually became black and ash-laden, fingers of lightning raked through it; and the thunderbolts, the rumble from the uprushing bombs, and the bangs when they crashed into the sea created a dramatic symphony. It is beyond the power of man, however, to describe in words a major volcanic eruption, especially when it is seen from a boat in a storm and high seas. Such elemental fury is beyond description.

On April 4, 1964, the eruption changed from a highly explosive to a purely effusive, lava-producing eruption of the Hawaiian type. The breach in the crater wall had been filled by a scoria wall thick enough to block the sea from the magma; and in the crater there was now a fountaining lava lake overflowing the lower part of the crater rim and running down its outer slopes toward the sea. In a few weeks the lava eruption built up a small shield-

volcano, the first of its kind in Iceland since the settlement of the country. The lava eruption, which lasted 13½ months, was at times a truly magnificent spectacle, especially at night. Sometimes broad golden streams of lava rushed down the slopes at tremendous speed straight to the sea, where columns of white steam rose to a height of many thousand feet. Often, however, the lava did not overflow the crater walls but found outlets through channels on the lava dome, flowing down the slopes and entering the sea in numerous rivulets.

No sooner had the outpouring of lava ceased than Syrtlingur began its explosive activity, which tourists could enjoy throughout the entire summer of 1965. . . .

The great interest earth scientists have taken in the Surtsey eruption is due at least partly to the attention paid the neovolcanic zone of Iceland ever since the discovery of the mid-Atlantic rift by geophysicists of the Lamont Observatory of Columbia University twelve years ago. Further studies have revealed that the mid-Atlantic rift is part of a worldwide system of rift zones, mainly submarine. The famous San Andreas fault in California belongs to this system. And the highly rifted neovolcanic zone of Iceland is part of the mid-Atlantic rift zone; being the only part of it that can be studied on dry land, it is far more accessible for various studies than its submerged parts. Surtsey is situated on this rift zone, and many of the problems concerning the Surtsey eruption may have a bearing on questions that have come to the fore in connection with the midoceanic rift systems. It may be added that Surtsey is also a fascinating object for geomorphologists, the students of land forms and their changes. Geomorphological forces, especially marine abrasion and wind erosion, work with astonishing speed; and it is of great value that at Surtsey it is possible to study their effect on a landscape of which the initial stage is known. Already now one finds on Surtsey landscape so varied and mature that it is almost beyond belief. There are sandy beaches and precipitous lava cliffs, there are gravel banks, lagoons, fault scarps, sand dunes, ravines, and screes. And this landscape is constantly and rapidly changing. (Sigurdur Thorarinsson, "Surtsey: Iceland's New Island," American Scandinavian Review, *54:2 (June 1966), pp. 117–125. Reprinted by permission of the publisher.)*

Earthquakes. Of the shallow earthquakes that occur in the earth's crust, many occur in belts roughly coinciding with the oceanic ridges. These shallow earthquakes all occur less than 70 km below the earth's surface. This contrasts with the foci of earthquakes beneath the trenches. The earth movements causing the earthquakes, therefore, are located near the earth's surface and indicate some instability there.

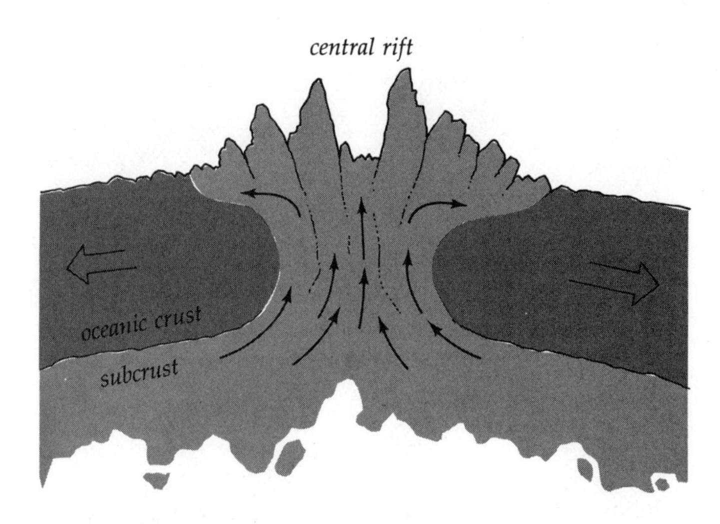

2–57 Rift valley, mid-ocean ridge, and possible directions of movement.

Composition. The mineral composition of all the mid-ocean ridges, regardless of their location, is surprisingly similar. The rocks on Iceland are more similar to rocks on the Pacific–Antarctic ridge than to rocks on Greenland. In general, the rocks are simatic and contain appreciable amounts of iron, magnesium, silicon, and oxygen. The common minerals would include magnetite, $FeO \cdot Fe_2O_3$, and olivine, $(Mg\ Fe)_2SiO_4$. Many other minerals, however, are also present.

Rift Valley. Another characteristic of most ridges is a rift valley (Fig. 2–57). The exception is the East Pacific rise. Why it has no rift valley is not known. A rift valley has relatively steep sides, a flat bottom, and is about 200 to 1,000 m deep. Valleys with similar topography are known in the Imperial Valley of California, the East African rift valley, and a number of other places on continents. They have been caused by faulting in which the central block drops down relative to the flanking blocks. This is probably caused by a

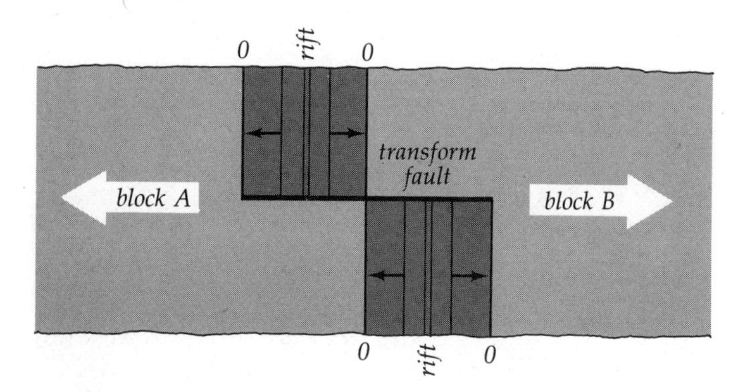

2–58 *Transform fault across the mid-ocean ridge.*

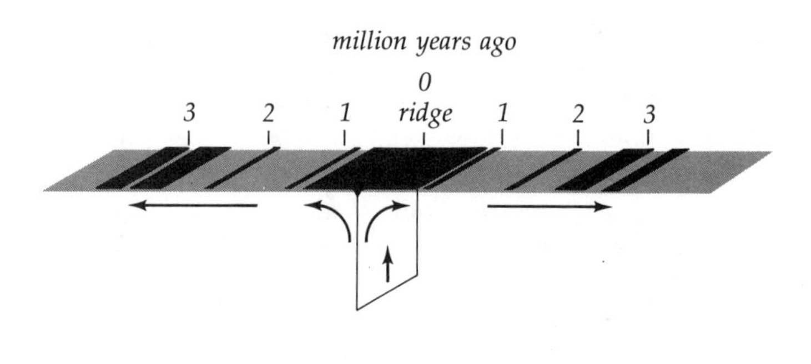

2–59 *Magnetic bands across the mid-ocean ridge.*

tensional, or pulling-apart, force. A study of the rift valley cutting across Iceland reveals a similar origin. Because Iceland's rift valley is an exposed section of the mid-Atlantic rift valley, the origin is assumed to be the same for the whole valley.

Sediment. Very little sediment has accumulated on the ridges or in the rift valleys. This seems to indicate that they formed in relatively recent times, during the last 20 million years. The sediment varies from only a few meters thick to a maximum of about 250 m thick. The sediment contains some volcanic ash, practically no sediment derived from continents, and moderately abundant shells of foraminifera, radiolaria, diatoms, pteropods, and coccolithophores. The shells simply rained down from surface waters.

Transform Faults. Slicing the mid-ocean ridges at approximately right angles are slightly curved transform faults (Fig. 2–58). The north and south sides of the transform faults indicate that sliding has occurred, because the sides are often displaced by 10 to more than 100 km. The mid-ocean ridges are broken by these faults, producing an irregular step-like pattern on a map. Transform faults are most evident on the mid-Atlantic ridge.

Magnetic Bands. When theories recently suggested that the sea floor might be spreading apart in the region of the mid-ocean ridges, measuring the magnetism of the rocks on both sides of the ridges was postulated as a test for the validity of the theories. This technique was picked because as magma cools below the Curie temperature (575°C) after emergence from the earth's crust, the magnetic material crystallizes and tends to align with the magnetic field of the earth. As the rest of the magma cools and crystallizes, the magnetic crystals such as magnetite are "frozen" in place. The relative amount of crystals aligned indicates the intensity of the earth's magnetic field at the time of crystallization and the direction of the alignment indicates the location of the poles. Because previous analysis of magnetic crystals in volcanic rocks on land had already supplied data on the intensity of the magnetic field and its polar locations at dated times in the past, the dates of the rocks flanking the ridges and the trends of movement could be identified from magnetometer readings taken over the ridges. The readings revealed astounding confirmation of the sea-floor spreading theories. North-south oriented bands of magnetic intensity exist, with the youngest closest to the axis of the ridge and each band progressively older away from the ridge. "Mirror images" of these bands exist on opposite sides of the ridges (Figs. 2–59 and 2–60). The rocks show that the magnetic poles have reversed in the past. These reversals are also mirrored on both sides of the ridges.

Gravity Anomalies. When interpreting crustal rocks, it is often helpful to compare seismic profiles, which provide sound-wave velocity data, with gravimeter measurements providing rock mass data. From such a comparison, it has been inferred that low density

2–60 Magnetic bands off the southern tip of Baja California, Mexico. (Data from Roger L. Larson.)

mantle-type materials are relatively near the surface and that rocks from the mantle may be mixing with rocks from the crust.

Age. Three dating techniques have been used in determining the age of the mid-ocean ridges. They are all in general agreement. The ridges are no older than 20 million years in any measured area and are still forming in others. These data are based on the radioactive decay of an isotope of potassium found in the rocks sampled from the ridges, the age of fossils in the sediment in rift valleys and draped on the ridges, and correlation of magnetic data with known dates for variations in the earth's magnetic field. The mid-ocean ridges contain relatively young rocks compared to the continents.

Heat Flow. The highest heat flow rates in the ocean basins are found in the rocks on the mid-ocean ridges. This is an almost certain indication that hot material is relatively near the surface in these areas.

Although the mid-Indian ridge cannot be clearly traced north of Cape Guardafui, a rift valley is present in the Red Sea. In this general region where a projected rift valley would intersect the East African rift valley at the bottom of the Red Sea, hot, mineral-rich brines occupy depressions. Heat flow is exceptionally high and the water temperature is over 50°C. This hot water stays on the bottom because of the high salt concentration. What happens to the East Pacific rise? It can be clearly traced northward for most of the distance of the Gulf. The Imperial Valley, just north of the Gulf, is a fault zone associated with the San Andreas fault complex.

Other Ridges and Rises. Not all the ridges and rises in the ocean basin have the same properties as the mid-ocean ridges. The Hawaiian Islands and the Chile rise are such exceptions. Hawaii is part of a chain of mid-Pacific volcanic mountains that do not have parallel magnetic bands or other indications of sea-floor spreading. They may be regarded as a stationary intrusion (hot spot theory) or as a northwesterly moving plate. The Chile rise extends from the East Pacific rise in a broad arc to the southern part of the Peru–Chile Trench. Its origin is presently unknown.

A third ridge that has not yet been thoroughly studied is the Lomonosov ridge, which stretches across the Arctic Ocean basin from the USSR to Baffin Island in northern Canada. This may be a mid-ocean type or a Hawaiian type ridge.

Global Plate Tectonics

A bold new model explaining the origin of ocean basins and many other geologic features of the earth's surface has emerged in the last ten years. This model, the global plate tectonic model, has caused a scientific revolution in geology. Before 1962, many geologists rejected the idea that the continents move and that the size of the present ocean basins has changed. F. B. Taylor from the United States and Alfred Wegener from Germany had separately evolved theories around 1910 to 1920 suggesting that such movements occurred. Their theories were based on the geographic similarities of continental coast lines. In 1963, a number of

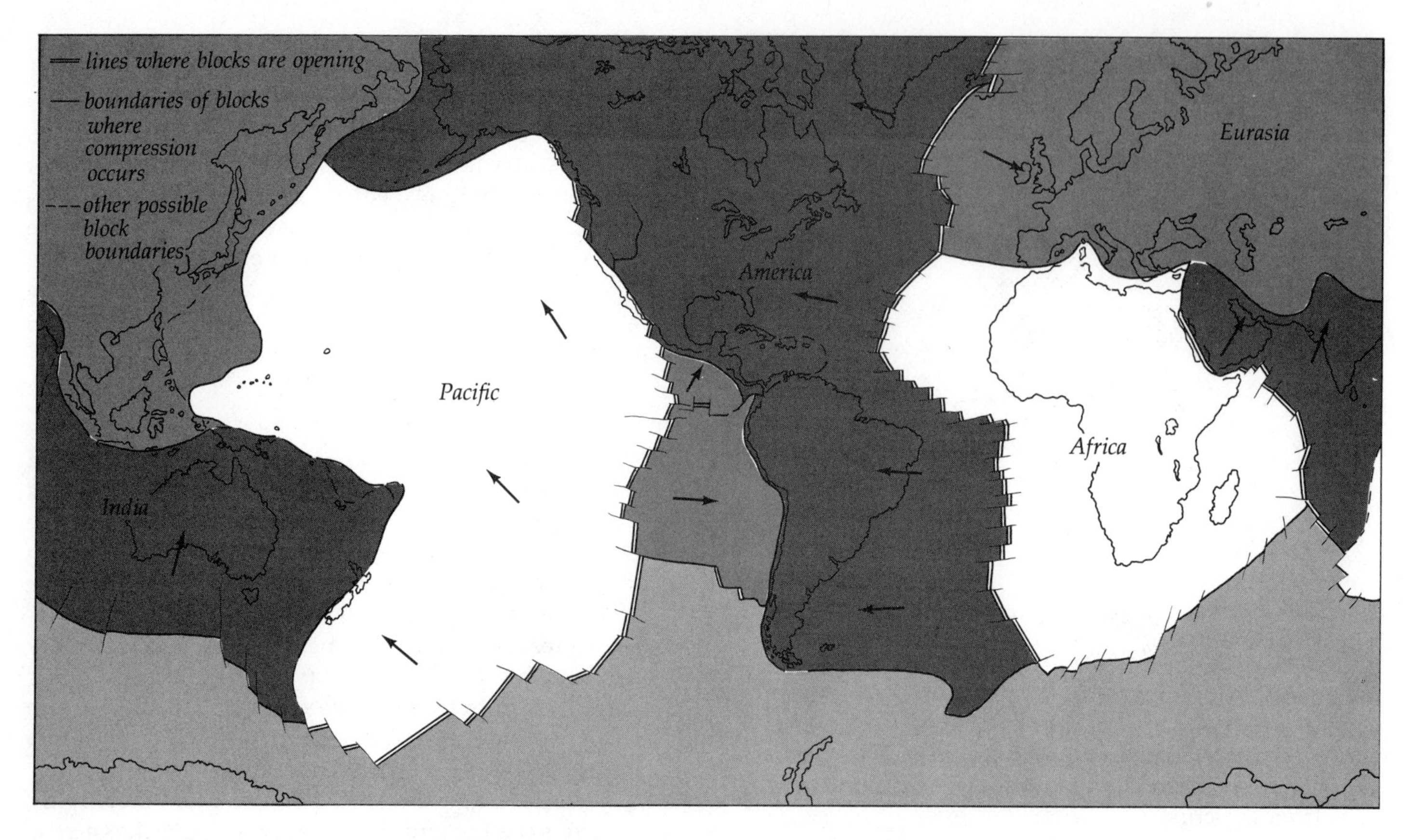

2–61 *Division of crustal blocks. (After Xavier Le Pichon,* Journal of Geophysical Research, *73, 3675, 1968.)*

scientists reintroduced their ideas. F. J. Vine and D. H. Matthews suggested that sea-floor spreading might be occurring and could be tested by analyzing magnetic properties of the mid-ocean ridges. J. T. Wilson described his theory of continental drift. Since then, marine geologists have acquired an amazing amount of data generally supporting the idea that the relative locations of continents and ocean basins have changed.

Hypothesis. The global plate tectonic model was suggested by Xavier Le Pichon (1968). The earth's crust is divided into six major plates: Eurasian, African, Indian, American, Pacific, and Antarctic (Fig. 2–61). These plates extend from 70 to 100 km into the earth. They each move about with respect to one another. In the region of the mid-ocean ridges, the plates are separating and new crustal rocks are moving to the surface. In the region of trenches, the plates are colliding, with one plate sinking down into the mantle (see Fig. 2–53). In California, two plates are sliding past one another along the San Andreas Fault.

At least three forces could allow these plates to move. One might be large convection cells in the mantle that rise where the mid-ocean ridges are and sink where the trenches are. The convection cells would be caused by heating of the mantle deep within the earth. A second possibility is that, as plates collide, one slides over the other. The lower plate is pushed down into the mantle, where the sinking plate cools the mantle causing a downward convection cell. The resulting displacement would cause the mantle to bulge elsewhere, namely under the mid-ocean ridges. A third idea is the exertion of some external force. Perhaps the combined effects of lunar and solar gravitational attraction are different for different plates de-

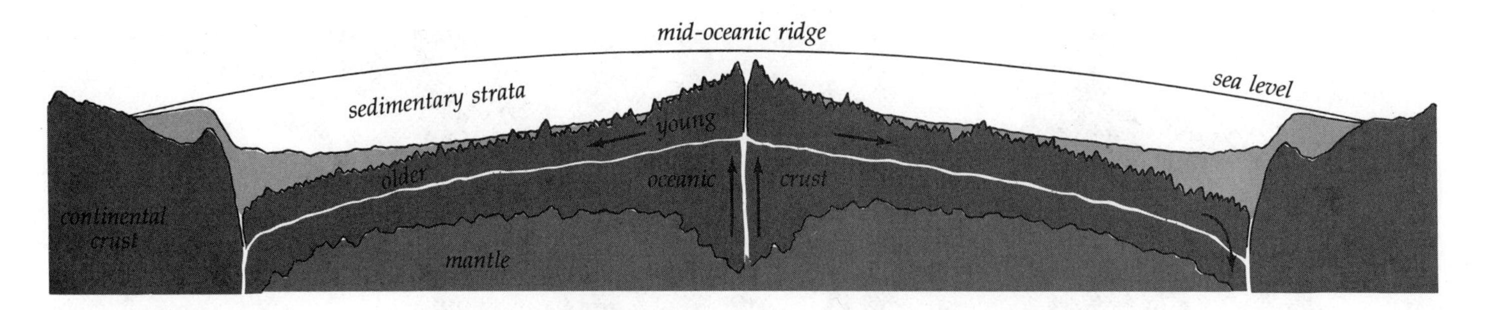

2–62 *Sea-floor spreading. (Courtesy of Scripps Institution of Oceanography, University of California, San Diego.)*

pending on their relative masses. This would move the plates. Where the plates separate, deep crustal material would move up like cold water when ice blocks separate on the surface of the Arctic Ocean. Where the plates collide, a plate might slide downward into the mantle or be crumpled into a mountain range or a combination of the two.

Similar Theories. The theories of continental drift and sea-floor spreading are not the same as the global plate tectonic model. Continental drift postulated that the continents move separately from the ocean basin crust. The original sea-floor spreading model postulated that the ocean basin crust moves separately from the continents (Fig. 2–62). The plates in the global plate tectonic model may be entirely ocean basin crust, like the Pacific plate, may contain ocean basin crust plus two continents, like the American plate, or may be predominantly continental crust, like the Eurasian plate (Fig. 2–63).

Supporting Facts

Continental coastline configuration. Although not sufficient alone to support the theory of global plate tectonics, the fact that the continental coasts and the configuration of the edges of the continental shelves fit surprisingly well (Fig. 2–64), especially for South America, Africa, and Antarctica, adds an important bit of evidence to the case.

Magnetic bands. Perhaps the most significant supporting evidence is the matching bands of magnetic intensity and polarity that parallel the mid-ocean ridges. They show relative spreading on opposite sides of the mid-ocean ridges and also the rates at which spreading occurs. Rates measured so far vary between 1 cm per year on the mid-Atlantic ridge to 8 cm per year on the East Pacific rise. At this rate the whole present Pacific Ocean floor could have formed in less than 200 million years.

Bedrock age. Cores of bedrock obtained in scattered parts of the ocean basins have produced no rocks 200 million years old or older. The oldest, about 175 million years old, have been found in the northwest corner of the Pacific plate, a location far from the actively spreading new crust at the East Pacific rise. The oldest rocks from any mid-ocean ridge are 20-million-year-old rocks found on the mid-Atlantic ridge just south of the equator.

Sediment age and thickness. The age of the sediments overlying the bedrock is always younger than the bedrock. Recent sediment is, of course, distributed all over the ocean basin. The thickness of the sediment and the time span it represents, based mainly on dating of microfossils, increase progressively with distance from the axis of the mid-ocean ridges.

Heat flow. The heat flow patterns for the ocean bottom indicate low conduction in the vicinity of trenches and exceptionally high conduction on the mid-ocean ridges. This would tend to confirm the idea

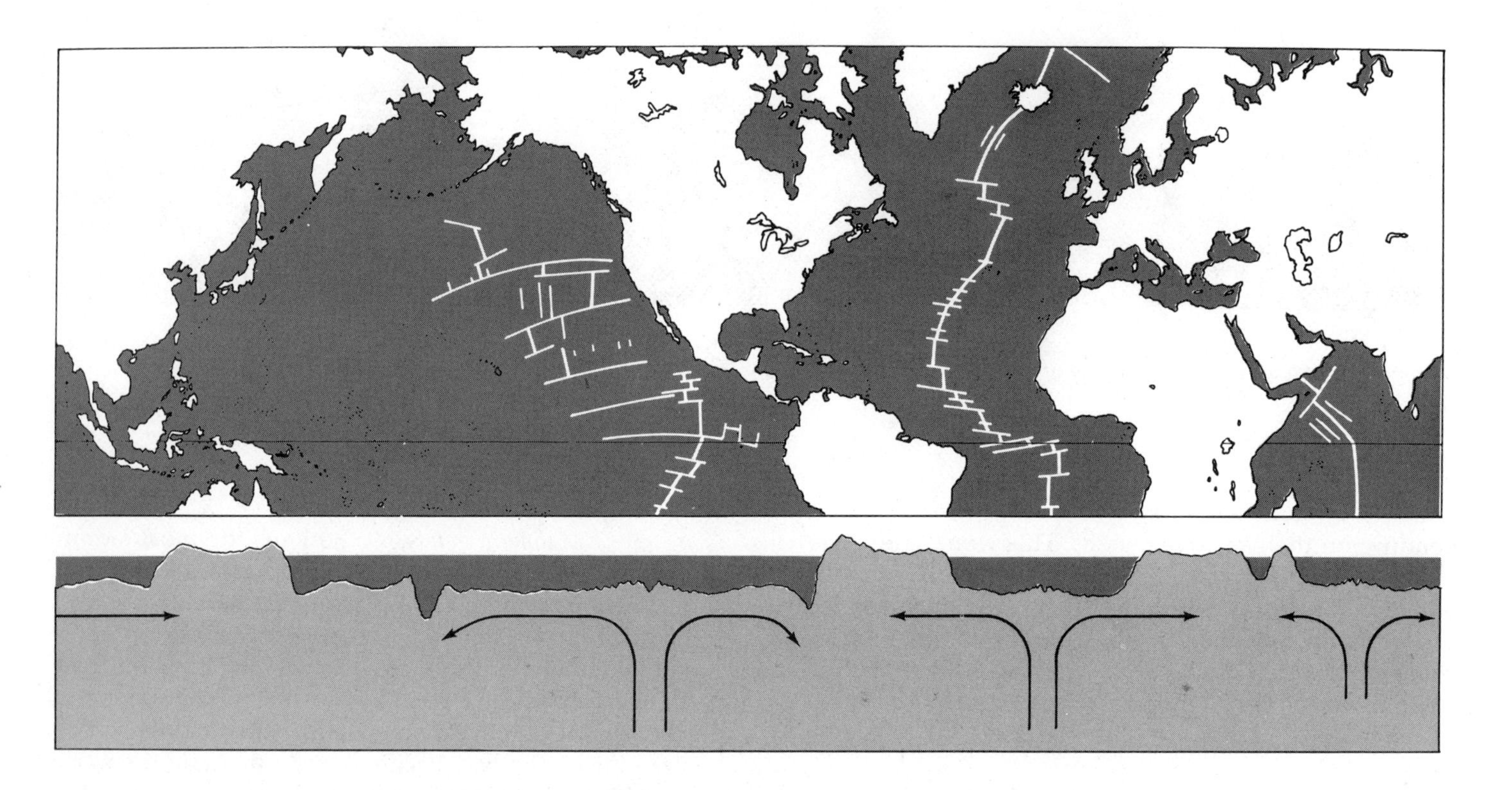

2–63 Sea-floor spreading takes place at several zones in the world oceans; from left to right, the East Pacific rise, the mid-Atlantic ridge, and the Carlsberg ridge. The Deep Sea Drilling Project has verified that this process takes place and has measured the rates at which it takes place. For example, the South Atlantic spreads at a rate of four cm per year from the center. The continents of Africa and South America move with the Atlantic spreading system. At certain places in the world, the oceanic crust apparently dives down into the interior of the earth; along these zones are many earthquakes, much volcanism, and deep trenches. (Courtesy of Scripps Institution of Oceanography, University of California, San Diego.)

of sinking crust or deep mantle at the trenches and a rising crust or shallow mantle at the ridges.

Earthquake locations. The distribution of earthquake foci in the earth's crust reveals deep foci at the trenches and shallow foci at the ridges, supporting sinking and rising materials respectively. Shallow earthquake foci also are located along the plate borders.

Antipodal point comparison. In one study (Harrison, 1966), antipodal locations on the surface of the earth were compared by computer. The frequency of finding a point in an ocean basin located antipodal to a point on a continent was so high that the probability of such an occurrence being random was 1 in 14. This supports the idea of a globally operating mechanism.

Lithologic correlation. The relative percentage of minerals and the detailed mapping of fractures in rocks in eastern South America have been shown to closely match rocks in western Africa. The chances that the same set of forces produced the same fracture pattern in rocks on separate continents are slim. Other such correlations between New England, Canada, and the British Isles have produced similar results.

Terrestrial fossil correlation. The famous Gondawana series of sedimentary rocks in Africa

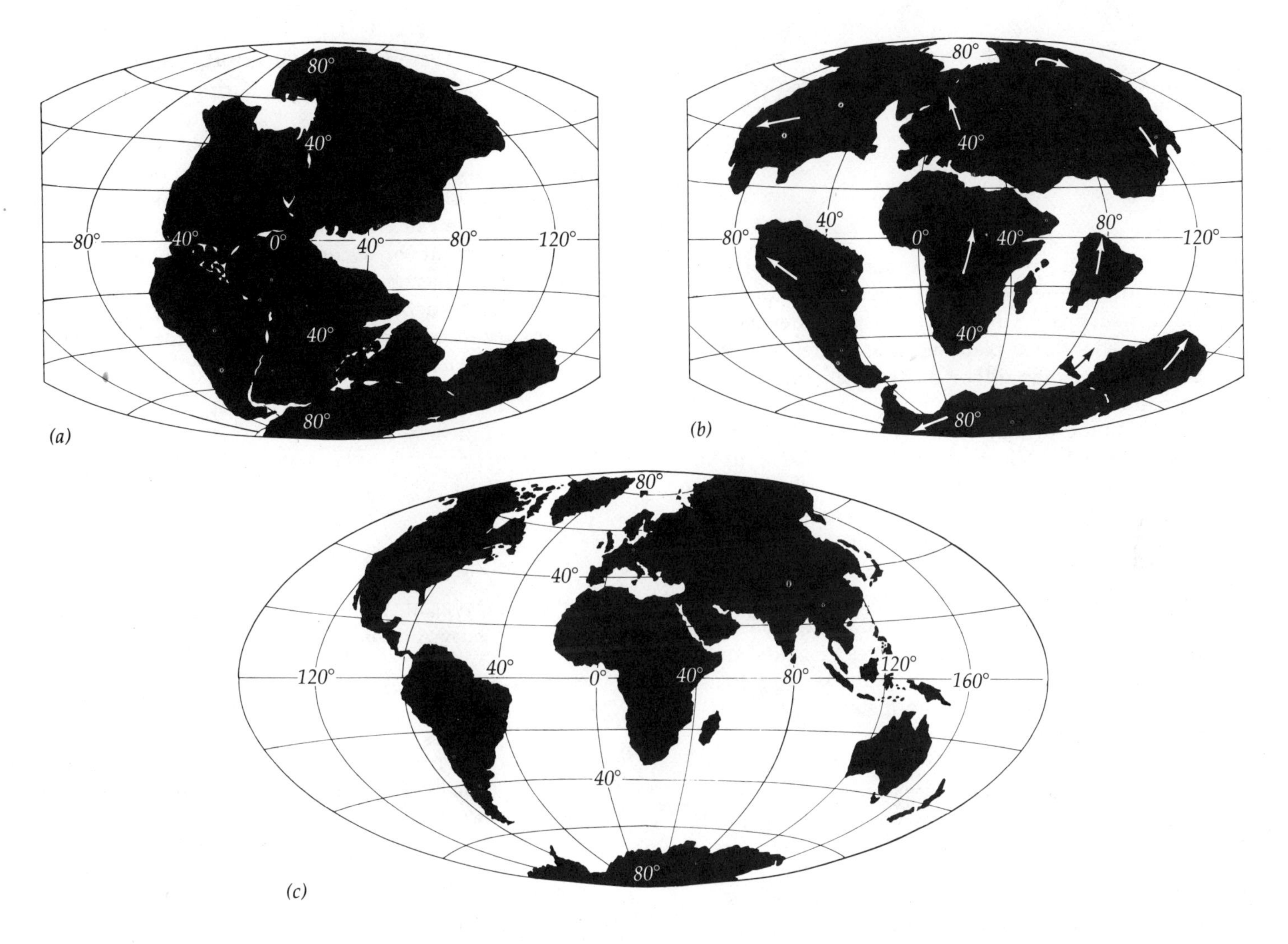

2–64 *Possible configuration of the continents (a) 150 to 200 million years ago; (b) 65 to 80 million years ago; (c) present configuration of the continents.*

contains fossils of ferns, amphibians, and other fresh-water swamp denizens. The fact that a number of the same fossils have also been found in Antarctica and South America is strong indication that a land bridge of some sort existed between the continents, because such organisms are not known to be capable of migrating across oceans.

Mathematical model. If the theory of global plate tectonics is correct, the movement of the plates should fit a mathematical model.

If two rigid plates on a sphere are spreading out on each side of a ridge that is crossed by fracture zones, the relative motion of the two plates must be a rotation around some point, termed the pole of spreading. The "axis of spreading," around which the rotation takes place, passes through this pole and the center of the earth. The existence of a pole of spreading and an axis of spreading is geometrically necessary, as was shown by Leonhard Euler in the eighteenth century. If the only motion on the fracture zones is the sliding of the two plates past each other, then the fracture zones must lie along circles of latitude with respect to the pole of spreading, and the rates of spreading at any point on the ridge must be proportional to the perpendicular distance from the point to the axis of spreading.

All this is well verified for the spreading that is going on today. (Bullard, 1969)

Summary. Thus, at least ten separate types of evidence support the present model. This is quite convincing, and it is surprising that no workable model existed before 1963. The global plate tectonic model seems to be a better model than either continental drift or seafloor spreading alone.

Until now, the primary activity of a geologist was observation and description of the earth. He used models to explain local phenomena. The global plate tectonic model is an important step toward enabling geologists to use a single, broad model to explain the origin and structure of continents and ocean basins. One major activity, then, of a modern geologist is testing this model. Such testing could lead to explanations for the formation of mountain ranges, islands, and earthquakes, the location of ore deposits, and generally to a reinterpretation of the history of the continents and ocean basins.

This scientific revolution is a prime example of the "process of science." Each area of science has experienced revolutions. Physics was revolutionized between 1890 and 1910 when Einstein and others produced models displacing Newtonian physics. In the 1950s and 1960s, microbiology also went through a revolution. The last revolution in geology was probably the establishment of the principle of uniformitarianism (present geologic processes have remained the same throughout the earth's history varying only in rate; often known as "the present is the key to the past") by Sir James Hutton (1729–1797), a famous Scottish geologist.

Some unanswered questions presently pursued by marine geologists are: Where is the division between the Eurasian and American plates in the vicinity of the Bering Sea? Where are the plate contacts in the Arctic Ocean basin? Why are there so few folded sediments in trenches? What is the origin of the Caribbean complex? Does spreading occur along the East African rift valley? Does the East Pacific rise extend northward under the western United States?

The Joint Oceanographic Institution for Deep Earth Sampling (JOIDES) Project of Scripps Institution of Oceanography at University of California in San Diego, the Lamont–Doherty Geological Observatory at Columbia University, the Rosenstiel School of Atmospheric and Oceanic Studies at University of Miami, the Department of Oceanography at Texas A & M, and the Department of Oceanography at University of Washington has been using the *Glomar Challenger* to test these theories since 1968. So far, much supporting evidence has been amassed. More questions, however, are arising.

Further Reading

Bascom, Willard, *Waves and Beaches.* Garden City, New York: Doubleday-Anchor, 1964.

Bullard, Sir Edward, "The Origin of the Oceans." *Scientific American,* September 1969, pp. 66–75.

Bullard, Sir Edward, "Reversals of the Earth's Magnetic Field." *Philosophical Transactions of the Royal Society of London:* series A, *263:*1143 (December 1968), pp. 481–524.

Dietz, R. S., and J. C. Holden, "The Breakup of Pangaea." *Scientific American,* October 1970, pp. 30–41.

Emery, K. O., "The Atlantic Continental Margin of the U.S. During the Past 70 Million Years." Geological Association of Canada, S.P. No. 4, *Geology of the Atlantic Region.* November 1967, pp. 53–70.

Emery, K. O., "Characteristics of Continental Shelves and Slopes," *American Association of Petroleum Geologists Bulletin, 49:9* (September 1965), pp. 1379–1384.

Emery, K. O., "The Continental Shelves." *Scientific American,* September 1969, pp. 39–52.

Harrison, C. G. A., "Antipodal Location of Continents and Oceans." *Science, 153* (September 1966), p. 1246.

Heirtzler, J. R., "Sea Floor Spreading." *Scientific American,* December 1968, pp. 60–70.

Holmes, Arthur, *Principles of Physical Geology*. Ontario, Canada: Nelson and Sons, Ltd., 1965.

Hurley, P. M., "The Confirmation of Continental Drift." *Scientific American*, April 1968, pp. 52–64.

Isacks, Bryan, et al., "Seismology and the New Global Plate Tectomics." *Journal of Geophysical Research, 73*:18 (September 1968), pp. 5855–5899.

Josephs, M. J., and H. J. Sanders, *Chemistry and the Environment: The Solid Earth, the Oceans, the Atmosphere*. Washington, D.C.: ACS Publications, 1967.

Keen, M. J., *An Introduction to Marine Geology*. New York: Pergamon Press, 1968.

Kummell, Bernhard, *History of the Earth: An Introduction to Historical Geology*. San Francisco: W. H. Freeman, 1961.

Le Pichon, Xavier, "Sea-Floor Spreading and Continental Drift." *Journal of Geophysical Research, 73*:12 (June 1968), pp. 3661–3697.

MacIntyre, Ferren, "Why the Sea Is Salt." *Scientific American*, November 1970, pp. 104–115.

Menard, H. W., "The Deep Ocean Floor." *Scientific American*, September 1969, pp. 53–63.

Milliman, J. D., and K. O. Emery, "Sea Levels during the Past 35,000 Years." *Science, 162* (December 1968), pp. 1121–1123.

Morgan, W. J., "Rises, Trenches, Great Faults, and Crustal Blocks." *Journal of Geophysical Research, 73*:6 (March 1968), pp. 1959–1982.

Phinney, R. A., *The History of the Earth's Crust*. Princeton, N.J.: Princeton University Press, 1968.

Rudman, A. J., "The Role of Geophysics in the New Global Plate Tectonics." *Science Teacher, 36*:7 (October 1969), pp. 21–26.

Shepard, F. P., *Submarine Geology*, 2nd ed. New York: Harper & Row, 1963.

Strahler, A. N., *A Geologist Looks at Cape Cod*. Garden City, N.Y.: Natural History Press, 1966.

Thorarinsson, Sigurdur, "Surtsey: Iceland's New Island." *American Scandinavian Review, 54*:2 (June 1966), pp. 117–125.

Thorarinsson, Sigurdur, *Surtsey: The New Island in the North Atlantic*. New York: Viking Press, 1967.

Vine, F. J., "Spreading of the Ocean Floor: New Evidence." *Science, 154* (December 1966), pp. 1405–1415.

Chemical and Physical Properties of Seawater

3

The fading sky
Cast patches of light,
And the mass of salt and water
Was a world inundated;

While an isle of clouds,
Streaming shrouds of rain,
Hung like a pall in a sea
Of a cosmos ending creation . . .

Edward L. Meyerson

No pencil has ever yet given
anything like the true effect of
an iceberg. In a picture,
they are huge, uncouth masses,
stuck in the sea, while their
chief beauty and grandeur—
their slow, stately motion,
the whirling of the snow about
their summits, and the fearful
groaning and cracking of their parts—
the picture cannot give.
This is the large iceberg,
while the small and distant islands,
floating on the smooth sea,
in the light of a clear day, look
like little floating fairy isles
of sapphire.

Richard Henry Dana

I. Water

II. Seawater

A. Composition

B. River water

C. Chemical and physical properties

III. Salinity

A. History

B. Definitions

C. Distribution

1. surface
2. depth

D. Marine life and water

IV. Oxygen

V. Sampling

VI. Temperature

A. Distribution

1. layering
2. temperature profiles

B. Measurement

VII. Pressure

VIII. Density

IX. Light

A. As it reaches the sea surface

1. nature of light
2. climate
3. weather
4. greenhouse effect

B. In the sea

1. absorption
2. attenuation

C. What it looks like in the sea

X. Sound

A. The nature of sound

B. Velocity

1. air
2. pure water
3. seawater

C. Sonar

1. active (depth recorders)
2. passive
3. submarines and fish
4. bottom contour

D. Sound properties

1. propagation loss
2. deep scattering layer
3. density layering

E. What it sounds like in the sea

XI. Ice in the Sea

A. Ice

B. Sea ice

C. Icebergs

1. Arctic
2. Antarctic

D. Ice islands

One encounters many surprises in the experimental study of nature. Contrary to many of the explanations of ancient and modern philosophers, nature is not intuitively obvious to man. Nature is, for example, left-handed (the DNA helix is a left-handed thread), whereas man tends to be right-handed. The motion of clocks, gauges, speedometers, and tachometers is clockwise, but nature is primarily counterclockwise. The orbital motions of the moon around the earth, of the earth around the sun, and of the earth on its own axis are examples.

Water

Life on this planet is based on water. No life encountered thus far can exist without it. Most scientists now believe that the evolution of life began in the water–salt solution, the sea. One of the most fascinating points to ponder concerning seawater is its remarkable similarity to blood. Both are mostly water. They have similar densities and contain similar percentages of dissolved salts. The *p*H of both fluids is almost identical, and both are very slightly basic. These similarities of the two fluids (as well as another vital body fluid, lymph) are put forth as a fragment of proof for the evolution of life from the sea.

The most remarkable thing about the waters of the sea is the water itself, and how much is still to be discovered about it. Water is by far the most abundant molecule on this planet's surface. Almost everyone can give the chemical formula for water, H_2O, but scientists are no longer sure of this formula. The ratio of atoms is correct, but there is evidence that the ratio may exist in varying multiples (such as H_4O_2, H_6O_3) depending on temperature and other parameters.

Properties. Chemists have shown that compounds produced from elements in the same vertical columns of the periodic table have similar properties (Fig. 3–1). Thus, NaCl, KCl, CsCl, and so forth, should be similar, as indeed they are. However, because the ionic radii (hence, the mutual attraction) of these elements are different, their compounds are expected to have differences. For example, the ionic radii of the elements increase as one reads down the table (Na, K, Cs). Compounds toward the bottom have bonds of greater strength (percent ionic character), giving these compounds lower melting points, and other characteristics. Accordingly, the hydrogen compounds of elements in the oxygen group should have properties that are reasonably close (Fig. 3–2). From the study of the other hydrogen compounds (such as H_2S, H_2Te), water should be a gas at normal temperatures. But because of the oxygen atom's extreme attraction for electrons, the bond formed between hydrogen and oxygen is different than would be expected, giving water molecules much greater attraction for one another and making water a liquid under normal conditions. Water exists naturally in all three states of matter—solid, liquid, and gas—but the vast bulk of it on earth is in the liquid form.

To really appreciate seawater, one should know something about the properties of water. Water is the superlative liquid. It has one of the greatest transparencies to light; it has the greatest heat capacity of all liquids except liquid ammonia; it conducts heat better than any other liquid; no other liquid has a greater surface tension. Because seawater is primarily water and therefore has properties almost identical to those of pure water, the following discussion of seawater will treat some of these characteristics (Table 3–1).

Seawater

Water forming the world's oceans accumulated in the basins of the earth (see Chapter 2). The vast preponderance of the water on this planet is in the oceans:

H																	He
Li	Be											B	C	N	O	F	Ne
Na	Mg											Al	Si	P	S	Cl	Ar
K	Ca	Sc	Ti	V	Cr	Mn	Fe	Co	Ni	Cu	Zn	Ga	Ge	As	Se	Br	Kr
Rb	Sr	Y	(Zr)	Nb	Mo	(Tc)	(Ru)	(Rb)	(Pd)	Ag	Cd	In	Sn	Sb	(Te)	I	Xe
Cs	Ba	*	(Hf)	(Ta)	W	(Re)	(Os)	(Ir)	(Pt)	Au	Hg	Tl	Pb	Bi	(Po)	(At)	Rn
(Fr)	Ra	**															

*La	Ce	Pr	Nd	(Pm)	Sm	Eu	Gd	(Tb)	Dy	Ho	Er	Tm	Yb	Lu
**(Ac)	Th	Pa	U											

3–1 A periodic table of the elements. Elements not in parentheses are present in seawater.

97 percent covers 70.8 percent of the earth's surface, with
2 percent occurring in rivers and lakes, and
1 percent in snow and glaciers, with less than .005 percent in the atmosphere.

Water has had a profound effect on the earth's surface. It is a primary erosive agent. If the earth had no water, its surface would be more irregular; it might look more like the surface of the moon.

Water is referred to as the universal solvent. It is the best solvent because it dissolves more substances more often and in greater amounts than any other liquid. Virtually everything dissolves in water, although some things dissolve only in very small amounts. The glass in a drinking glass is slightly soluble in water. If a water-filled sealed soft glass bottle were allowed to sit for a long time, such as a year or two, the water would become noticeably cloudy because of the presence of dissolved glass. The rate at which some things dissolve may be very slow. Oil has an extremely low solubility in water or seawater. It takes the sea a long time to rid itself of an oil spill, which is one reason large oil leaks or spillages can be so disastrous.

Composition. Anyone who has ever come in contact with the ocean knows that it is salty. One hundred pounds of seawater contains about 3.5 lb of salts. In other words, the salt content of the world's ocean is 3.5 percent. Oceanologists find the kilogram (1000 g or approximately 2.2 lb) of seawater a convenient unit to use so the salt content, referred to as salinity, would be 35 parts per thousand (or 35‰).

How did the salt get into the oceans? The classical answer to this question is that rivers supplied the salts. As the rivers flowed over the earth they dissolved salts from rocks and carried the salts into the ocean. This answer is too simple.

The salt in seawater is not just regular table salt (sodium chloride, NaCl) but a mixture of a number of salts. If regular salt is dissolved in water to make a salt water solution, the dissolved salt is not present as salt particles but breaks up into charged particles called ions.

$$NaCl \xrightarrow{\text{water}} Na^+ + Cl^-$$

Of course, the net average charge is zero, because the number of positive and negative charges is equal. The presence of these ions accounts for the fact that seawater is an excellent conductor of electricity, whereas pure water is a poor conductor. The ions conduct the current. The elements, in ion form, that make up the salt content in seawater are shown in Table 3–2. These

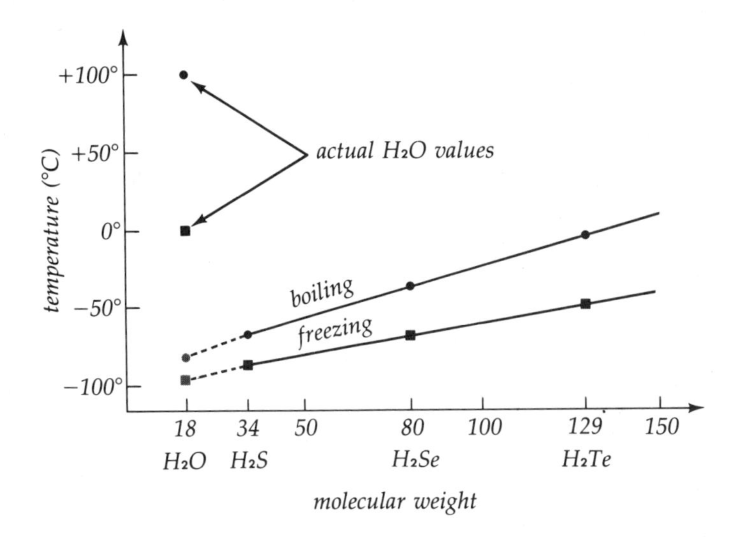

3–2 Some properties of hydrogen compounds similar in structure to water. The theoretical values for H_2O show what the compound's properties would be if water conformed to the pattern established by other hydrogen compounds in the oxygen group.

comprise 99.9 percent of the total salt content in seawater. The remaining 0.1 percent includes most of the other natural elements except about 20 (present in such minor quantities that they generally need not be mentioned). Sometimes the idea that salts do not exist as particles of salt in solution is hard for nonscientists to understand.

River Water. Table 3–2 shows that, in seawater, chlorine (or the chloride ion) is the most abundant material next to the water itself. If river water supplied most of the salts to the sea, it would seem reasonable to expect river water to be similar in composition to seawater, though less concentrated. River water should be proportionally high in chlorine concentration. Such is not the case. Although the amount of total dissolved solids varies from river to river, their compositions are relatively similar (Table 3–3). For all practical purposes, tap water may be regarded as representative of river water in terms of dissolved salts because much tap water comes from rivers. This is especially true in areas where the water is very hard. Evaporation of a seawater sample shows that sodium chloride is the primary salt constituent, but in river water the major residue is calcium bicarbonate. Seawater cannot then be concentrated river water, so the notion that rivers alone account for the total salt content of the ocean is incorrect.

Chemical and Physical Properties. Much of the salt in the sea did and does come from river runoff, and there may now be a very slight annual increase in the salt content of seawater as the world's rivers continue to denude the continents. However, this build-up of salts in the ocean is so slow that it would take about 6 million years to increase the salt content by 1 percent. The early ocean was probably always salty, the salts having been dissolved out of volcanic rocks in the ocean basins. The early atmosphere (see Chapter 1) also contributed in all probability to the composition of the ocean. The primordial atmosphere seems to have been rich in chlorine and hydrogen chloride. These gases, resulting from volcanic activity, are quite water soluble and were in part flushed out of the atmosphere by rain. Undoubtedly many chemical reactions occur between constituents in the ocean to give its present composition. One such reaction is the formation of calcium carbonate, which precipitates out of seawater when Ca^{+2} and CO_3^{-2} react. This precipitate contributes to ocean sediments (although the calcium carbonate in sediments comes chiefly from biological shell debris).

As the precipitation of calcium carbonate indicates, the ions of each element do not always stay in the ocean. Some ions such as sodium and potassium remain in the ocean for extremely long periods of time, perhaps 250 million years for a sodium ion. Salt material in the sea is removed in a number of ways. One method is by ocean spray. When the wind blows across the tops of waves or bubbles burst or a wave breaks or a raindrop strikes the surface of the water, some water and a small amount of dissolved salt is blown into the air. The increased rate at which an automobile bumper will rust or aluminum house trim or screens will corrode near the sea is evidence of salt in the air. The means by which the bulk of the salt material in the sea is removed, however, is through sedimentation and biological processes. The length of time the ions of each constituent will remain in the ocean, known as its residence time, is a function of its reactivity. Sodium, for example, will stay in the sea for a very long time, while aluminum may remain only about 1000 years because it is very reactive to inclusion in sediments.

Table 3–1 The Properties of Water

Property	Compared to other substances
Physical state: occurs in all three states—gas (or vapor), liquid, solid	The only substance occurring naturally in the three states on the earth's surface
Quantity present on the earth's surface	Three times as abundant as all other substances combined
Dissolving ability	Dissolves more substances in greater quantities than any other common liquid
Density: mass per unit volume (g/cm³ or g/ml)	Density determined by (1) temperature, (2) salinity, (3) pressure, in that order. The temperature of maximum density for pure water is 4°C. For seawater the freezing point decreases with increasing salinity. For water with salinities greater than 24.70‰ the temperature of maximum density is below that of the initial freezing point.
Specific gravity: the ratio of the density of a substance to the density of pure water at 4°C (in the cgs system, the two terms have the same numerical value)	
Surface tension	The highest of all common liquids
Conduction of heat	The highest of all common liquids, except mercury
Heat capacity: the quantity of heat required to raise the temperature of 1 g of a substance 1°C (cal/g/°C)	The highest of all common solids and liquids
Latent heat of fusion: the quantity of heat gained or lost per unit mass by a substance changing from a solid to a liquid or liquid to solid phase without an accompanying rise in temperature (cal/g)	The highest of all common liquids and most solids
Latent heat of vaporization: the quantity of heat gained or lost per unit mass by a substance changing from a liquid to a gas or gas to liquid phase without an increase in temperature (cal/g)	The highest of all common substances
Viscosity	Relatively low viscosity for a liquid (decreases with increasing temperature)
Refractive index	Increases with increasing amounts of salts and decreases with increasing temperature
Transparency	Relatively great for visible light
Sound transmission	Transmits sound well, compared to other fluids
Compressibility	Only slightly compressible

The salt content of seawater obviously has some effect on its properties. Pure water freezes at 0°C, but seawater, the salt content of which makes it act like an antifreeze, freezes at a lower temperature. Because the amount of salt in water varies, the freezing point of seawater also varies. There is a specific freezing point for each salinity (Table 3–4). Water of 35‰ freezes at −1.91°C.

One of the unusual and lesser known effects of salt content on water deals with its ability to foam (Fig. 3–3). Foam in water is simply the coalescing of a multitude of tiny bubbles. Unlike fresh water, which allows the bubbles to come together, the salt in seawater causes the bubbles formed by churning to bounce off one another so that true foam generally does not form in the sea.

Salinity

Ninety-nine percent of the solid inorganic matter in seawater is chlorine, sodium, magnesium, sulfur (as sulfate), calcium, and potassium. With this in mind it should be relatively simple to determine the salt content of a seawater sample using the advanced techniques of modern chemistry. Of the arbitrary branches into which oceanology is divided, the chemical study of the sea is the oldest. Yet the problem of determining the total amount of salt, let alone the separation and identification of the individual components, has always been a difficult one.

Table 3–2 Dissolved Constituents of 1 kilogram of Seawater (chlorinity = 19‰)

Constituent	g/kg seawater	Percent by weight
Chloride, Cl^-	18.980	55.04
Sodium, Na^+	10.556	30.61
Sulfate, SO_4^{-2}	2.649	7.68
Magnesium, Mg^{+2}	1.272	3.69
Calcium, Ca^{+2}	0.400	1.16
Potassium, K^+	0.380	1.10
Bicarbonate, HCO_3^-	0.140	0.41
Bromide, Br^-	0.065	0.19
Boric acid, H_3BO_3	0.026	0.07
Strontium, Sr^{+2}	0.013	0.04
Fluoride, F^-	0.001	0.00
Total	34.482‰	99.99%

History. As early as 1674 Robert Boyle (1627–1691), the great English chemist, decided that simply evaporating the water from a pound of seawater and weighing the salt residue did not yield values that were consistent or reproducible even with larger water samples. A comparison of the results obtained 100 years later by the renowned chemists Antoine Lavoisier (1743–1794) and Torbern Bergman (1735–1784) shows definite variances not only in weighed values but in the specific salts discovered. Although 11 constituents comprise 99.9 percent of the total salt content, this seeming simplicity is deceptive. If one were to dissolve 10 g of sodium chloride and 10 g of potassium bromide in 1 liter of pure water and then evaporate the solution to dryness, 20 g of solid material would remain, but not all of it would be the two original salts. Depending on factors such as the rate at which the mixture was evaporated, one might find trace amounts of sodium bromide and potassium chloride. Recall that, once dissolved, these two salts exist as ions and on approaching dryness randomly combine with other ions to form some different salts. With a total of 11 ions, the situation becomes much more complex. The problem with

Table 3–3 Dissolved Constituents in River Water (average)

Constituent	Percent by weight
Carbonate, CO_3^{-2}	35.15
Calcium, Ca^{+2}	20.39
Sulfate, SO_4^{-2}	12.14
Silicon dioxide, SiO_2	11.67
Sodium, Na^+	5.79
Chloride, Cl^-	5.68
Magnesium, Mg^{+2}	3.41
Oxides (Fe, Al)$_2$O$_3$	2.75
Potassium, K^+	2.12
Nitrate, NO_3^-	0.90
Total	100.00%

Table 3–4 The Freezing Point of Seawater as a Function of Salinity

Salinity (‰)	Freezing point (°C)
0	0
10	−0.53
20	−1.08
30	−1.63
35	−1.91

seawater is further compounded by the fact that not even the total solid residue obtained is the same each time. Substances such as magnesium chloride (which gives seawater its slightly bitter taste, as well as making table salt cake) may partially decompose from the heat used to evaporate the sample. The English chemists Alexander Marcet (1770–1822) and John Murray (d. 1820), working independently, decided about 1820 that the best way to determine salinity would be to measure the separate components (the ions—but at that time the ion concept was not known in chemistry) and then add together the discovered values to infer their original composition. Even with the definite work of these two men it was not until 1850 that J. Usiglio, an Italian chemist, proved that evaporation was a poor analytical technique for seawater. The reliance of chemists on evaporation as a technique, which arose from the analysis of the much more dilute mineral waters, probably hindered seawater analysis.

But the analysis of these individual constituents is a long, complicated, and exacting procedure, much too long for the vast number of salinity determinations needed in the study of the sea. It was known by 1820 that the salt content of waters varied from place to place. Inshore waters, especially near river mouths, might easily be less saline than water from the open ocean. Marcet suggested in 1819 that although salt content varied it seemed that the ratio of the constituents was nearly constant. The great Danish chemist Johann Georg Forchhammer (1794–1865), in a work published in 1865 after 20 years of study, suggested that because these ratios were so constant, and because the chlorine content could be determined accurately and rapidly, the relationship between chlorinity and salinity could be experimentally determined. Then salinity could be calculated by the rapid, accurate measurement of chlorine and multiplication by a correction coefficient. The

3–3 Much foam is present on the water even though there is little surf activity. Seawater does not foam. Air bubbles from surf and propellers do find their way into seawater giving the appearance of foam, but it is short-lived, as the bubbles do not coalesce into foam. Here foam is caused by the exudate (alginic acid) from large amounts of brown algae or kelp (Macrocystis). *(Photo by William J. Wallace.)*

Challenger expedition of 1872–76 collected 77 water samples, specimens from all of the world's oceans on the surface and at depth (Fig. 3–4). Wilhelm Dittmar (1833–1894), the chemist who analyzed the samples, published his results in 1884. These results, which seemed proof of Forchhammer's earlier conclusion, showed the chlorinity–salinity ratio to be almost constant. Dittmar calculated a coefficient of 1.806 (as compared to Forchhammer's 1.812).

Definitions. The International Council for the Exploration of the Sea was founded in Denmark in 1902 as an outgrowth of an international conference held at Stockholm early in 1899. The Council was formed to encourage all research connected with sea exploration. The problem that seemed most important at the first conference was the establishment of certain standards. A committee of experts was appointed to investigate the problem of salinity and its possible determination and definition. In 1902 this committee, led by the Danish oceanographer Martin Knudsen, gave a long, detailed gravimetric definition of salinity, but

3–4 The chemical laboratory aboard H.M.S. Challenger.

recommended the calculation of salinity by an equation from the actual determination of chloride by titration. The equation was: salinity = 1.805 chlorinity + 0.030. This equation has been used in oceanography until very recently.

The problem with the equation, however, is that the long definition of salinity as determined by chloride is not the same as the actual salt content. A kilogram of seawater must contain a certain amount of dissolved salts. But the amounts of salt described by the equation and by the titration are slightly different. The problem is several-fold. The original equation contained the constant 0.030. This would mean, then, that the salinity of distilled water should be 0.030 (but the actual salinity of distilled water must be zero). Determining salinity by chlorinity is based on the assumption that the ratios of the ionic constituents in seawater are constant, but this is now known to be only an approximation. Very recently a new definition, again based on chloride ion determination, has defined salinity as: salinity = 1.80655 Cl‰. Although this relationship removed the troublesome constant 0.030, it still does not exactly describe the salt content of seawater. The difference in values between these two equations is not large, but it is important to oceanologists.

More recently, a new technique has been developed to measure salinity. Seawater is a conductor of electricity, and the degree of conductance is a function of the number of ions. Thus, determining salt content by this property of seawater is not only feasible, but now may be even more commonly used than chloride determination. Modern conductivity meters, or salinometers as they are generally called, are capable of the same accuracy as chloride titrations measured in the laboratory —about 20 parts per million. Salinometers have the additional value that they can be run *in situ*. The water sample is not taken out of the water for testing; the salinity is determined as the sensing coil of the salinometer is lowered into the water. Salinometers function to a depth of more than 6,000 m.

The concept of constant ionic ratios has been and still is a useful one in oceanology. The fact that salinity as defined by oceanologists is not exactly the same as the total salt content formerly made little difference; the important thing was the definition. In 1902 it was important that oceanologists agree to a particular def-

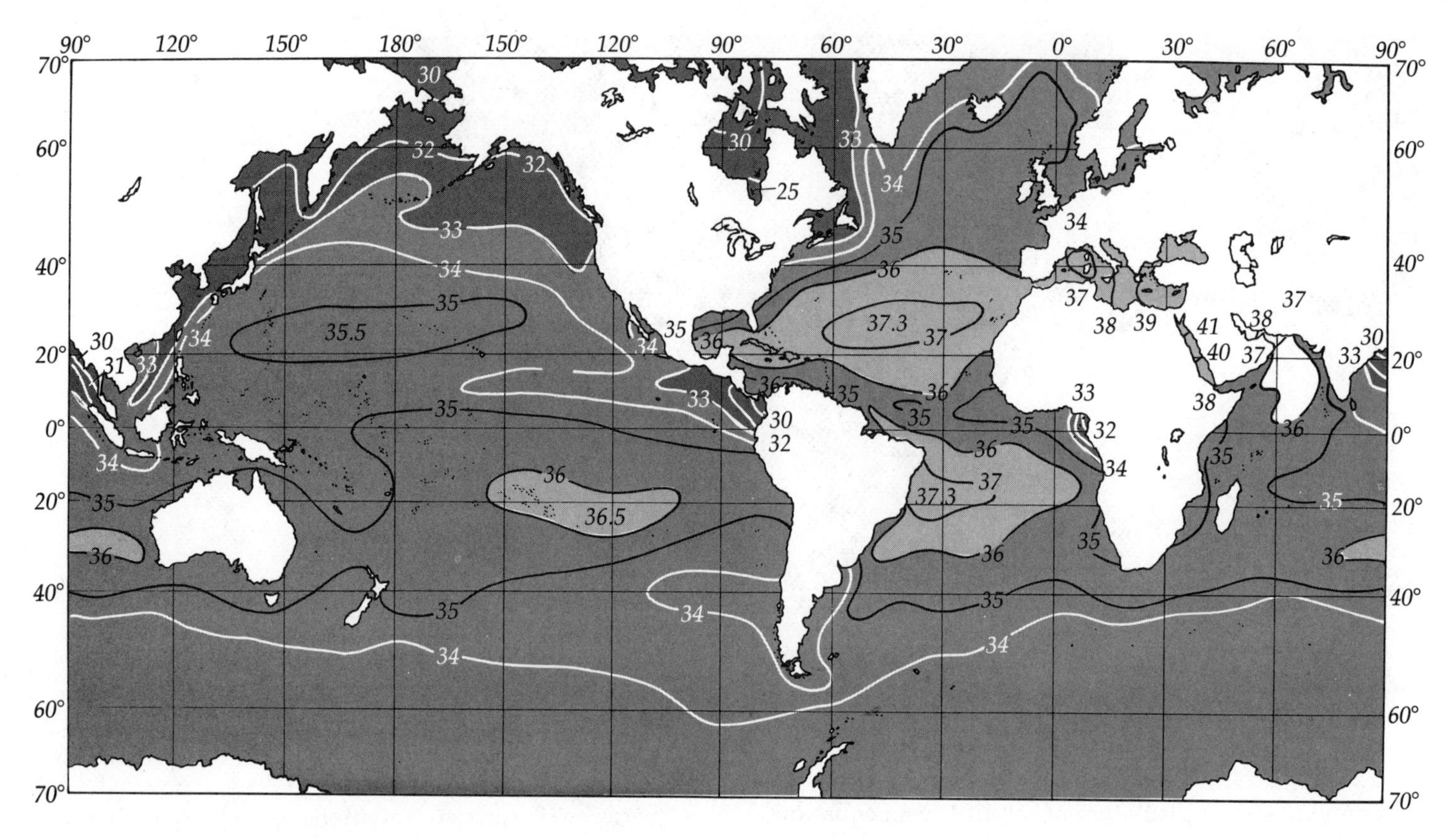

3–5 *Oceanic surface salinities (‰) in August.*

inition, even though the definition they chose did not coincide with the real physical properties of seawater, than to have an exact, accurate measure of total salts. In other words, the term *salinity* was as it was defined. In all probability, the term *salinity* will remain in use by oceanologists as a convenience. As a chemical concept, however, the idea of salinity is defunct, because most salinity determinations will probably be done by electrical conductivity. This is an example of a scientific definition changing as technology improves the means of measurement.

Distribution. If one takes a bottle of seawater, adds some salt, and shakes it, the salt will dissolve. Seawater, then, is not saturated with salt. Therefore, salinity could vary from place to place. The open oceans are fairly well mixed, so salinity there is reasonably consistent (Fig. 3–5). There are small but measurable differences in various parts of the ocean as well as slight variations with season.

Processes such as evaporation, freezing, salt into solution, currents, and mixing, which add salt or remove water, increase the salinity. Processes that remove salt or add water, including rain or snow, runoff from rivers, melting of ice, currents, mixing, and precipitation out of solution, decrease the salinity. Of these, evaporation and precipitation are the most important factors. Regions of highest salinity occur where evaporation is greater than precipitation. Conversely, where precipitation exceeds evaporation, regions of lower salinity occur. In regions where these factors are balanced, salinity is constant. In the open ocean, these variances are not too great (Table 3–5).

In almost all places in the ocean, the salinity lies between 34‰ and 37‰. The Baltic Sea, with abundant precipitation and runoff from rivers, has the lowest salinity of any sea, 12‰. The Red Sea, with little entering water and high evaporation, has the greatest salinity, 40 to 42‰. In regions such as the Gulf of Bothnia and the waters off Finland the salinity may be as low as 5‰. During the International Geophysical Year (1958–

Table 3–5 Average Values of Salinity (S), *Evaporation* (E), *and Precipitation* (P), *and the Difference* (E − P) *for every Fifth Parallel of Latitude between 40°N and 50°S*

Latitude	Atlantic Ocean				Indian Ocean			
	S (‰)	E (cm/yr)	P (cm/yr)	E − P (cm/yr)	S (‰)	E (cm/yr)	P (cm/yr)	E − P (cm/yr)
40°N	35.80	94	76	18				
35	36.46	107	64	43				
30	36.79	121	54	67				
25	36.87	140	42	98				
20	36.47	149	40	110	(35.05)	(125)	(74)	(51)
15	35.92	145	62	83	(35.07)	(125)	(73)	(52)
10	35.62	132	101	31	(34.92)	(125)	(88)	(37)
5	34.98	105	144	−39	(34.82)	(125)	(107)	(18)
0	35.67	116	96	20	35.14	125	131	− 6
5°S	35.77	141	42	99	34.93	121	167	−46
10	36.45	143	22	121	34.57	99	156	−57
15	36.79	138	19	119	34.75	121	83	38
20	36.54	132	30	102	35.15	143	59	84
25	36.20	124	40	84	35.45	145	46	99
30	35.72	116	45	71	35.89	134	58	76
35	35.35	99	55	44	35.60	121	60	61
40	34.65	81	72	9	35.10	83	73	10
45	34.19	64	73	− 9	34.25	64	79	−15
50	33.94	43	72	−29	33.87	43	79	−36

Latitude	Pacific Ocean				All oceans			
	S (‰)	E (cm/yr)	P (cm/yr)	E − P (cm/yr)	S (‰)	E (cm/yr)	P (cm/yr)	E − P (cm/yr)
40°N	33.64	94	93	1	34.54	94	93	1
35	34.10	106	79	27	35.05	106	79	27
30	34.77	116	65	51	35.56	120	65	55
25	35.00	127	55	72	35.79	129	55	74
20	34.88	130	62	68	35.44	133	65	68
15	34.67	128	82	46	35.09	130	82	48
10	34.29	123	127	− 4	34.72	129	127	2
5	34.29	102	(177)	(−75)	34.54	110	177	−67
0	34.85	116	98	18	35.08	119	102	17
5°S	35.11	131	91	40	35.20	124	91	33
10	35.38	131	96	35	35.34	130	96	34
15	35.57	125	85	40	35.54	134	85	49
20	35.70	121	70	51	35.69	134	70	64
25	35.62	116	61	55	35.69	124	62	62
30	35.40	110	64	46	35.62	111	64	47
35	35.00	97	64	33	35.32	99	64	35
40	34.61	81	84	− 3	34.79	81	84	− 3
45	34.32	64	85	−21	34.14	64	85	−21
50	34.16	43	84	−41	33.99	43	84	−41

Parentheses indicate limited data.
From H. U. Sverdrup, Martin W. Johnson, and Richard H. Fleming, *The Oceans: Their Physics, Chemistry, and General Biology*, © 1942, renewed 1970. By permission of Prentice-Hall, Inc., Englewood Cliffs, New Jersey.

59), local pockets of deep (below 2,000 m) very salty water were found in the Red Sea with salinities greater than 250‰. This is very close to the saturation point for salt in water. The saltiest open ocean is in the subtropical North Atlantic where the salinity is 37.5‰. The Pacific is less salty than the Atlantic because it is less affected by dry winds.

Only near the poles does the formation of ice and its effect on salinity become important. Salt water under forming ice has a higher salinity, because salt tends to separate out of the water as freezing occurs. This helps to account for the fact that waters near the poles, particularly at the Antarctic, are more saline than perhaps would be expected. When the ice melts, the salinity of adjacent water decreases. The net effect on freezing and melting of seawater probably averages out over the years so that the net change is zero.

Most of the comments made thus far concerning salinity and changes in this parameter have referred to the surface. Of the processes given above for the change in salinity, only one, mixing, can change the salinity of subsurface waters. The vertical distribution of salinity in seawater is not as simple as one might expect. Salinity does not simply increase with depth. In the equatorial, tropical, and subtropical regions, there is a minimum salinity region between 600 and 1,000 m. Beyond this, salinity increases to a depth of about 2,000 m, after which it decreases slightly in the Atlantic and Pacific; there is a very slight increase with depth beyond this point.

For all practical purposes, the salinity of a large volume of ocean water below the surface with a specific temperature and salinity range changes very little if at all with time. Salinity, then, is a conservative property of seawater. The nitrogen concentration and the concentration of rare gases like helium and neon in seawater are also classified as conservative properties as they, too, change very little if at all at depth. Other materials, such as dissolved oxygen or carbon dioxide, change concentration appreciably, especially because these gases are part of a number of biological processes. These are considered nonconservative properties.

Marine Life and Water. Life in the sea is very different from that on land; but, of course, there are similarities. As on land, there are complex food webs. Plants are grazed on by herbivores, which in turn may be food for carnivores. The owl, the mouse it kills, and the seed eaten by the mouse are all a part of the web. When plants become scarce, so do the animals that eat them and the predators who prey on these. So it is in the sea. Sea creatures, however, never feel the shortage of water that may exist on land. The grass of the sea consists of countless numbers of tiny free-floating plants called phytoplankton. But in the sea there are also limitations in food. The plankton takes nutrients from the water itself, silicates for the cell walls in the case of diatoms, nitrates and phosphates for protoplasm, and proteins. Note that in Table 3–2 these chemical constituents of seawater are not mentioned; they are not abundant enough. Life in the sea depends on the presence of these very minor amounts of nutrients. Where they are available, there are also phytoplankton and zooplankton. And where these exist, there are diverse forms of sea life. Unless an exchange with the deep nutrient-rich waters occurs or nutrients are introduced from the land, the surface waters are soon stripped of these nutrients. The sea equivalent of a desert is the result (see *Upwelling* in Chapter 4 and *The Carbonate Cycle* in Chapter 6).

For marine plant life, light must be a limiting factor in some locales. The Arctic Circle (67°N) is the limit of light on the shortest day of the year—December 21. Beyond 67°N there is no daylight on December 21. Because light reaching the sea's surface reflects primarily for angles of incidence less than 21° (97 percent of the light is reflected), on December 21, 46°N is the theoretical limit of light in the water ($67° - 21° = 46°$). Yet large amounts of kelp grow north of 46° all year round. The coast of Norway abounds with algae.

Salinity cannot be much of a problem for most marine plants and animals, but some are oriented to brackish water and must follow it upriver in the summer and fall and downriver in spring, when there is greater runoff. For creatures living near the sea's edge, however, salinity variances can be dangerous. During rain, tidepools can become quite diluted, and during hot weather the pools may become much more saline.

Even temperature, which is very uniform in the ocean, can be a problem at the sea's edge. Normal ocean temperature variance is quite small, much less than on land. Siberia, for example, can easily vary by 100°F or more from summer to winter, yet the sea at the same latitude changes only a few degrees centigrade sea-

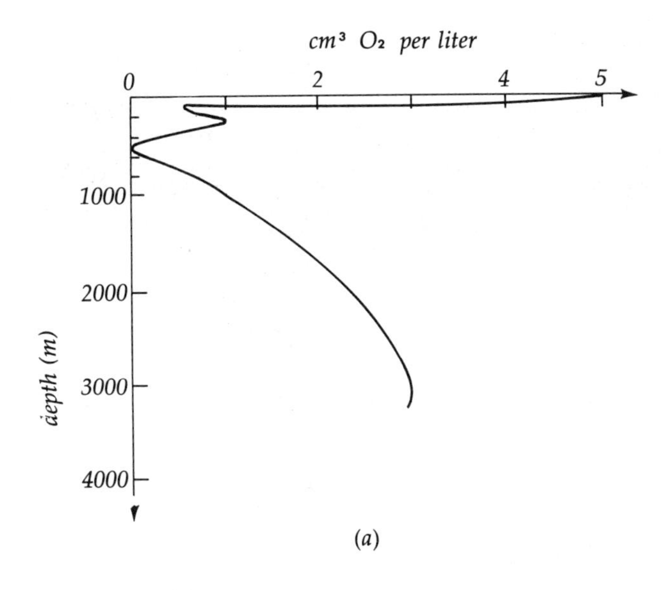

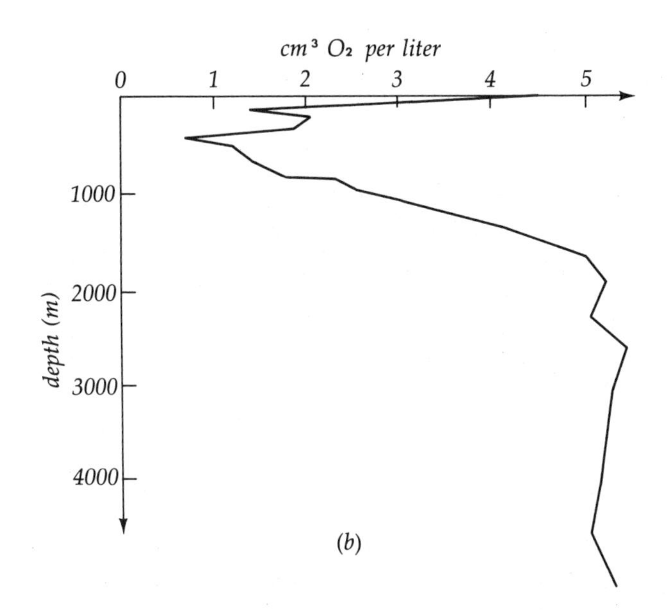

3–6 Change in the amount of dissolved oxygen with depth (a) in the eastern tropical Pacific; (b) in the equatorial North Atlantic.

sonally. In a tidepool temperature varies much more often and within a greater range. In addition, many organisms at the sea's edge (in tide pools and on pilings, for example) are often left exposed by the tides. They not only have to endure this (although some seem to require it) but the often associated great temperature extremes. Any organism in the sea, like those on land, has had to adapt to the physical parameters that control and are its environment.

Oxygen

Aside from food, most animals in the sea need oxygen, which many must take from water through gills or similar structures. Although nitrogen is the most abundant gas in the atmosphere, oxygen is more abundant in seawater because it is more soluble. A representative value for the amount of oxygen in seawater would be between 1 and 8 ml per liter, with 6 ml per liter being average. The oxygen in seawater can come from the atmosphere or from plants by photosynthesis. Both of these are essentially surface phenomena. Proceeding downward from the surface, the oxygen content decreases, being consumed by creatures at all depths and being used in the decomposition of dead vegetable and animal matter.

The vertical profile of the oxygen content (Fig. 3–6) shows a maximum at the surface, where there is actually an overproduction of oxygen in the upper 10 to 20 m. One would expect this to reach a certain minimum at depth as the light penetration to plants decreases. One characteristic of the vertical oxygen distribution in the open ocean is the existence of such an oxygen minimum, usually somewhere between 700 to 1000 m below the surface, but there is a wide variance. Below these values, the oxygen content increases somewhat, although the values are still less than those at the surface. Currently, no theory satisfactorily explains the existence of the oxygen minimum. Because oxygen is more soluble in cold water than warm, more oxygen is dissolved in the surface waters at the higher latitudes. Seawater at 0°C may contain 50 percent more oxygen than at 20°C. Because the circulation of the deep waters in the Pacific is more sluggish than in the Atlantic, deep waters in the Pacific contain less dissolved oxygen than those in the Atlantic.

Sampling

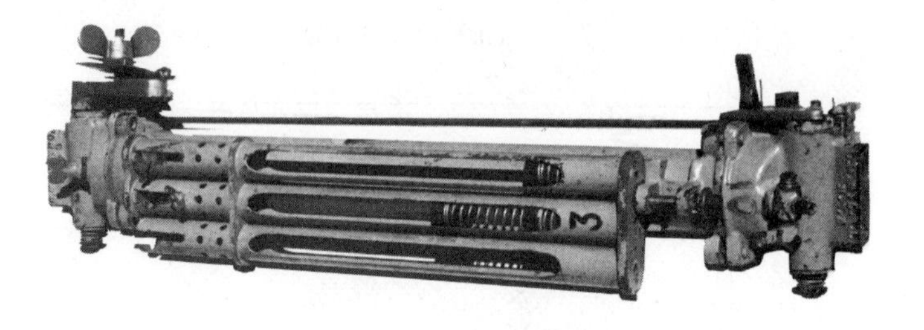

3–7 A Nansen bottle. Note the externally attached tubes for housing the protected and unprotected thermometers during a hydrocast. (Photo by William J. Wallace.)

So far, little has been said about how oceanologists collect the seawater samples they work with. It's fairly simple to throw a bucket over the side, making sure it is away from the hot water effluent of the ship's engines or bilge pumps, and haul some water back in. Sampling water below the surface is more of a problem. Today, most subsurface water samples are taken with a Nansen bottle (Fig. 3–7). This is a brass tube with valves on both ends. The bottle is lowered vertically by a cable with both valves open so the water can pass through as it descends. When the desired depth is reached a collar-like brass weight, called a messenger, is dropped along the wire, tripping the Nansen bottle, which reverses and closes its valves. A number of bottles are lowered at the same time on the cable to specific intervals, usually at the following depths for a 1 km cast: surface, 50, 100, 130, 160, 190, 220, 300, 350, 400, 450, 500, 550, 600, 650, 700, 800, 900, 1000 m, and after that at 500 m intervals for deeper casts. As the cable is played out, it is momentarily stopped by the winch operator at the desired intervals (in reverse since the deepest bottles are of course going down first). An operator in a hydrocage fixes each bottle to the cable (Fig. 3–8). Once the bottles are at the desired depth, a certain time is allowed for them to come to equilibrium, and the messenger is released at the surface. As the messenger trips a particular Nansen bottle, it releases a subsequent messenger, which drops to the next Nansen bottle, tripping them in sequence. While the operator is fixing each bottle to the wire as it goes over the side, he must be careful not to prematurely release a messenger, which will trigger all the bottles already on the line. When all the bottles in the series have been tripped, the total hydrowire with all the bottles (referred to as a hydrocast) is brought up. As each bottle slowly approaches the cage, the winch operator stops the wire, and the hydrocage operator removes each Nansen bottle from the cable, placing it in a special rack adjacent to the cage (Fig. 3–9). Tests are run on the water at that point, or the water samples may be stored in containers and frozen for later study. Normally, the samples are analyzed in a laboratory immediately adjacent to the hydrocage, sometimes called a wet lab—and literally it very often is.

The hydrowire is a primary tool by which man collects data about the ocean. Oceanologists attach all kinds of instruments to it. The hydrowire of today is a twisted or braided steel cable varying in size depending on the equipment that will be tied to it. Before the cable, piano wire was used and, even earlier, hemp rope. The time involved for a 5000 m hydrocast is between 5 and 6 hours. The hydrowire itself presents many problems. Care must always be exercised to keep it away from the ship's propellers. Sometimes bottles don't trip. The bottles and the wire must constantly be checked for corrosion. And there is always the chance that in deep samples especially the wire might tangle. Add to this the fact that the ship is not a steady platform on which to work, the bottles and the wire together are very heavy, and often it is cold and wet. But the biggest shortcoming of the hydrowire is its inability to support heavy loads. A wire with a diameter of 4 mm has a rated breaking point of approximately 3200 lb. The weight attached to make the cable sink weighs about 100 lb, and the weight of 5000 m of the wire itself brings the total to over 650 lb. Commonly, the wire is loaded to only about one-third of its breaking strength, or about 1050 lb. Subtracting the 650 lb for the wire and its weight leaves a payload of about 400 lb for instruments. Because a Nansen bottle weighs approximately 7.5 lb in water, only 50 bottles can be accommodated for a 5000 m cast.

Considering the high cost of operating a research vessel at sea, including the salaries of the crew but not those of the scientists, a representative price for a seawater sample would probably be about $12 a fifth.

The constant attempt to improve methods for taking water samples continues, and there is a need for new

3–8 A marine technician attaches a Nansen bottle to the hydrographic wire. Activated by sliding weights called messengers, the bottles on a line reverse and collect a sample of seawater at a specified depth. Thermometers on the side of the bottle record temperature in situ. *(Courtesy of Scripps Institution of Oceanography, University of California, San Diego.)*

3–9 Cores such as these, taken from the sediments at the bottom of the sea, may record millions of years of earth history; these were obtained on the Capricorn *expedition in the southwestern Pacific in 1952–53. In the background is the rack in which Nansen bottles are placed in order on their return to the surface after a hydrocast. (Courtesy of Scripps Institution of Oceanography, University of California, San Diego.)*

engineering ideas. Some of the most promising procedures appear to be the use of pumps. Submersible pumps and especially pumps located just below the surface of the water are being considered. If these pumps do become feasible, and there is every indication that they may, oceanologists would be freed from the use of sample bottles and would, in principle, be able to take much larger samples.

As the cost of manned research vessels is extremely high, other methods of collecting oceanologic data are also being constantly developed. Among these are automatic buoys designed to sit off the coast, able to withstand large storms and constantly take and transmit data on salinity, pressure, wind velocity, and temperature. The use of aircraft, especially to collect surface data such as temperature, is by no means new and, of course, a number of offshore towers have been built in continental shelf regions. The use of satellites is especially promising in the collection of surface data concerning the ocean. With each passing day there is more and more interest in scuba diving, and one can anticipate further development of underwater laboratories and research submersibles.

Temperature

Distribution. Of the parameters or properties of seawater that oceanologists measure, easily the most important single one is temperature. Contrary to what the walrus in *Through the Looking Glass* said, the sea is not boiling hot. The ocean may conveniently be seen as consisting of three layers (Fig. 3–10). In the deep layer, below 200 to 1,000 m, the water temperature of the world's oceans is very uniform and quite cold. No other place on the planet possesses such a narrow temperature range. The total ocean, for example, has a temperature range of −2 to 30°C, a much smaller range than that of the atmosphere, but at depth even this temperature range is extremely narrowed. The hottest surface temperature of any water of any ocean is in the Persian Gulf, as high as 36°C (96.8°F) nearshore in the summer. The hot, deep, salty regions of the Red Sea (see p. 80) have been found to have a temperature as high as 56°C (132.8°F) at a depth of 2,000 m, when the expected temperature is about 20°C (68°F). Geothermal solutions emanating from fractures along ocean rifts apparently cause the phenomenon.

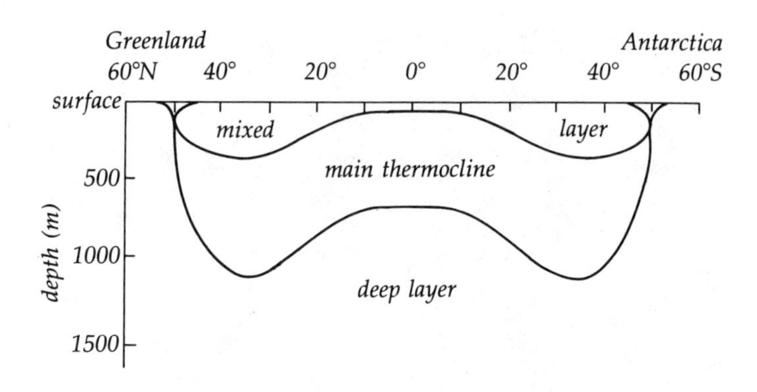

3–10 *A three-layered ocean.*

Water has a very high specific heat. Because of this, it warms and cools very slowly. Ocean water and the massive amount of heat it receives and contains from sunlight moderate and control most of the earth's climate. A diver knows that it gets colder as one goes down and this, of course, is a general situation over the world's oceans. Fig. 3–11 shows the general vertical temperature of the world's oceans. A temperature gradient that changes abruptly with depth is called a thermocline; there are both permanent and seasonal thermoclines. The sea receives most of its heat at the surface from the sun (Fig. 3–12 and Table 3–6; see also p. 97). Because the light from the sun and, therefore, the heat striking an area varies with the season, so does the temperature of the surface waters (Fig. 3–13). A vertical temperature profile for regions with notable seasons (that is, the mid-latitudes) is shown in Fig. 3–14a. Notice the warmer summer waters. The equator does not experience much change in heating over the year so a temperature profile there looks the same all year round (Fig. 3–14b).

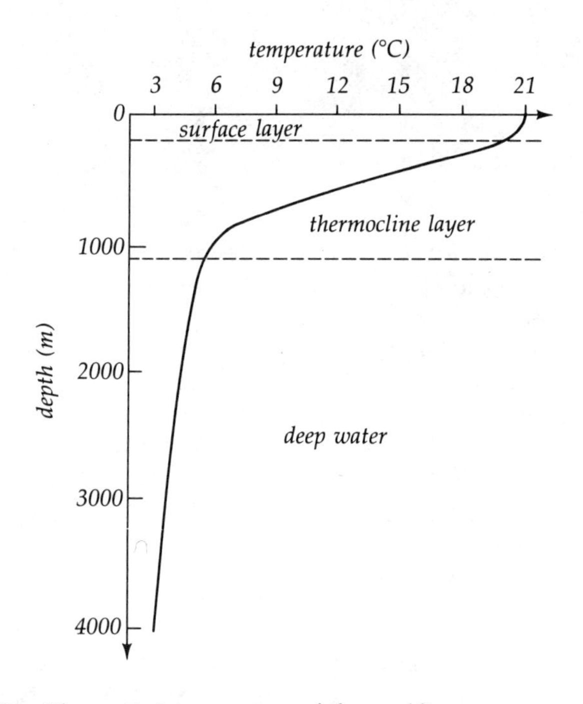

3–11 *The vertical temperature of the world's oceans.*

Although the areas near the poles (the high latitudes) are quite different in actual temperature from areas on the equator, the regions are similar in profile (Fig. 3–14c). Temperature layering in the ocean is obvious (Fig. 3–15). Very recent studies by a number of scientists have shown that this layering is much more extensive and detailed than had been previously thought. Variations in temperature might now be referred to as microlayering or microstratification because many layers vary only a fraction of a degree from one to another.

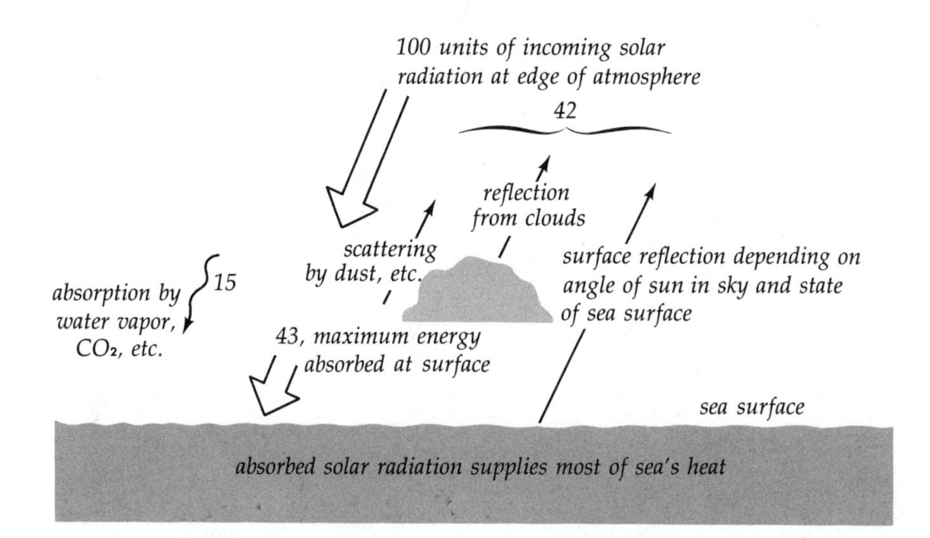

3–12 Solar radiation from the upper atmosphere to the sea.

Differences in surface water temperature and air temperature cause a number of phenomena. Fog is one of these; a somewhat more unusual phenomenon is sea smoke (Fig. 3–16), which is caused when water temperature is greater than that of the air.

Measurement. When a hydrocast is made, the temperature of the water column at each interval is also recorded. Each Nansen bottle has two reversing thermometers attached to the outside of its cylindrical body. Because the temperature registered by a thermometer will usually decrease on the way down or increase on the way up, these thermometers have the capability of fixing a temperature recorded at a desired depth. The cast is left at a depth for approximately 10 minutes so that the thermometer can come to the temperature of that particular depth. When the Nansen bottle flips as the messenger triggers it, so do the thermometers. Reversing thermometers have a small capillary constriction so that when the bottle flips, part of the mercury column is separated, making a further temperature change by the thermometers impossible (Fig. 3–17a).

Two thermometers are attached to the Nansen bottle housing. One is shielded against pressure and the other is unprotected (Figure 3–17b). When a hydrocast is down, there is no way to tell with certainty that the cable is vertical or that it has no kink or bend in it, which would give a false recording. The two thermometers give the only true depth values for a sample depth. Pressure forcing in on the unprotected thermometer causes it to show a higher temperature reading than the protected one. The glass of the unprotected thermometer is squeezed in, and the capillary size decreased, causing the mercury level to rise higher. From a cross check of the different temperature readings using known empirical relationships, the pressure to which the thermometers were subjected is calculated and from the pressure, the depth may be determined. To an oceanologist, pairs of these thermometers that have proved reliable and accurate are treasured almost as friends and cared for accordingly. They are the most important part of a hydrocast.

Other methods are available to measure temperature. The one that has been most used in the past is the bathythermograph (Figs. 3–18, 3–19, and 3–20). The temperature readings from the bathythermograph (or BT) are not as accurate as those from a reversing thermometer. Generally the BT is capable of registering only to a depth of about 270 m, which is adequate for a number of purposes. A BT can be used while the ship is moving, although the maximum towing speed is approximately 10 knots. More recently an expendable bathythermograph, the XBT, has come into use (Fig. 3–21). It contains a temperature sensor and a spool of

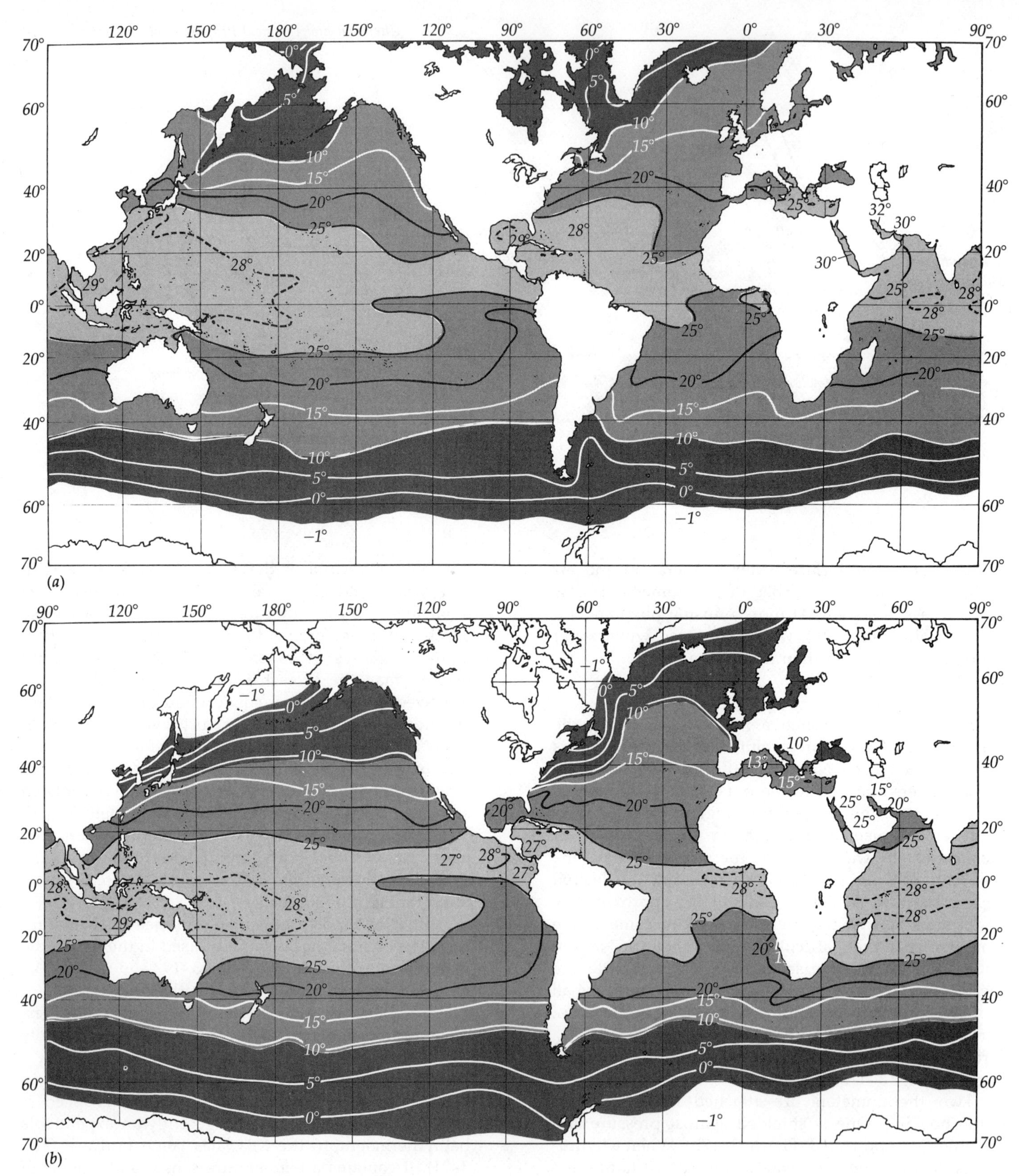

3–13 *Surface temperature of the oceans (a) in August, (b) in February.*

Table 3–6 Average Surface Temperature of the Oceans between Parallels of Latitude

Latitude	Atlantic Ocean	Indian Ocean	Pacific Ocean
70°–60°N	5.60	—	—
60 –50	8.66	—	5.74
50 –40	13.16	—	9.99
40 –30	20.40	—	18.62
30 –20	24.16	26.14	23.38
20 –10	25.81	27.23	26.42
10 – 0	26.66	27.88	27.20
70°–60°S	− 1.30	− 1.50	− 1.30
60 –50	1.76	1.63	5.00
50 –40	8.68	8.67	11.16
40 –30	16.90	17.00	16.98
30 –20	21.20	22.53	21.53
20 –10	23.16	25.85	25.11
10 – 0	25.18	27.41	26.01

From H. U. Sverdrup, Martin W. Johnson, and Richard H. Fleming, *The Oceans: Their Physics, Chemistry, and General Biology,* © 1942, renewed 1970. By permission of Prentice-Hall, Inc., Englewood Cliffs, New Jersey.

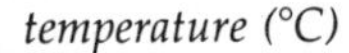

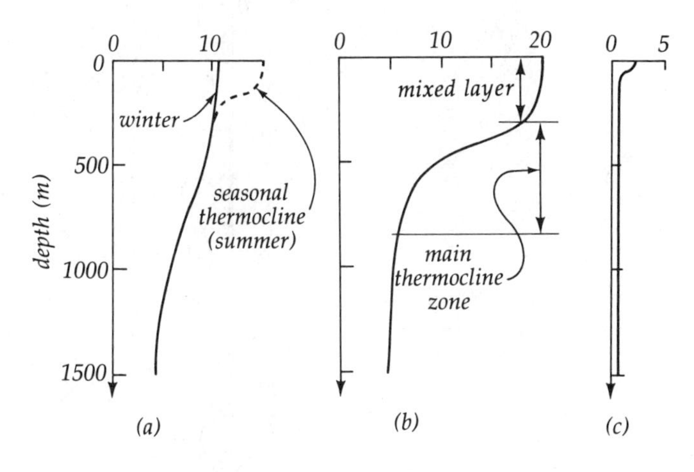

3–14 Average temperature profiles for the open ocean. (a) Mid-latitudes; (b) low latitudes; (c) high latitudes.

wire. After it is thrown over the side, the data is relayed electrically by wire to the ship. The depth is determined by its known rate of sinking.

An even more recent method of measuring temperature is the thermistor chain. A number of electrical temperature sensors called thermistors, which work on the thermocouple principle, are attached to a cable. The cable is towed and a constant reading of temperature throughout the depths is recorded (Fig. 3–22).

Pressure

Little need be said about pressure. Of all the parameters of the sea, the only predictable one is pressure. As one might figure, with an increase in depth the pressure increases. For each 10 m of depth, the pressure increases by one atmosphere. The term *atmosphere* refers to the pressure (force/area) exerted on a surface at sea level by a column of air extending to the edge of the atmosphere, a distance of about 160 km. At sea level, the weight of such a column of air with a cross section of one square inch is 14.7 lb. It is equal in weight to a column of mercury 76 cm high or a column of water approximately 10 m high. Thus, the pressure on the bottom of each container in Fig. 3–23 is the same. Pressure is a function of the height of the water column on a unit area at the bottom (the unit area usually chosen is 1 in.^{2}). The weight is the force that gravity exerts on the water. Then what is the pressure inside a large blob of water existing in space? From place to place for a particular depth, the variation in pressure is extremely slight. These small differences result from the fact that the density of seawater in the column may not be uniform.

Because the deepest part of the ocean is 7 mi (11 km) deep the pressure there is almost 1000 times atmospheric pressure (almost 7 ton/in.^{2}). These pressures are tremendous. A pressure of 1,350 lb/in.^{2} (the pressure at 1000 m) would compress a block of wood to about one-half its volume, so much so that it would sink. Pressure is the biggest reason man has not explored more of the deep ocean in submarines. Man has, however, reached the great depths of the ocean. On January 23, 1960, the U.S. Navy's *Trieste* descended to a depth of 10,920 m, to the bottom of the Challenger Deep in the Mariana Trench. The trip down took 4 hours 48 minutes. The *Trieste,* however, is not capable of sustained

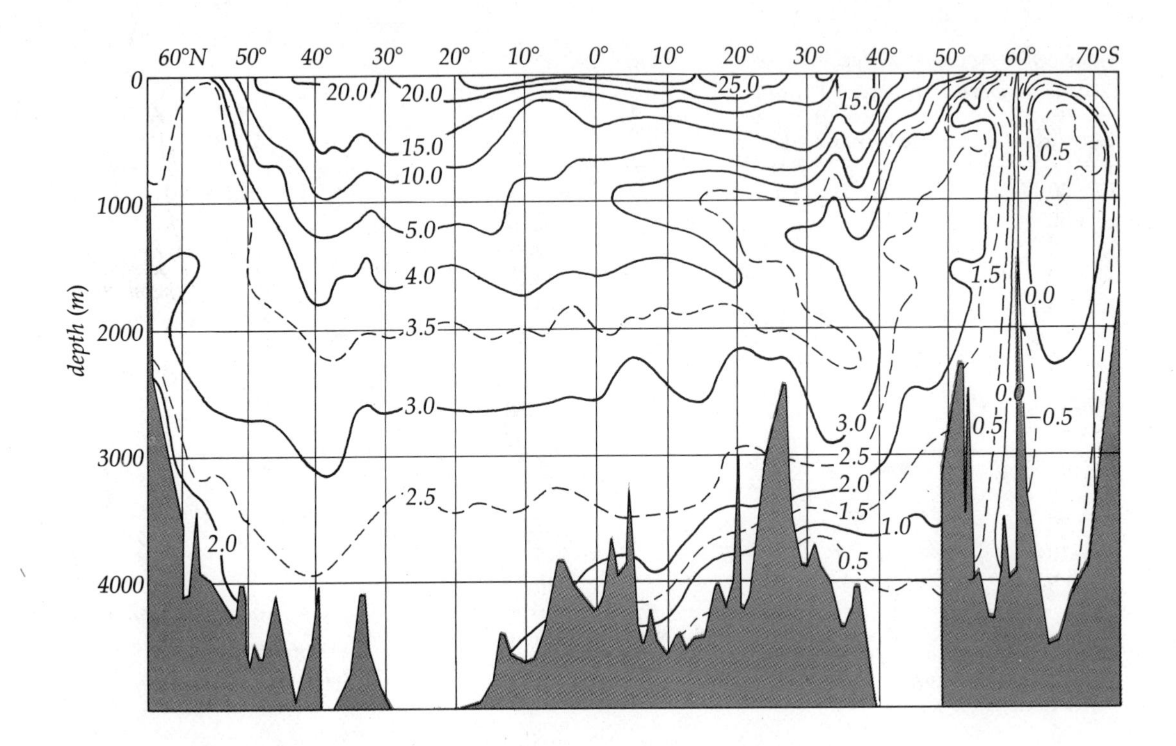

3–15 *North–south temperature (in °C) profile of the Atlantic Ocean.*

operation or many dives. It is slow and is purely a research vessel. While scuba divers can presently descend to 180 m, for all practical purposes they are limited to a working depth of about 60 m (because of the restriction imposed by air supply). A helmeted or hard-hat diver can normally work at 60 m for 3 or 4 hours. He must then make a very slow return to the surface. World War II submarines had very shallow diving depths, seldom exceeding 60 to 100 m. The more modern nuclear submarines can dive three times as deep.

Often associated with deep dives by hard-hat and scuba divers is an affliction known as "the bends." At the higher pressures associated with depth, more nitrogen dissolves in a diver's tissue and in his blood. If the rate of ascent is too fast to allow for the normal return or loss of nitrogen, it will collect as bubbles in his blood, joints, bone marrow, and tissues, causing extreme pain. Usually it is not fatal, unless very deep dives are involved; then, if normal decompression procedures are not followed, bubbles will collect in central nervous system tissue such as the brain and the spinal cord, which might cause death. Should a diver for one reason or another fail to adequately decompress on the way up, he is put into an air pressure chamber to simulate the pressure of the ocean depth, and the pressure is slowly decreased. In scuba diving, decompression is not necessary unless one goes deeper than 10 m (1 atmosphere). A good rule to follow for shallow dives when decompression tables are not needed is that one should never ascend faster than his own air bubbles. In snorkling, one normally does not go deep enough to worry about pressure. Air of a different pressure is not taken into the system in such a case, so bends cannot occur. Whales and other diving aquatic mammals, like a free diver, cannot get the bends.

Density

The terms *density* and *viscosity* are often incorrectly related and, worse, interchanged. Of the substances that exist in nature as liquids water, especially seawater, has one of the highest densities yet a low viscosity. Oil, on the other hand, is less dense than water (it will float on water), yet it has a higher viscosity than water.

Viscosity is the measure of a fluid's ability to flow. If a fluid flows easily, it is said to have a low viscosity. If it flows slowly, it possesses a high viscosity. This property is not a function of density but rather of the internal structure of the particles and interparticle

3–16 *Sea smoke. (Official United States Navy Photograph.)*

attraction. Temperature is often a factor, and one need only recall the proverbial comment about the rate at which cold molasses moves. An old pane of glass is not optically clear like a new one is; it shows some distortion. An accurate measurement with a micrometer of the thickness of the pane at the top and the bottom would reveal that the bottom is thicker than the top. Glass, technically, is classified by scientists as a supercooled liquid. As such, it has an extremely high viscosity and, because of gravity, moves very, very slowly downhill toward the center of the earth, making the bottom of the pane thicker.

The high density and low viscosity of water make the sea's surface easily traveled by shipping. Heavily loaded ships still float high in water because of the water's high density; because of seawater's low viscosity, the ships can move easily through the water. If seawater had a much higher viscosity, the job of pushing a ship through the water would be much more expensive and difficult. These two factors, high density and low viscosity, also produce the primary danger of the sea to ships and shore. Because the water has such a low viscosity, waves can build up to very large heights, exerting potentially crushing forces on ships, piers, and coasts. If the oceans were as viscous or syrupy as molasses, there would be virtually no waves, even at the shoreline.

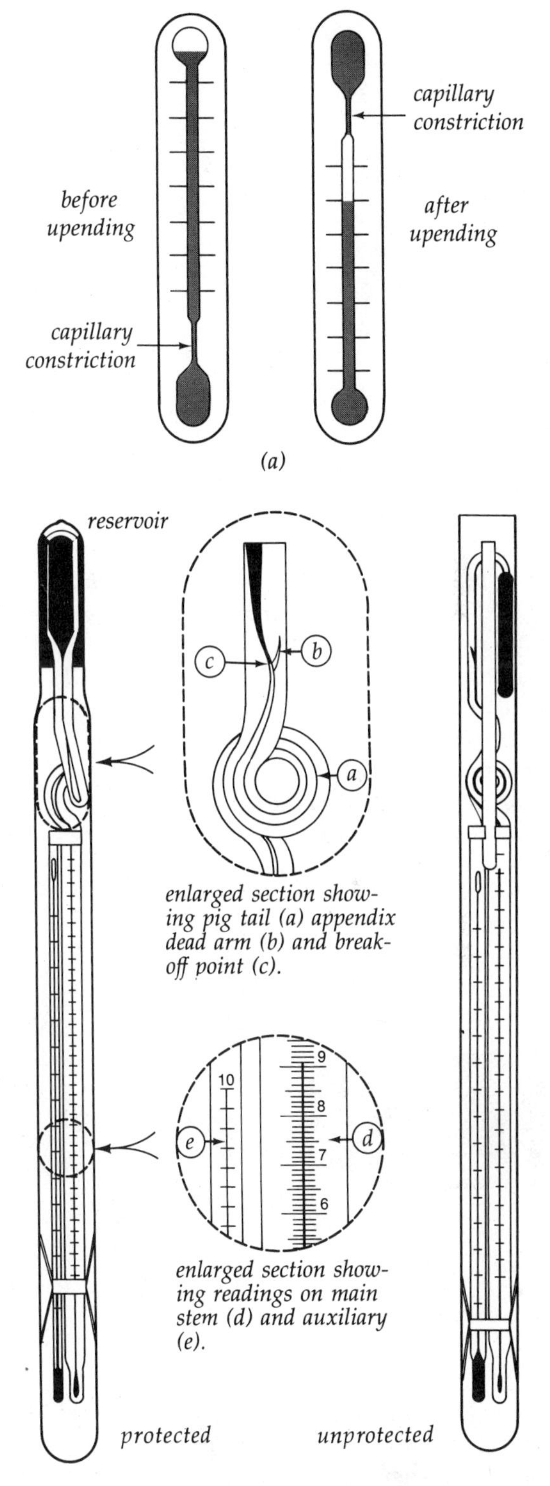

3–17 *(a) The reversing thermometer (simplified). (b) Actual reversing thermometers, protected and unprotected. (From the United States Naval Oceanographic Office, Washington, D.C.)*

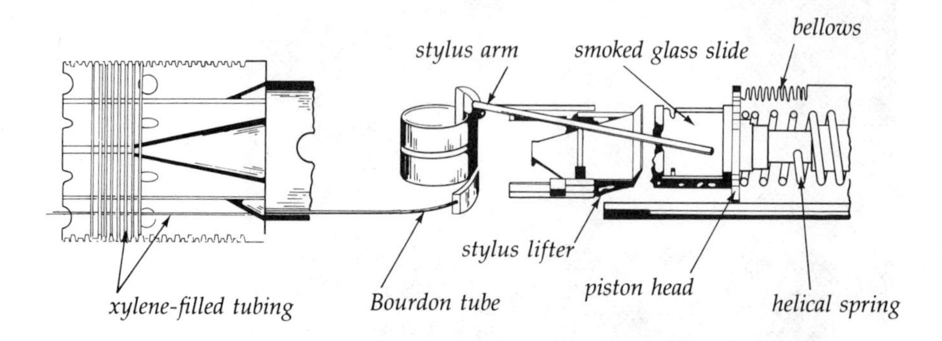

3–18 Cross section showing the bathythermograph mechanism.

(*a*)

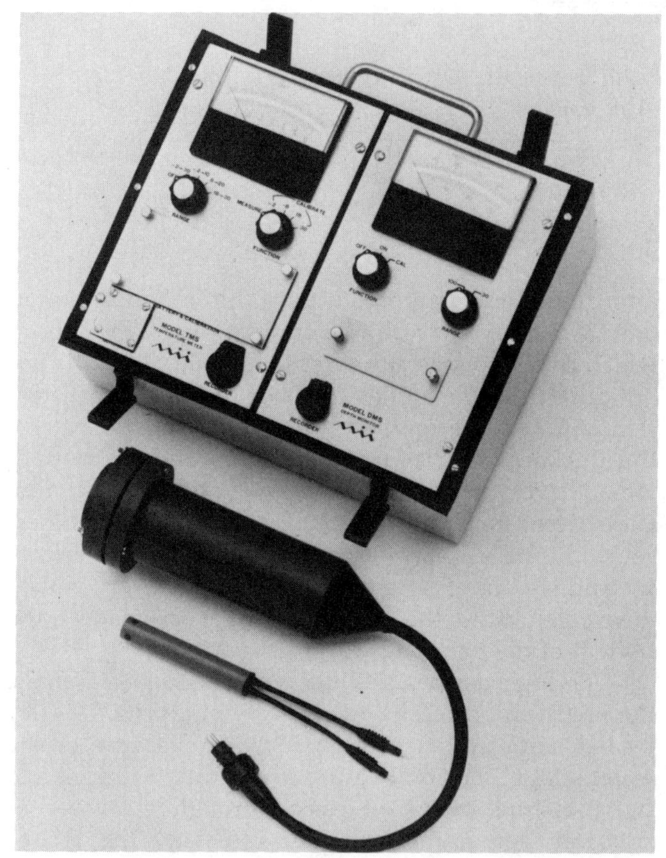

(*b*)

3–19 (a) A mechanical bathythermograph. A temperature profile is sketched on a smoked or coated glass slide inserted in the rectangular opening in the side of the cylindrical middle section. The thin brass collar-like sleeve is moved to the left (in this photo) over the hole to protect the mechanism somewhat. The slides, called BTs, are shown in Fig. 3–20. (Photo by William J. Wallace.) (b) An example of some of the new light portable equipment now appearing. This electronic bathythermograph (EBT) couples temperature and depth sensors. (Courtesy of Martek Instruments, Inc.)

The high density of water in a gravitational field is responsible for the tremendous pressures at considerable depths. Density is defined as mass per unit volume or $D = m/v$. The terms *weight* and *mass* are commonly interchanged. Although scientists knew the distinction, it was not until the advent of satellites and space travel that it became really necessary to clearly differentiate between the two. Any object, such as a half dollar, is composed of a certain number of molecules or atoms. The sum of these particles is its mass. The half dollar has weight. You can feel it in your hand. If the half dollar were orbiting earth, 200 miles from the

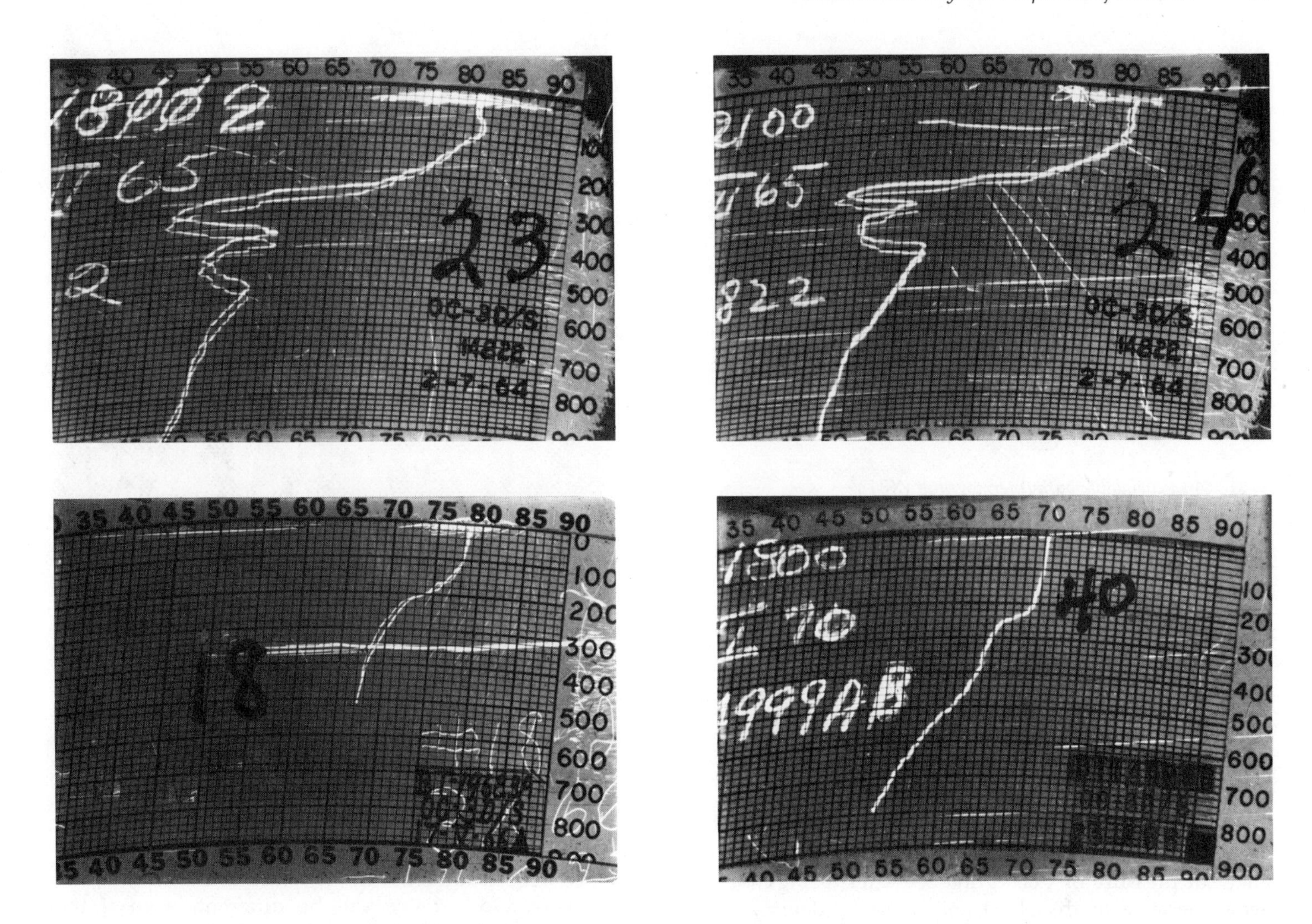

3–20 Representative BTs from the world's oceans, made on various National Oceanographic Data Center cruises. # 23, made 8/28/65, 1800 hours, at 37°30′N 71°02′W (Hudson Canyon). # 24, made 8/28/65, 2100 hours, at 37°N 71°30′W (Hudson Canyon). # 18, made 10/4/66, 0010 hours, at 35°10′N 44°13′W (central Atlantic west of the mid-Atlantic ridge). # 40, made 11/9/70, 1800 hours, at 30°N 140°W (near a Pacific fracture zone half way between Honolulu and San Francisco). Coordinates are depth in feet and temperature in Fahrenheit degrees. (Courtesy of Scripps Institution of Oceanography, University of California, San Diego.)

earth's surface, it would be weightless. But it would still be made of the same number of particles, so its mass would not have changed. Weight, then, is only a measure of the attraction a gravitational field like that of the earth has on matter. The same half dollar would weigh one-sixth as much on the moon as it does on earth because the moon's gravitational attraction is only one-sixth of earth's. The value for gravity and acceleration, normally referred to as 980 cm/sec^2, varies from place to place on the earth's surface, but only slightly. If they are at the same locale, the old question, "Which weighs more—a ton of feathers or a ton of

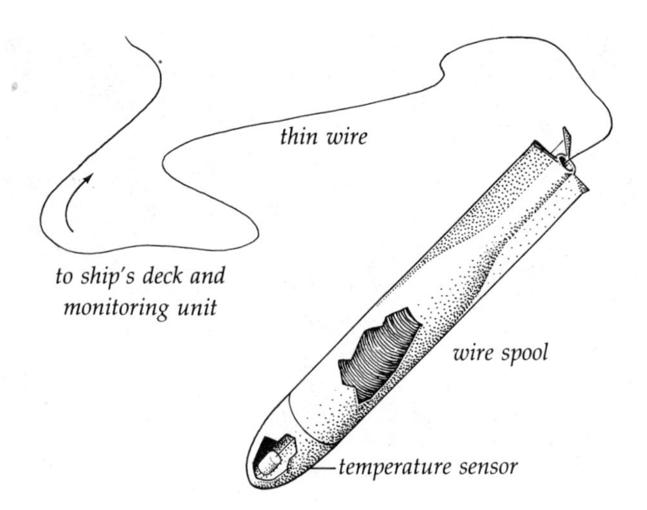

3–21 *An expendable bathythermograph (XBT).*

3–22 *Thermistor chain on a winch mounted on the after deck of the U.S. Navy ship* S. P. Lee. *(Official United States Navy Photograph.)*

lead?" is silly. The mass is the same, but the volume occupied by the feathers would be the size of a room and that of the lead might be roughly the size of a desk. The density of feathers, then, is much less than that of lead. Density is simply a measure of the amount of matter in a particular volume.

The density of seawater is expressed by oceanologists in the metric units grams per milliliter. By definition, pure water has a density at 4°C and 1 atmosphere of pressure of 1.000 g/ml. Seawater, being more dense because of dissolved solids, has a density of between 1.02 and 1.03 g/ml. This is not a great variation. The density of pure water is a function of temperature and pressure, but in salt water there is the additional effect of the salinity. These three factors, temperature, salinity, and pressure, in order of decreasing relative significance, determine the density of seawater. Though water is not very compressible, at depth pressure is still a factor to consider.

The question arises, "How might density be measured?" The simplest method is to throw a bucket over the side, bring up some water, weigh accurately the particular volume, and calculate the density at some known temperature. Because water, like other substances, changes its volume as the temperature changes, its density also changes as the temperature changes. To be meaningful, density must be expressed at a certain temperature. This method will work for surface waters. But what about subsurface waters? Once the subsurface water is raised to the surface, the pressure and density will be different, even if the temperature could be kept constant. Only the salinity of the sample would remain unchanged. Consider the following situation. Suppose one weighs a flask with the stopper sitting on the balance pan so it too is weighed (Fig. 3–24). The stopper is then put into the bottle to seal it. Is there a change in weight? Assume the mass of an aircraft can somehow be determined in flight from the ground. If a bird sits on the wing of the flying aircraft, does the weight of the plane increase? If a bird were sitting on a chair inside a closed pressurized jet, would the plane weigh more or less than if the bird were flying around in the cabin? All of these questions are similar. In the first case, there is no change in the mass of bottle and stopper. The reason is simply that you can't weigh air in air. By using an air pump to evacuate the flask one could weigh the absence of air in air; similarly, even if you used a scale that could function in deep water, you cannot weigh

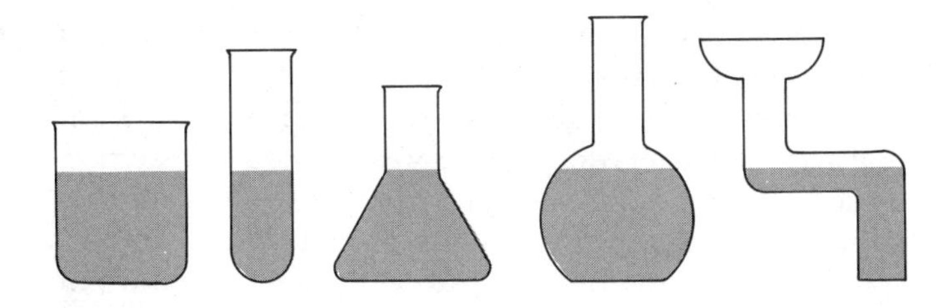

3–23 The pressure per square centimeter (or any square unit) is the same at the bottom of each container.

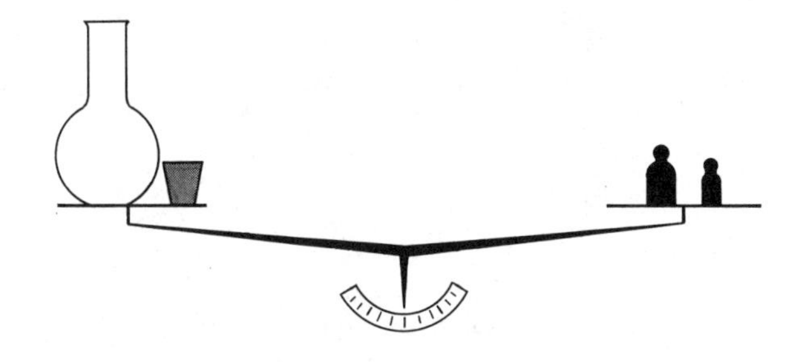

3–24 Can air be weighed in air?

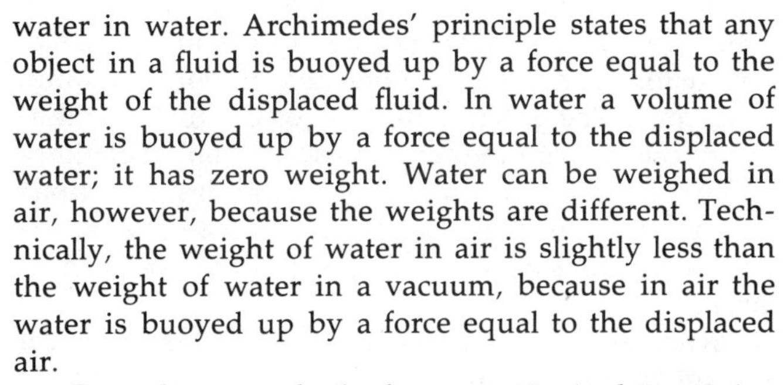

water in water. Archimedes' principle states that any object in a fluid is buoyed up by a force equal to the weight of the displaced fluid. In water a volume of water is buoyed up by a force equal to the displaced water; it has zero weight. Water can be weighed in air, however, because the weights are different. Technically, the weight of water in air is slightly less than the weight of water in a vacuum, because in air the water is buoyed up by a force equal to the displaced air.

Several new methods show promise in determining the actual density of water as it is in the ocean. But the most commonly used method is the calculation of density. To do this, one must know the temperature, salinity, and depth or pressure. Tables have been constructed to aid in this determination (see Appendix IV).

The question might be asked, "Why is it necessary to determine density when it is fairly obvious that density must generally increase with depth?" Density does increase with depth, although not necessarily uniformly. The heaviest or most dense waters are found at the bottom of the ocean. The lightest or least dense waters are found at sea level. In some regions of the deep sea the density is constant for some depth. In other areas, called pycnoclines, the density rapidly increases with a small change in depth. Regions such as these are very stable. A water column is stable, or has little tendency to mix vertically, if the lighter fluid is above the heavier fluid and there is a sharp gradation in density. In some cases lighter water may be below the heavy water, and in this case the water column is unstable. When density is the same over a wide region, the area has neutral stability. The primary reason for determining density and stratification is to study currents. In deep waters, gravity is the motive power behind these deep currents. The waters sink until they reach layers of equal density. Water masses move along these density discontinuities. The study of subsurface currents is based largely on the determination of the densities of the deep waters.

Two water masses having the same density may have different temperatures and salinities, as varying values of each can result in the same density. As a rule, temperature is more important than salinity in density determination. Because temperature and salinity can vary, it is not too unusual to find warmer waters below cold or less saline waters below more saline waters, but still have a stable water structure. A region with a clearly defined pycnocline is the most stable (Fig. 3–25).

Light

When the English poet Lord Byron said, "Roll on, thou deep and dark blue ocean!" either he was referring to a very small part of the ocean or he hadn't seen a great deal of it. The world's oceans and seas have a variety of colors. Surprisingly, waters of a clear, transparent azure blue, although they are generally found in the warmer regions, do not contain particularly abundant marine life. The vast bulk of marine life occurs in deep, colder, and almost opaque dark blue-green waters, only about 10 percent of the ocean.

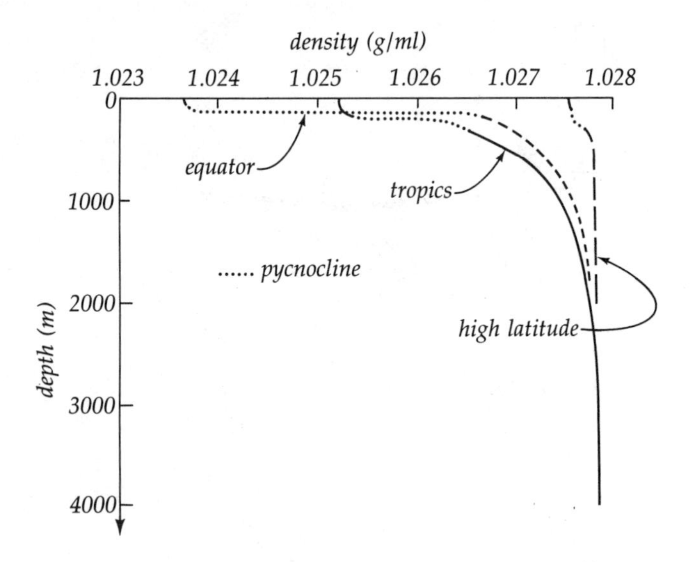

3–25 *Typical density profiles.*

The names of many of the world's seas refer to color: the Black Sea looks dark because its bottom is covered with black sediment; the Yellow Sea off northern Siberia contains an appreciable amount of river-borne yellow mud, especially during flood season; the Red Sea and the Vermilion Sea (the Gulf of California) appear red because of the presence of blue-green algae that *in toto* have a reddish coloration; the White Sea owes its name to the fact that it is frozen solid more than 200 days of each year; the Kuroshio Current, sometimes called the Japanese Current (not altogether correctly), is dark or colorless—hence the name, which in Japanese means "black stream."

As It Reaches the Sea Surface. Color is a function of the light spectrum. White light is composed of red, orange, yellow, green, blue, and violet. The color of a particular area of seawater may change as one views it because a cloud is passing overhead or because the angle of the sun changes, for example, toward sunset. The reason for the sea's general blue color is the same as that for the blue sky. Blue light, which possesses a short wavelength (compared to red), is most easily scattered by water particles and the microscopic matter that water contains; in fact, the general color of the sea is the function of light scattering through suspended particles, the reflection of the sky's color, and the nature of the suspended and dissolved matter in the water. All of this light originates in the sun; the light one sees, however, does not represent the full spectrum of solar radiation. Pure light from the sun would be lethal to all forms of life on this planet. To best understand why, it would be well to consider the nature of light.

Light, unlike sound or water waves, needs no medium for its transfer. Thus, it can travel through a vacuum such as space where, for example, sound cannot. Light and the term *electromagnetic energy* are often used interchangeably. Technically, only the portion of the electromagnetic spectrum that can be seen is known as light. The entire energy spectrum breakdown is shown in Fig. 3–26. To the right of the visible spectrum on the chart is the region called infrared, which one feels as heat (as from a heat lamp). Left of the visible spectrum, the energy of light becomes dangerous. The intense heat of black light or ultraviolet causes tanning and burning and sometimes sunstroke and death, especially in very bright regions such as the Sahara or Death Valley. Sunlight can be more dangerous to human beings than most people think.

Scientists have not yet developed a concise model to explain the nature of light. Very often, light is treated as waves, but it may also be treated as small discrete bundles of energy called photons. Experiments can "prove" either that light exists as waves or that it exists as photons. Perhaps it is best to think of light as neither a wave nor a photon phenomenon, but something that has the properties of both. Some generalizations about light may be helpful here. Wavelength is the length of a wave of light. Red light has a longer wavelength than blue light, but it also has a lower frequency than blue light and is less energetic.

The sun's light as emitted at its surface contains a large portion of the spectrum shown in Fig. 3–26. Most of this energy range never reaches the earth's surface. In fact, only a minute fraction of the sun's light ever approaches the vicinity of the earth (Fig. 3–27). Much of the light that approaches the earth is reflected either by the earth's atmosphere or by its surface. Also, a large percentage of the light that hits the earth's atmosphere is absorbed as energy by the particles that comprise the atmosphere. The net effect is a shift in the wavelength of light as it reaches

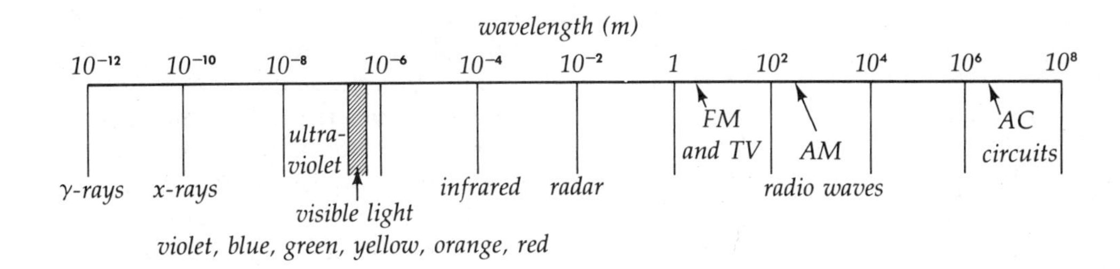

3–26 The electromagnetic spectrum.

the earth's surface. Approximately 65 percent of the light reaching the earth's surface is infrared; about 10 percent of the light is ultraviolet; the remaining 25 percent is visible. The nature of the earth's surface to a large extent determines the degree of reflection. In areas toward the poles, where there are large amounts of snow and ice, the degree of reflection is much higher than in forests, plains, or grassy areas. The surface of the sea is also a fairly good reflector of sunlight. When would more light energy penetrate into the ocean, when its surface is smooth or rough?

Electromagnetic radiation, obviously, causes the temperature structure of the sea. Interaction with water causes a shift in the wavelength of light from ultraviolet to infrared. All of the absorbed energy either is turned into heat, chemical, or kinetic energy or is lost by radiation (especially to the atmosphere), convection, or conduction.

From year to year the heat content of the oceans varies little. Although the bulk of the oceans' heat comes directly from the sun, other processes such as water condensation at the sea surface and conduction of heat from the atmosphere add solar heat indirectly. Other processes, such as evaporation of water to the atmosphere, are sources of heat loss.

Since the relative angle of the earth's attitude with respect to the sun changes during the year, the amount of the light striking the earth varies as one moves away from the equator (Fig. 3–28). This change in the amount of light falling on a square meter of the earth as one moves away from the poles is responsible for seasons. In winter, the northern hemisphere is 3 million miles closer to the sun than in mid-summer, yet winter is colder. This difference in temperature cannot result from closeness but must be caused by the change in attitude of the earth with respect to the sun. During the summer the northern hemisphere is tilted toward the sun; and, because of this, each square meter receives more light than in winter. Near the equator, the attitude of the earth with respect to the sun changes very little; as a consequence, the climate in equatorial regions is essentially the same all year round. The regular change in the sun's light and heat reaching an area creates the seasons and determines the area's climate. Local changes, like clouding and wind patterns, cause the daily changes in weather, which averaged over a long period of time like a year, are called climate. All weather—winds, snow, rain, or anything—results from sunlight heating the earth's surface. As mentioned, the nature of the earth's surface helps to determine the average degree of light absorption. A forest, for example, has an entirely different light absorption characteristic than a beach area.

It is impossible to overemphasize the importance of light for the earth. Aside from the fact that all weather is generated by light, the light reaching the surface of the earth and the sea is largely responsible for all life. Light is life. All the energy expended on this planet (with the exception of the extremely small fraction of nuclear energy) comes from the sun. The burning of fossil fuel like coal and oil releases solar energy that was stored for thousands of years as chemical energy in the remnants of plants and animals. A normally active person uses energy at about the same rate as a 100 watt light bulb. This energy, too, comes from the sun through food that could not have been produced without sunlight.

The existence of life on this planet depends on the small amount of light the planet receives from the sun. This energy is necessary for heating and for photosynthesis, the basic process by which food for higher organisms is generated. If the average temperature of the earth and its atmosphere is relatively constant,

3–27 *Solar radiation received by the earth.*

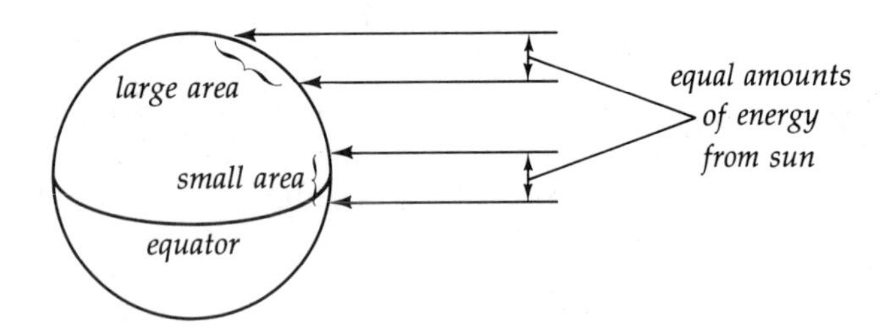

3–28 *Unequal heating by the sun at different latitudes.*

then the earth must be losing energy to space at the same rate that energy is received from the sun.

During certain periods in the earth's history, other factors have affected the amount of light the planet's surface received. Ice caps, for instance, reflected more light in the past, but changes in the nature of the protective umbrella of the atmosphere have primarily determined the amount of light received on the earth's surface. A recent increase in carbon dioxide in the air concerns scientists. As sunlight penetrates the atmosphere, some is absorbed by particles of certain gases in the atmosphere and reradiated as infrared or heat. Carbon dioxide and water vapor are largely responsible for this transfer. The same effect occurs when an automobile is parked in the sun with the windows rolled up. The temperature in the automobile becomes much greater than that of the outside air. The glass allows sunlight to pass through it. As it hits the interior of the car the sunlight is absorbed by the material and transformed into infrared or heat. The glass does not allow this radiation to pass out of the car at a same rate and, consequently, the temperature inside the car increases. This phenomenon, referred to as the "greenhouse" effect because greenhouses receive their energy in the same manner, largely determines the temperature of the earth's atmosphere. Because of the tremendous increase in burning of fossil fuels, which adds carbon dioxide to the air, scientists have been worried about the possibility of an undue increase in the earth's temperature. Varying carbon dioxide content in the atmosphere is a possible explanation for the occurrence of ice ages.

Much more recently, another factor has become a source of alarm. Particulate matter in the atmosphere can and does reflect sunlight. The volcanic dust and ash pumped into the northern hemisphere in 1815 by a volcanic eruption of Mount Pambora on Sumbawa, Indonesia, canceled summer in the northern hemisphere in 1816 simply because this large amount of airborne matter reflected much more of the sunlight back into space than was normal. Man, by his burning of fossil fuels, is rapidly increasing the amount of particulate matter in the atmosphere. One need only watch most jets taking off to see the dark trails of particulate carbon being emitted from their engines. Although the most recent data tend to show a marked increase in turbidity in the atmosphere in general, perhaps as much as 30 percent in some areas, few particulates have been unleashed into the very high upper atmosphere. Commercial jet aircraft now fly roughly between the altitudes of 6 to 12 km (20,000 to 40,000 ft). The supersonic transport (or SST) would fly at much higher altitudes and, with present and proposed engines, would constantly dump thousands of tons of exhaust emissions into the upper atmosphere. Unlike the lower atmosphere, the upper regions have little or no vertical circulation. Any matter dumped in that region will probably stay there for a much longer period of time, and the build-up would continue. There is also an additional effect, first noted during World War II when B-17s and B-24s flying at high altitudes created trails of condensing water vapor or contrails. It is quite possible that the contrails of high-flying SSTs would enormously increase the cloud cover of the high atmosphere.

The net effect of one or both of these phenomena would be a decrease in the light hitting the earth's surface, causing the average temperature of the atmosphere and the whole earth's surface to go down. Quite frankly, scientists don't know enough yet to determine what is really happening to the atmosphere. If the earth's atmospheric temperature increases markedly, then more polar ice would melt and sea levels would increase. If a greater amount of sunlight is reflected, then atmospheric temperature would drop and polar ice would increase in size. Apparently, most people are hoping for a balance of the greenhouse and SST effects. Perhaps the only positive thing that can be said about air pollution at low or high altitudes is that the sunsets get prettier, when you can see the sun.

In the Sea. Water is transparent to light. Although a glass of water seems to allow all the light to come through it, most people do know that as one goes deeper in the ocean it (eventually) gets dark. The water in the glass does absorb some light, but the light intensity around the glass is so great that the amount of light absorbed is too small to be noticed. In the ocean, with so much water available, the abundance of light energy absorbed by the water becomes important and ultimately at depth there is darkness. The water's clarity and lack of turbidity are primary factors in determining light penetration in different regions. Thus, it is difficult to generalize exactly at what depth the water becomes dark. An excellent indication of light penetration is the depth to which microscopic plants exist in the ocean. Because most marine plants, like most land plants, depend on photosynthesis they need light. The vertical region of the ocean where light exists is called the euphotic zone and extends from the surface to the depth where about only 1 percent of the light that enters the surface will penetrate. In regions of extreme water clarity, such as the Mediterranean and Caribbean, the euphotic zone may be as deep as 100 to 160 m. The maximum penetration may possibly be as low as 15 m near some coasts. Even though coastal waters are more turbid, most plant life in the ocean occurs there because of the availability of nutrients. Because of man, coastal waters are becoming more turbid. Marine life there may be in danger of decreasing just because of man's impact, not to mention the more direct poisoning effect of shoreline and estuarine pollution.

As one begins to descend into the ocean, the world of white light changes rather abruptly to one of blue-green. If one slowly descends he may note more yellow at first, which rapidly yields to the blue-green. The red end of the spectrum is absorbed so fast that it is not noticeable below the surface. The question is sometimes asked why a brilliant red or orange fish such as a garibaldi can still be seen at depths greater than a meter or two, where no red or orange light ever penetrates. However, the red light or orange light emitted by these fish does not come from the surface. Pigments in the scales of the fish absorb the blue-green light present at that particular depth and re-emit it as red or orange light. The distance at which one may see these brilliantly colored fish underwater is not particularly great, however. This phenomenon is not unusual. A red shirt in effect does the same thing. The whiteness advertised by detergent manufacturers is created by compounds that make the fabrics appear whiter by translating some of the invisible ultraviolet light spectrum into visible light.

Table 3–7 Percentage of Light Passing through the Sea Surface That Penetrates to Specific Depths

Depth (m)	Clearest ocean water	Turbid coastal water
0	100%	100%
1	45	18
2	39	8
10	22	0
50	5	0
100	0.5	0

What It Looks Like in the Sea. At a depth of 33 m, no color distinction is possible. There are no shadows; lights seem to be coming from all around rather than from any single source. At 100 m, water visibility is limited to a few feet. By 305 m, it is quite dark. Normally, light intensity decreases steadily with depth (Table 3–7 and Fig. 3–29). The deeper one goes, the shorter the horizontal visibility is. In some regions, however, because of higher amounts of suspended matter in the waters near the surface the horizontal visibility may actually be better at depth. Light intensity still decreases with depth but, because the water is more turbid near the surface, the suspended matter lessens the horizontal visibility at the surface. Where waters are turbid, visibility decreases vertically for awhile, then increases below the area of turbidity,

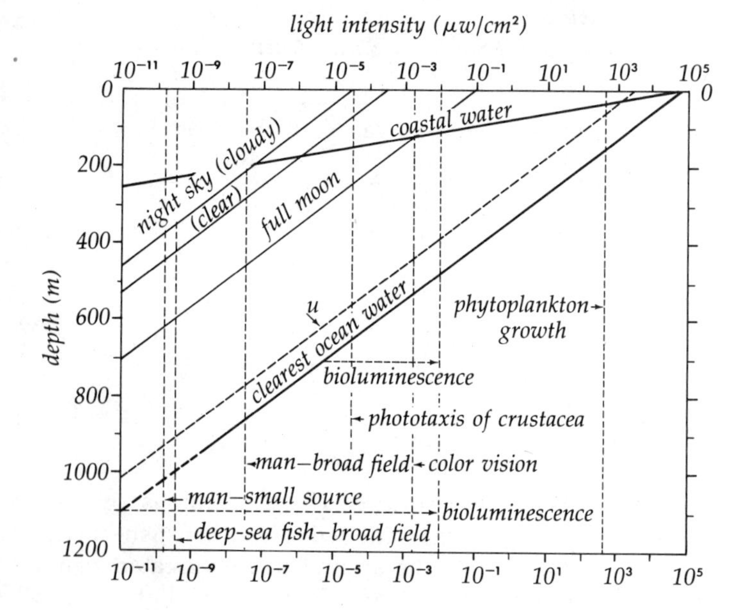

3–29 *Light in the sea.* (*After George L. Clarke.*)

and slowly decreases until, at about 300 m, normal lighting is barely sufficient to distinguish objects only a short distance away.

Very recently scientists have discovered that green light, or at least certain narrow wavelengths of green light, may be much more transparent in ocean water than had previously been thought. This effect, called the "green window," is currently under intensive investigation. If this green light is truly much more transparent in seawater than other forms of light, the development of an underwater radar might well be possible.

Several unusual visual effects occur underwater. When a fish is viewed through an aquarium window, it actually is somewhat magnified or appears closer than it really is. This effect becomes much more pronounced when looking at the fish through the edges of one's field of vision (peripheral vision). When one is swimming underwater, objects look larger or closer than they really are. This effect is noticeable through goggles or through a swim mask but also with the naked eye. The explanation for this is based on the index of refraction of water. The speed of light in water is less than that in air. A spoon or a soda straw placed in a glass of water appears to be bent at the air–water interface. The bending of light rays makes an object appear closer than it actually is. The object is often somewhat distorted, and this distortion increases as the object gets closer to one's outer edge of vision. This refractive index makes it difficult to see into water. Much of the light available at the air–sea interface is reflected from the surface of the water, also making it very difficult to see into the water. Some fishermen wear polarized sunglasses, which cut down some of the glare of reflected light, making them more apt to see fish in the water.

If the water surface were extremely placid and a person held himself just below the surface and looked straight up, what would he see? There are several possible answers. First of all, he might see a reflection, because the surface can be considered to have two sides. Light reflects off the top of the surface, but it can also reflect off the bottom of the surface, much like a mirror. So a person underwater looking straight up will see the light bounce off the bottom of the surface and go back down. He may see himself, much as he does in a mirror. If he is only a few feet below the surface, he will see a circle of light at the surface, very much as if he were looking out of a manhole. In that circle he will actually have a 360° distorted panorama of the total outside area, much as if it were photographed through a fish-eye camera lens. A boat might be seen coming across the surface of the water but, since it would be at the edge of the picture or the circle, there would be much distortion. The effect is seldom seen because the water surface is not smooth enough except in pools or small ponds, where wind and wave action may be small enough to allow the water surface to become almost mirror-like. Going down from the surface, this effect would close in and ultimately vanish.

Sound

The Nature of Sound. The transfer of sound through air can be thought of as similar to that of surface waves through water. As in the case of waves through water, there is no net transfer of air with the sound. Sound is a form of energy. This energy is transported from particle to particle of the air. Unlike light, sound depends on the presence of a medium of transmission. This can be illustrated by placing an electronic device such as a buzzer in a bell jar (so named because of its shape) connected to a vacuum pump. The level of the sound of the buzzer will remain essentially the same until low pressure levels are reached; then the sound will noticeably diminish and finally will vanish because there is not enough air for the sound to be transmitted, even though the buzzer is still emitting the mechanical energy. Thus, in a science fiction movie, when a spaceship whooshes by, it is disconcerting because there could not really be such a whoosh in space. A drumhead (like the human eardrum), when struck, vibrates back and forth, constantly striking billions upon billions of air particles (about 98 percent nitrogen and oxygen molecules). These in turn bounce into other particles, causing sound energy to be transmitted. As the drumhead vibrates back and forth, small pressure variations in the air are created as the drumhead pushes into the air particles increasing the density and as it moves back decreasing the density. The smallest sound in air that could be heard by the average human ear is about 2×10^{-10} atmospheres of pressure. The loudest sound, without pain, that the human ear could experience is about 2×10^{-3} atmospheres.

Velocity. Because the type and number of particles in a volume of matter vary, sound moves with different velocities in different media. In dry air at 20°C the speed of sound is 344 m/sec. Would sound move faster in wet air or dry air? In aluminum, which is much more dense than water, sound moves at 5104 m/sec. The speed of sound in water is between 1400 and 1550 m/sec. Because seawater is slightly more dense than fresh water, the speed of sound should be slightly greater in seawater. The velocity of sound in fresh water increases with its density. This is only approximately so with seawater because the presence of some salts, especially magnesium sulfate, affects the velocity of sound. In addition, the constituent salts in the vertical water column are not exactly the same throughout. The amount of calcium ion, for instance, is slightly greater in deep water than at the surface. The speed of sound in seawater depends on salinity, temperature, and pressure. For water with a salinity of 34.85‰ and a temperature of 0°C, the speed of sound is 1445 m/sec. Increasing the salinity by 1 percent will increase the speed by 1.5 m/sec; raising the temperature 1°C will increase the speed of sound in seawater by 4 m/sec; an increase in depth of 1000 m will increase the velocity of sound by about 18 m/sec. Because the temperature and salinity vary in a water column, it is not easy to generalize about sound velocity with depth. Knowledge of all three factors is necessary to specify the speed of sound at a particular depth (Fig. 3–30). Further, both salinity and temperature are affected by seasonal changes and latitude, so that the velocity of sound is different at different latitudes and various times of year (Fig. 3–31). Though all three factors vary in the water column, with the most consistent being pressure, the average value of 1450 m/sec is generally taken as representative of the velocity of sound in seawater. While most sonar equipment has the ability to set a specific speed of sound on the equipment, this average value is used on most sonar gear.

Sonar. The great bulk of data that scientists possess concerning the topography of the ocean bottom has been gathered by sonar (*so*und *na*vigation and *r*anging). Sonar is a collective term and may be either active or passive. Passive sonar is an equipment system for listening to noises in the sea. It can determine the presence and relative direction of a sound source, be it a shrimp or a submarine. Active sonar, originally called an echo sounder, was invented by an American named August Hayes in the early 1920s. It is the source of the pinging noise most people have seen the sweating submariners in movies listen to as they sit on the bottom. The operation of active sonar is like that of radar in principle, but uses sound rather than radio waves. The sonar gear (sometimes called for accurate equipment a PDR, precision depth recorder) sends out a sound impulse or ping. The sound travels three-dimensionally with a hemispherical front through the seawater until it hits an object or the bottom. It reflects and travels back to the vessel, where it is picked up on listening gear. Because this sound travels at an average speed of 1500 m/sec, the depth is equal to one-half the time for the impulse to come and go multiplied by 1500 m/sec. The location of a submarine or school of fish can be pinpointed by active sonar. The people in a submarine can actually hear the sound impulse or ping bounce off the pressure hull.

Ships can trace out a continuous ocean bottom profile on a chart with sonar as the ship proceeds normally on its course. Countless profiles have been recorded by oceanologic, navy, and merchant vessels,

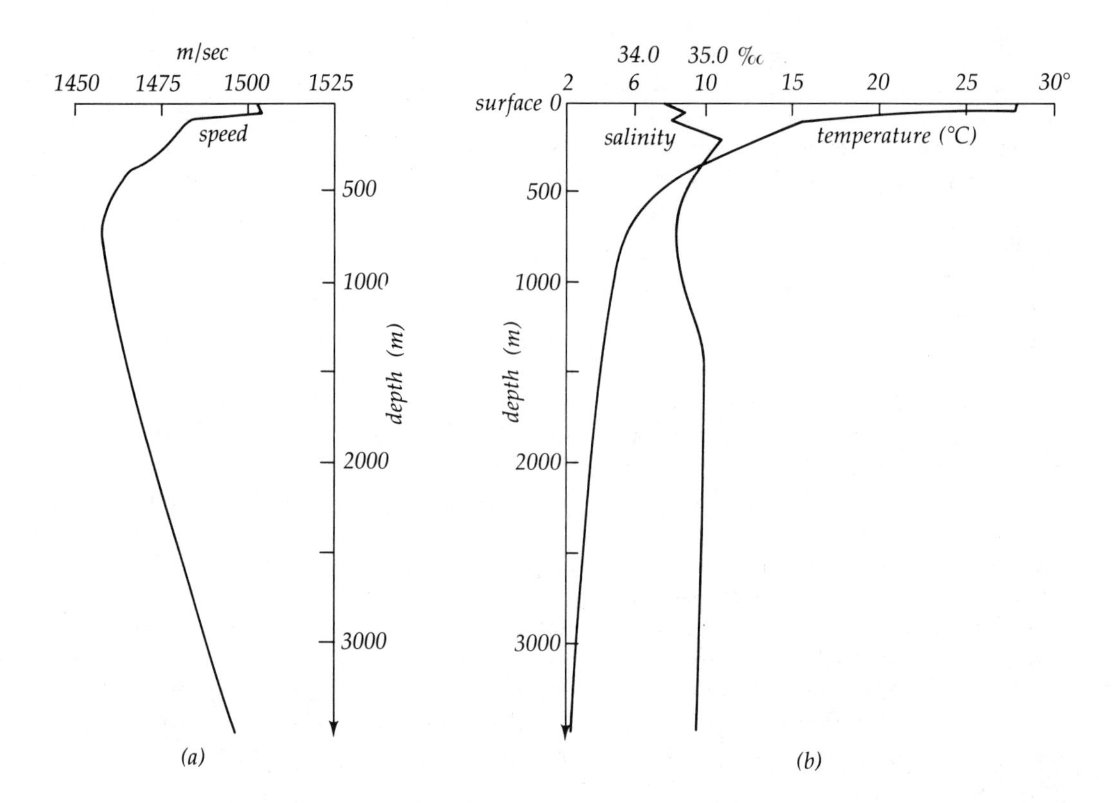

3–30 (a) A graph of the speed of sound versus depth, compared with (b) a graph of salinity and temperature with depth.

giving an approximation of the depth of the ocean throughout much of the world, although this is by no means a totally adequate picture of most of the ocean's bottom. Without sonar, water depth data would be sparse and tedious. One sounding with a weighted rope in water several kilometers deep takes three to four hours and involves handling tons of wire or wet hemp rope. With only a few such soundings taken before the invention of sonar, charts of bottom contour were generally misleading (Figs. 3–32 and 3–33). With sonar, some entirely different-looking charts have been produced.

Sound Properties. On the basis of comments thus far, seemingly the only problem with the use of sonar is the fact that the speed of sound might be different than the calibrated speed at which the sonar gear is set. Only in very precise depth measurement is this of sufficient concern that corrections need be applied. As it moves away from its source, however, sound energy naturally decreases in intensity. This decrease, known as propagation loss, is a cumulative loss of energy because of absorption, scattering, and spreading. Absorption is the actual absorbing of some of the sound energy by water particles. Because of its dissolved salt content, seawater absorbs more sound energy than pure water. Losses by spreading occur as the sound spreads out, usually in a spherical front or sometimes in a cylindrical front. The amount of energy at any square surface area is less as it moves away from the source because the same amount of energy must cover a wider front. The energy is spreading out or decreasing in intensity. Scattering occurs when anything in the water that has different sonic properties reflects the sound energy. This may be caused by fish, suspended inorganic particles, or especially bubbles of gases such as carbon dioxide or air.

Most of the world's oceans, with the exception of the Arctic, the Antarctic, and possibly parts of the central Pacific, have a region at depth that reflects or scatters sound. This is often known as the deep scattering layer (DSL), although some regions may have more than one. On sonar equipment, the DSL may be sufficient to produce a false bottom. The DSL is a layer of

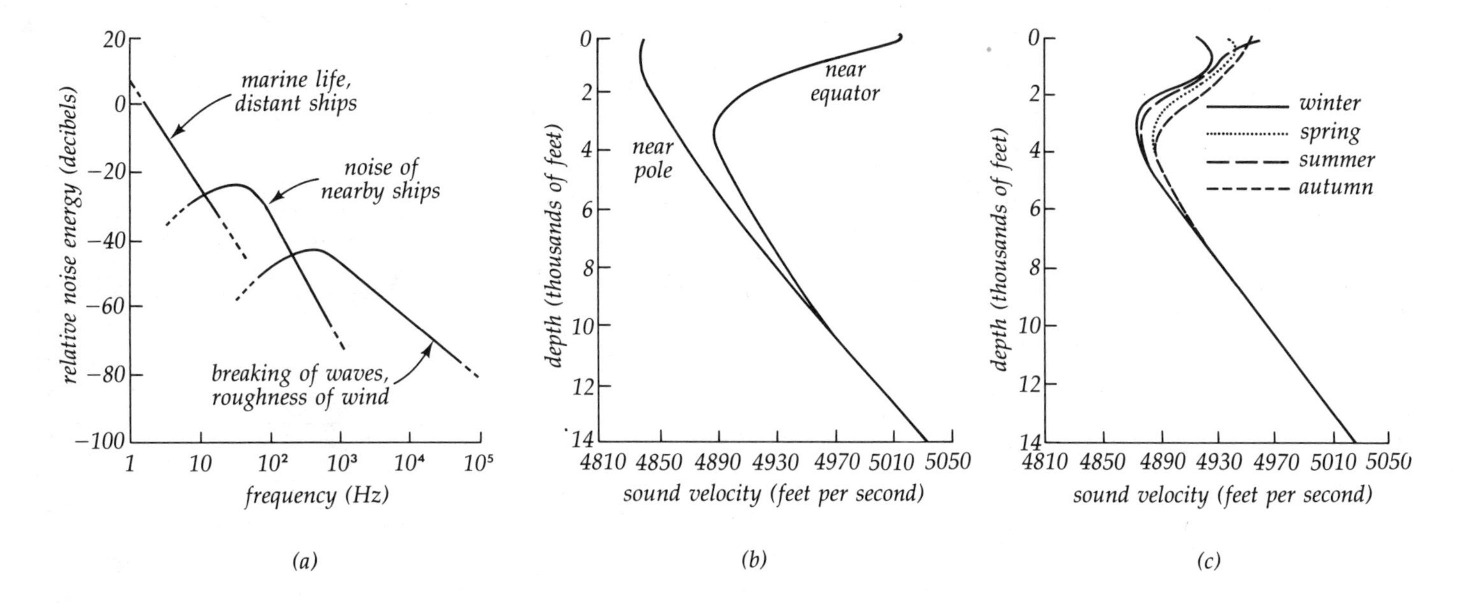

3–31 (a) The main sources of noise in undersea communication channels may be grouped into three classes according to their frequency ranges. Ships passing at a distance and marine life are the two significant factors below 10 Hz (or cycles per second). Nearby shipping greatly affects the range from 10 to about 300 Hz. Above this value, waves are the major noise sources, which, of course, depend on the wind. (b) In the ocean, the speed of sound near the surface critically depends on the water temperature. Near the equator (near 19°N) the minimum speed is attained at a much greater depth than near the polar regions (near 61° latitude). From the minimum value, however, both curves show that the speed increases almost linearly with depth, regardless of the geographical location. (c) Seasonal speed fluctuations for an area in the North Atlantic Ocean are more dominant near the surface. Smaller speed changes can also occur daily and even hourly, depending on local weather. (Copyright 1969, Bell Laboratories, Inc. Reprinted by permission, Editor, Bell Laboratories Record.)

living organisms, ranging from almost microscopic zooplankton to copepods, shrimp, and squid, that prey on one another. This layer migrates vertically in the water column, dependent on light intensity, and occurs at depths between 230 and 800 m during the day (although most concentrate at 310 to 460 m) and close to the surface at night. The greatest depth of the deep scattering layer would be at noon on a cloudless day. The DSL may be a potential new food source for man.

Another hindrance to smooth sonar operation is the existence of different density layers in the water column. When sound impulses strike these layers, some of the energy bends, penetrating into the layer, and some reflects off the layer (Fig. 3–34). For the sonar operator, the net result is a shadow zone, where little or no sound penetrates. It is quite possible for a submarine to hide in these regions. A knowledge of temperature, salinity, and the density structure of the water is strategically important for submarine warfare tactics. A submarine in, above, or below one or more of these density layers may also notice the same effect on its own sonar.

The existence of these density layers shows definite promise for use in communication. Fig. 3–30a shows a region of sound minimum at the bottom of the thermocline. Sound emitted or generated in this region has a strong tendency to remain in the layer; the sound has a greater tendency to bounce back into the layer than to penetrate out of it (Fig. 3–35). Sound, then, travels mostly in a horizontal direction in this layer and is affected only by absorption and scattering rather than by vertical scattering also. As a consequence, energy

PACIFIC OCEAN

Longitudinal Temperature Section . Meangis I.s to Admiralty I.s

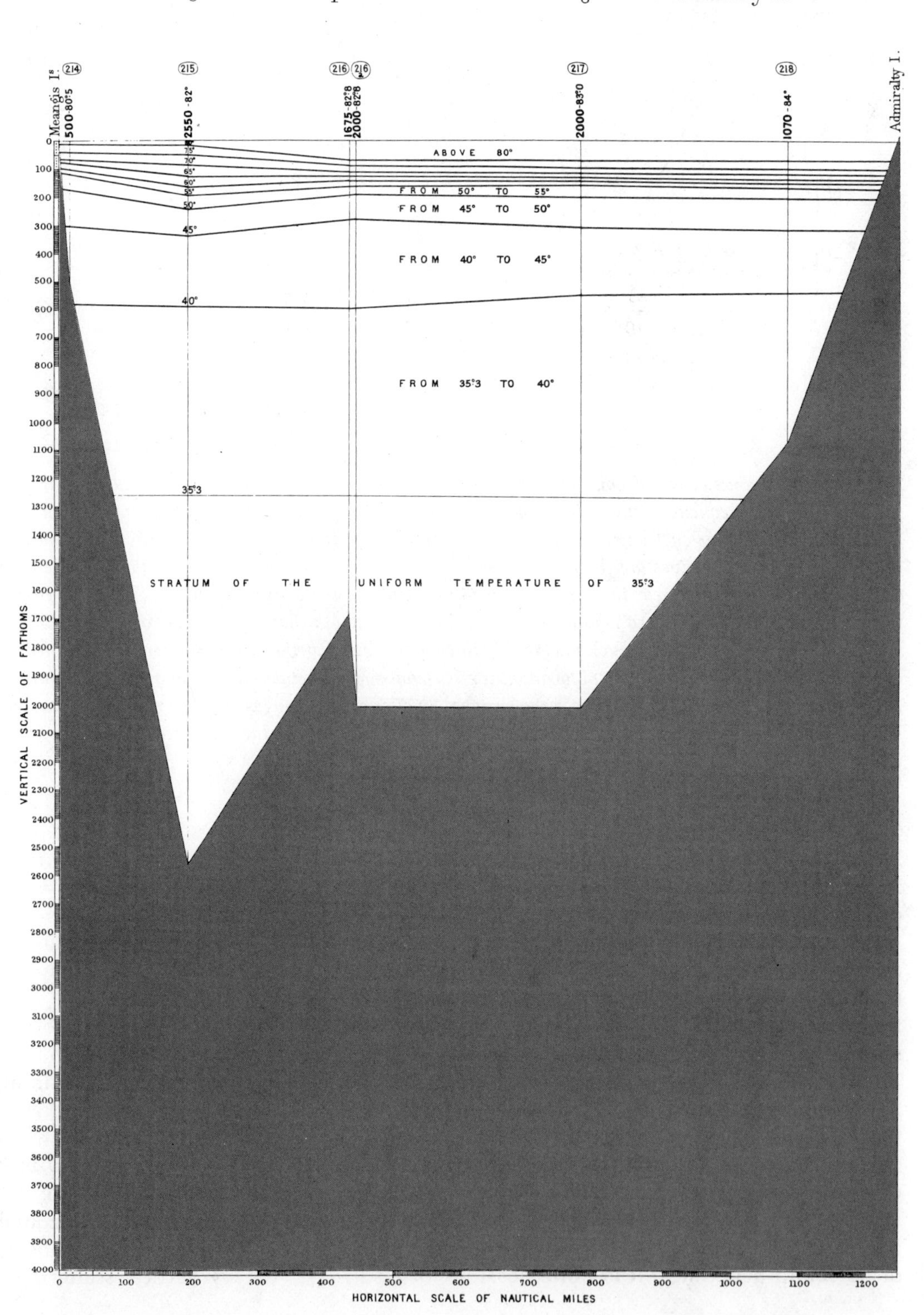

3–32 A lead-line bottom profile of a segment of the Pacific as determined by H.M.S. Challenger. *Because the soundings were long and tedious, they could only be made periodically. Notice that this does not give a good idea of the bottom typography.*

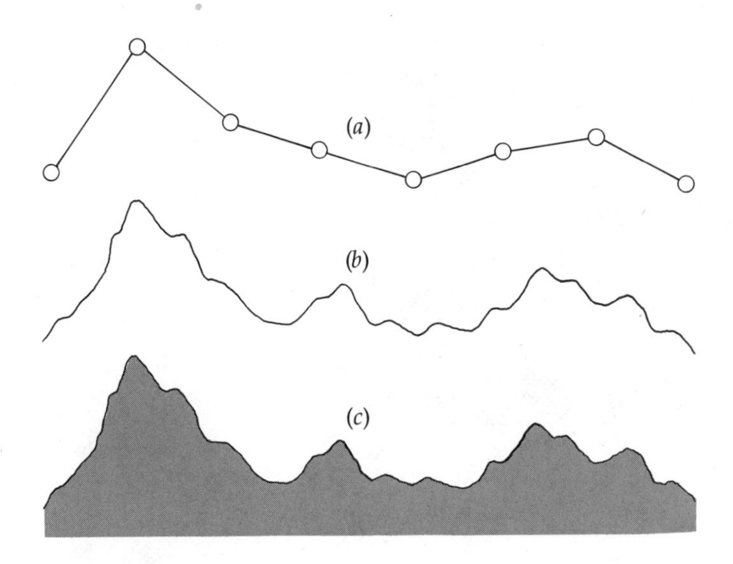

3–33 *Bottom contour. (a) Lead-line sounding; (b) sonar (PDR); (c) actual contour.*

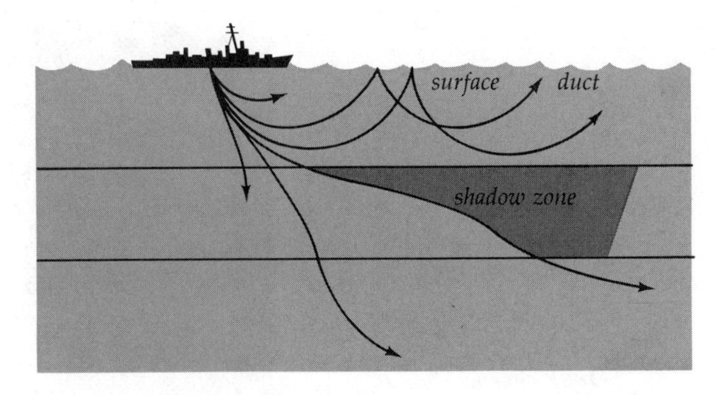

3–34 *A sound surface duct and shadow zone caused by the bending and reflection of sound fronts at layers of different densities.*

losses in this layer or sofar channel (*so*und *f*ixing *a*nd *r*anging) are extremely small, and the sound remains coherent or may be picked up long distances away—as far as 25,000 km. A small explosion, from a one-pound charge of dynamite, can normally be heard in air for a distance of about 1 km. Underwater the same one-pound charge may easily be heard for several miles, or perhaps as much as a 160 km. The same sound in the sofar channel may be heard 20,000 km away. As yet, underwater telephones do not use the sofar channel; and the range of such equipment is limited, at best, to short distances, generally much less than a mile.

The regions of slight pressure variation caused by sound exert pressure on a human's ear, setting the ear-drum in motion; then one hears. The intensity of sound waves is proportional to the square of the pressure. It is necessary to compare intensity and pressure to understand the difference of sound in air and water. In air versus water, these two factors vary appreciably, since air and water are themselves considerably different. If the pressure is equal, the sound intensity in air is 3000 times greater than that in water. But if the intensity is equal, the acoustic pressure in water is about 60 times greater than that in air. This makes it a bit difficult to compare sounds in these two media directly. A few comparisons will probably be helpful. If one were within a few dozen feet of a 25 horsepower outboard motor with his head out of the water, the sound level would not be too bad, but underwater the sound would be just at the level of discomfort. In air, standing about 100 feet from a jet aircraft taking off, one would have about the same sound level as being underwater about 100 yards from an underwater dynamite explosion. The average level of a television set is about equal to the sound level generated by a fairly calm sea surface. In air, the intensity of sound can be quite a bit greater than that in water but, because water is a much more dense medium, the pressure generated by an equivalent sound level is much greater. Therefore, being relatively close to the same high sound (intensity), a large decibel (loudness or pressure) level in air is less dangerous than the same sound underwater. Since audible sound is nondirectional in water, a diver cannot tell from which direction a sound comes.

What It Sounds Like in the Sea. What sounds can be heard in the ocean? To a sonar operator, the sea is full of background noises. Penetration of sound waves through temperature and density layers bends, distorts, and muffles his signals, but other natural phenomena such as earthquakes and volcanoes give the ocean a dull background rumble. Superimposed on this are the sounds of the sea's many inhabitants. These noises from sea creatures may come from within the animals' bodies. Fish, for example, produce sounds in the operation of their swim bladders (a gas sac that they inflate or deflate to change depth). Often drums, sea bass, and catfish sound like fog horns. The crustaceans

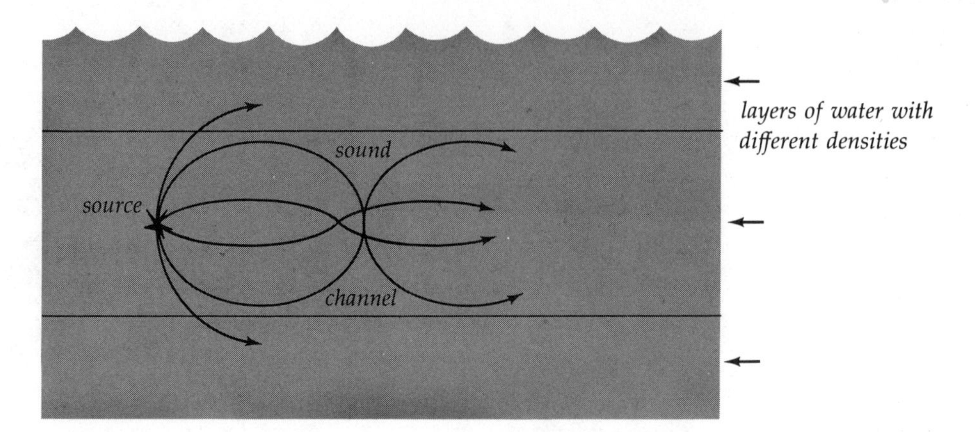

3–35 SOFAR channel.

—shrimp, crabs, lobsters—make audible clicking sounds as they move external parts such as claws, or as they snap legs together. Sea mammals, too, make noises. Of all the creatures in the sea, the porpoises may be the most fascinating to man. Perhaps it is because porpoises are also mammals, or perhaps it is because in an environment as seemingly hostile as the sea they are friends. Porpoises may emit either a clicking sound or a whistle. The clicking sound is the audio signal for a sophisticated natural sonar navigational system. The whistle or squeal is apparently for communication with other members of his species, and perhaps now in some cases for man.

If a person puts his head underwater, what does he hear? Nothing? In all probability, the first thing he would hear would be himself splashing. Even with scuba gear the only thing heard would be sounds of breathing. As usual, man makes so much noise that he overrides the sounds of the environment about him. There is a tremendous amount of sound in the sea, but most men never hear it. Underwater, around rocks, it is not unusual to hear shrimp. A precise passive sonar receptor or hydrophone placed in the water can pick up a tremendous amount of noise: the snapping of shrimp, the clicking of crabs, the bubbling of fish, even the noise perhaps of some distant surf. There are very few places if any in the sea where it is totally quiet, but the level of sound intensity is generally lower than man can pick up or gets around to listening to. Contrary to popular belief, some nuclear submarines, for instance, though a thousand miles away, at certain depths might sound on sonar gear like an express train in one's own backyard.

Ice in the Sea

If a piece of ice were chipped from the frozen ice cover of a harbor on the Maine coast during the winter and tasted, it would be salty—not as salty as seawater but salt-tasting nevertheless. A similar piece of ice taken from an iceberg, well out to sea in the North Atlantic, would have no salt taste. Ice with some salt content, correctly termed *sea ice,* accounts for the majority of ice in the oceans. An iceberg, however, is not sea ice but glacial ice in the sea.

Ice. Pure water freezes at 0°C. If any other substance, such as salt or clay, is mixed into the water, the freezing point is lowered. The freezing point lowering is entirely a colligative or collective property, depending only on the total concentration of particles in the water. These particles normally are dissolved (in solution), but muddy water containing suspended silt or clay does freeze at a slightly lower temperature than the pure water. However, not nearly as much suspended matter can exist in water as can dissolve. This is the principle used when antifreeze is added to an automobile radiator to prevent the water from freezing. Unlike most substances, water expands when it solidifies generating tremendous pressures, perhaps up to 30,000 lb/in.^{2}, which may crack an automobile block. As the freezing point of water depends on the amount of dissolved matter, a specific freezing point for seawater cannot be given unless the salinity is specified (Fig. 3–36).

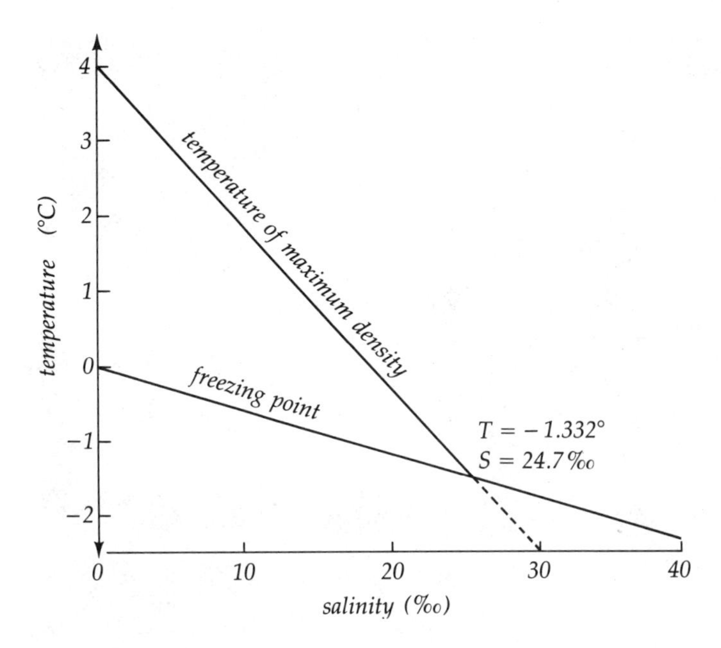

3–36 The relationship between the temperature of maximum density and the freezing point of seawaters with different salinities.

Because water expands as it freezes, the density of ice is less than the density of water. If this were not the case, icebergs would not float, and lakes would freeze from the bottom up. Sea ice varies somewhat in density because it contains salt from the water, but it is always less dense than the seawater itself.

Fresh water has a density maximum at 4°C. As the water temperature in a lake, for example, drops, its density increases until the temperature reaches 4°C. The water that is now at 4°C begins to sink in the lake, and warmer water rises to take its place; extensive mixing occurs until the entire body of water is at 4°C. Only then, as the surface waters cool down to 0°C can freezing begin at the surface. It would be possible to take advantage of this property to keep some lakes such as the Great Lakes, which freeze during the winter, open all year round or at least for longer periods during the winter. An artificial heat source that constantly heats the water to 5°C could be placed on the bottom to keep the water constantly circulating and above freezing temperature. While this has not been tried, there is no reason to believe that such a small increase in temperature, especially during the winter, would have any noticeable effect on the environment, with the possible exception of making the climate in that area slightly less severe during the hard winters.

Seawater, too, will undergo this overturn and mixing if the surface is exposed to cold air. Unless the water is shallow or the surface waters are underlain by a region of high salinity (approximately 35‰) not far from the surface (150 to 250 m plus), seawater will not freeze. Because the density maximum of seawater is below its freezing point, it may seem that seawater in shallow regions would freeze faster than lake water. Such is seldom the case. Lakes are in constant contact with the surrounding land, which is a poor insulator and therefore loses heat rapidly. This causes a more rapid cooling of the lake than is generally possible for seawater.

Sea Ice. · Sea ice is a collective term for ice with a number of different forms. As sea ice begins to form, the surface first appears oily or greasy, indicating the formation of ice crystals. These crystals are no more than a few centimeters in length. They are flat, thin, pointed, and form oblong plates. As the formation of these ice crystals increases, the sea surface becomes slushy. It is quieter with few ripples and has a slightly grayish tinge. There is almost no noise when a ship passes through this slush. Unlike fresh water lakes and ponds, the ice does not form into a clear ice sheet. In seawater, the closest equivalent to the fresh water ice sheet occurs in fjords. In the formation of slush individual ice crystals do not coalesce. As freezing continues, the crystals form a hard ice called ice rind, usually less than 5 cm thick. If the surface of the sea is in motion as the slush coalesces, the ice formed bumps and grinds together, causing disk-like cakes of ice known as pancake ice to form, usually 0.5 to 1 m in diameter. Ice rind subjected to a swell sufficient to break it up may then become pancake ice. These cakes form into a continuous ice sheet or flow (floe) ice on further freezing. These ice floes range in size from 10 m to about 8 km (Fig. 3–37). Ice floes are found in the northern harbors during winter and in the Greenland straits. It is not uncommon to find a variety of these types present simultaneously in close proximity (Fig. 3–38).

The ice in the Arctic Ocean has several different names. There is an extensive polar cap ice, which covers about 70 percent of the Arctic Ocean. It is not a

3–37 An Adélie penguin silhouetted against the sun in the frozen pack ice of the Weddell Sea from the Coast Guard icebreaker Glacier *during her oceanographic expedition as part of Operation Deep Freeze 1970. Note the unevenness of the ice, caused by the pressure of expansion when the water froze, as well as some trapped small icebergs. The "halo" appearance of the sun is called "sun dogs" and is caused by light hitting ice crystals in the air. (Official United States Coast Guard Photograph.)*

smooth surface but very hilly or ridged because of the constant collection of thick ice, which creates tremendous pressure and causes hillocks of ice extending perhaps 10 m above sea level and 40 to 50 m below sea level to form (Figs. 3–37 and 3–39). There may be considerable rafting or tenting as the ice is pushed together, some pieces sliding over and against others (Fig. 3–40). The thickness of ice varies from place to place and from time to time. The average winter thickness is 3 to 3.5 m (with an extreme known of 9 m), which decreases in summer to about 2.5 m. Local clearings of open water known as polynyas (a Russian-Eskimo word that may also mean regions of very thin ice) occur occasionally. Polynyas are much more common in summer than in winter. A nuclear submarine surfacing in polar waters would, of course, choose to do so in a polynya.

Around the outside 25 percent perimeter of the cap ice lies pack ice, sometimes known as drift ice. Its limits vary greatly with season, and it is this ice, not the cap ice, that can be penetrated by icebreakers (Fig. 3–41). Pack ice seldom exceeds a thickness of 2 m, and it usually melts entirely in summer months. Pack ice that is attached to the shore rather than to cap ice and extends seaward is called fast ice (see Fig. 3–39).

A normal ship lacks the capability of making its way through ice. Ice is an extremely hard substance and can easily slash a vessel's hull. Ships designed to penetrate ice have an entirely different hull configuration (Figs. 3–42 and 3–43). The hull design of an icebreaker, aside from being very rugged, is extremely rounded and dish-shaped. The icebreaker does not cut its way through the ice like a knife blade but uses its weight to make its way through the ice. The icebreaker's bow is constantly moving up and onto the ice, and its weight crushes the ice below it. Occasionally, the icebreaker can make no more forward velocity. At such times, it is commonly backed up for another running start at the ice, the propellers being carefully protected at all times from chunks of ice. Most of the icebreakers owned by the U.S. government are under

3–38 Icebergs in Arthur Harbor, Antarctica, seen from the Coast Guard icebreaker Glacier *during Operation Deep Freeze 1970. A tabular berg appears in the middle background and a pinnacle berg on the right. The pack ice has broken up "releasing" the icebergs. By grinding into one another, pieces of pack ice become almost circular. (Official United States Coast Guard Photograph.)*

the control of the U.S. Coast Guard. The majority of these were built during World War II, in approximately 1944. Few of these vessels are in the best condition. Clearly, the Coast Guard could use better and more recent equipment.

The salinity of sea ice varies considerably. As the ice forms from needle-like crystals of pure water, the salt tends to separate from them, not as pure salt but as particles of trapped, more concentrated salt solution. It is difficult to generalize about salinity, but as a rule for fairly new ice, the faster the ice was formed, the more saline it is (Table 3–8). Technically, sea ice is not totally solid, because liquid water is trapped in the hexagonal holes of the water crystal. Ice age is a primary factor in the salinity of the ice. The density of sea ice also depends largely on its age. The tiny brine deposits in the ice structure tend to migrate downward and out of the ice. Old ice (sea ice seldom has an age of more than 5 or 6 years) that has lost most of its salt appears mottled and honeycombed and is called rotten. Because air takes the place of the deposited brine, the resulting density of the ice may be as low as 0.85 g/ml, compared to that of 0.925 g/ml for new ice and 0.917 g/ml for pure water ice. Because ice more than a year old contains less salt, it is harder than the younger ice. Ice less than one year old is sometimes called winter ice. New ice may have a salinity of 5.5 to 10.2‰, depending on conditions, while the salinity of old ice is usually less than 2‰. Old ice and the puddles of melted water on its surface have a low enough salinity for human consumption, and more than once sea ice has been used by explorers who no longer had drinking water.

Although the cap ice in the Arctic is always present, it is not always the same ice. Some ice melts each summer, and some is constantly being reformed. Imbedded in the cap ice are large, rather flat pieces of ice with no salinity, called ice islands (see p. 114).

Icebergs. During spring in the northern hemisphere, appreciable amounts of sea ice float out of the

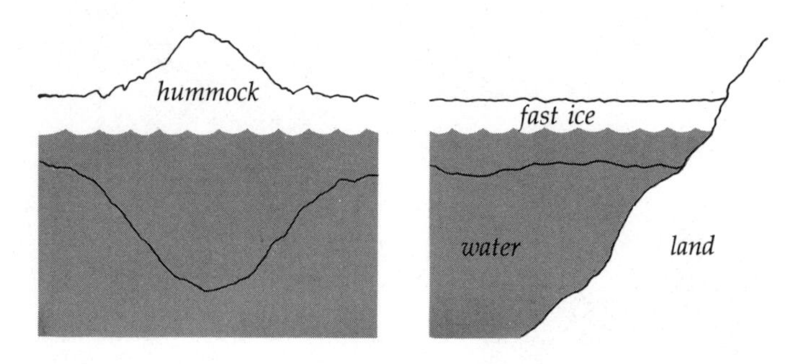

3–39 *Hillocks or hummocks and fast ice.*

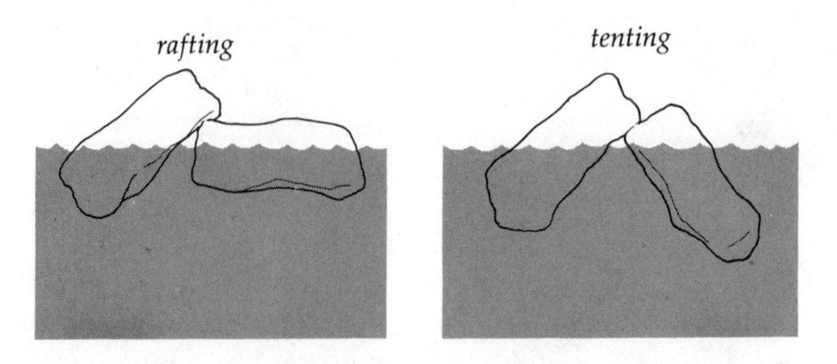

3–40 *Rafting and tenting of sea ice pieces.*

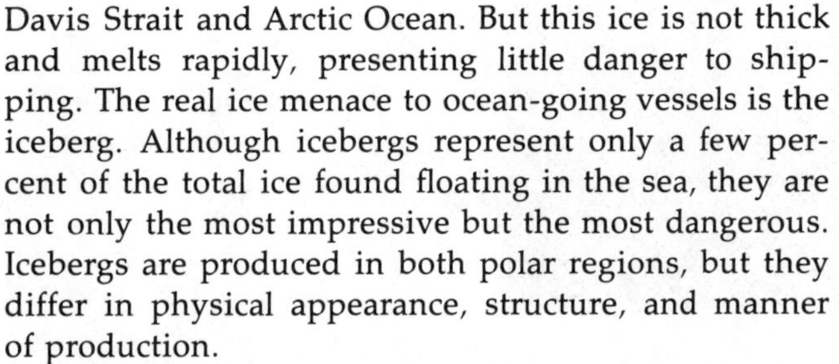

Davis Strait and Arctic Ocean. But this ice is not thick and melts rapidly, presenting little danger to shipping. The real ice menace to ocean-going vessels is the iceberg. Although icebergs represent only a few percent of the total ice found floating in the sea, they are not only the most impressive but the most dangerous. Icebergs are produced in both polar regions, but they differ in physical appearance, structure, and manner of production.

For a long time icebergs were thought to be broken or fragmented pieces of the polar cap ice that simply floated into the world's oceans. To contrast their origin with that of sea ice, icebergs might better be called land ice. While sea ice, as stated, is produced in the sea from seawater, icebergs are a land phenomenon. Icebergs are formed on land from fresh water, which explains their lack of salt. The only other fresh water ice found in the ocean is the small amount of river ice in some of the more northern rivers that finds its way into the ocean.

The massive glaciers covering most of Greenland are the source of most of the icebergs in the North Atlantic. Some of the bergs break off Greenland's east coast but by far the greatest number are produced on the western coast in the Davis Strait (Fig. 3–44). Glaciers, though composed of solid ice, may best be viewed as rivers of ice, formed from compacted snow, which flow steadily and constantly downhill. The average rate of flow of Greenland glaciers is about 10 m per day, with a value of 20 m per day being exceptional. As the front of the glacier reaches the water, it simply slides into the water and breaks off simultaneously. This breaking process is termed *calving*. When local water conditions permit, along with the breaking pack ice, currents carry the icebergs to the sea.

Men have always seemed compelled to name things. The extensive terminology associated with shoreline processes is one example. Because of long-term contact over well-traveled shipping lanes in the North Atlantic an extensive terminology has developed for icebergs in this area. North Atlantic icebergs are irregular and, because peaks are common, they are often called pinnacle bergs. A fairly small iceberg, about the size of a house, is known as a "bergy bit." Smaller icebergs that make a fair amount of noise as they bob in the water and melt are "growlers." Because most of an iceberg is below the surface of the water, the exposed segment is no indication of what lies below (Fig. 3–45). Ships of any size do not venture near icebergs for this reason. Often there is a projecting underwater segment, called a ram, that could easily split a hull. Probably, between four-fifths and six-sevenths of an iceberg's total volume lies below the surface of the water, with seven-eighths being a high

Table 3–8 The Salt Content of Sea Ice Is a Function of Several Factors, Such as the Air Temperature at Which the Ice Was Formed and the Time It Takes for Freezing to Occur

Salt content of sea ice (‰)	5.6	8.0	8.8	10.2
Air temperature at which ice was formed (C°)	−16	−26	−30	−40

3–41 The 269-foot U.S. Coast Guard icebreaker Northwind *making an unprecedented early summer west to east voyage of 2431 miles alone through the Northwest Passage in 1969. (Official United States Coast Guard Photograph.)*

value. North Atlantic bergs normally are no more than a few hundred meters long and 50 m high, but 80 m is not uncommonly high, and occasionally 100 m high icebergs have been sighted. In 1882, the largest North Atlantic berg ever sighted had a length of 7 miles and a width of over 3 miles.

The number of bergs produced in the North Atlantic is by no means constant from year to year. Most icebergs are found in April, May, June, and July. Slightly over 1000 icebergs were sighted in 1912, but in 1924 only 11 of those seen were big enough to worry about. In 1929, 1350 were sighted and tracked. When the *Titanic* hit an iceberg off Newfoundland on April 14, 1912, 395 icebergs had been sighted that spring, making to that time a particularly large number of known bergs. Actually, 15,000 icebergs might be produced in a year, but most of them are not very large and will melt after a short time. The loss of the *Titanic* with 1500 lives tragically indicated the need for more effective tracking of icebergs. The International Ice Patrol was formed by international treaty in 1913. The U.S. Coast Guard conducts a surveillance of icebergs using reconnaissance ships and aircraft, as well as data sent in on iceberg sightings from any vessels in the area. Seventeen nations contribute funds to maintain the Ice Patrol.

An iceberg calved on the west coast of Greenland may take months to reach the North Atlantic shipping lanes. Unlike free-floating sea ice, which is moved by the wind, icebergs move with ocean currents (Figs. 3–46, 3–47, and 3–48). In occasional instances, ships have moored to an iceberg to be towed through pack ice. As a chart of oceanic surface currents in the North Atlantic would show, most icebergs float directly east across the Atlantic. Some penetrate as far south as 30°N off the coast of the British Isles. They may have a total life span of 2 years. An average iceberg with a mass of about 1 million tons near the region where it was formed, say around Baffin Bay, will be approximately eight times smaller by the time it reaches southern Newfoundland. Icebergs travel at a rate of from 9 to 15 nautical mi a day. An iceberg may remain frozen in pack ice well over a year after it calves off the glacier. Once in the open ocean, however, its rate of melting is fairly fast, especially if it moves anywhere near the Gulf Stream, which it will do sooner or later if it comes from Greenland. In the warmer waters of the Gulf Stream, the melting rate is much faster, perhaps 3 to 3.5 m of loss in height per day. While the values do not seem particularly large, the difference in these two figures represents a tremendous difference in the amount of ice melted. Seldom do icebergs

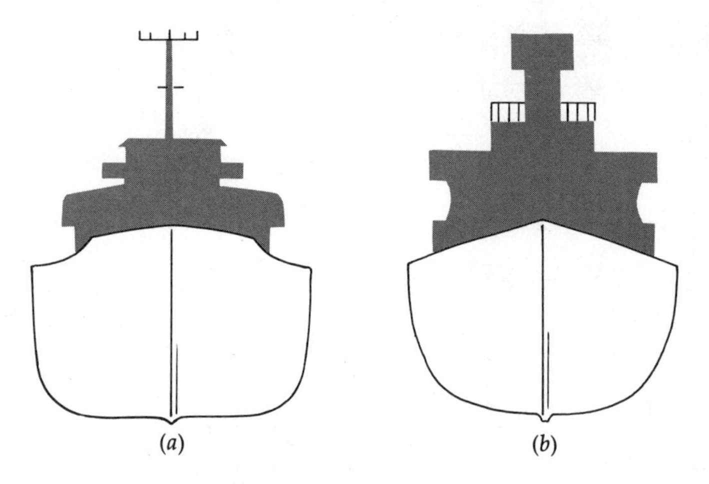

3–42 *Hull designs for (a) a conventional ship; (b) an icebreaker.*

3–43 *The* Burton Island, *a 270-foot Coast Guard icebreaker moving through thin pack ice at the Antarctic. Note the curved hull design. (Official United States Coast Guard Photograph.)*

proceed very far down in the Atlantic on its western side. In June 1926, however, an iceberg was sighted near Bermuda. Normally, once in the open ocean, the life span of the iceberg is probably less than 3 months. An iceberg may become smaller naturally by breaking into parts or calving within itself, often more than once. Wind, rain, and waves may greatly erode icebergs, especially causing scouring, but melting is the reason for the ultimate demise of the icebergs. The heat that melts icebergs comes from the water and from the heat in the air and the rays of the sun, which can cause appreciable melting of the upper part. It is not unusual to see small waterfalls cascading down an iceberg as the air temperature and the light of the sun cause the top to melt. There is little to do with an iceberg except track it and perhaps mark it with some type of bright coloring to make it more readily identifiable. Certainly, one cannot shoot and sink them. This has been tried, on the premise that the smaller fragments will melt faster, but the results were not particularly successful. Black material, such as carbon black, has been sprayed over icebergs in an attempt to make them absorb more sunlight and therefore melt faster. Although this does increase the rate of melting enormously, the results apparently are not worth the effort involved. Like the snow removal problem in many cities in the United States, ultimately nature must remove the problem.

The North Pacific is a region of heavy ship traffic. This is especially so nowadays with so much trade with Japan. Why is there no iceberg danger there? The answer is as simple as it seems; there are very few icebergs. The narrowness of the Bering Straits gives the North Pacific little access to the Arctic Sea, the only possible region of heavy icing. Further, the spring current moves south to north in the Bering Straits. The lack of major glaciers on the land adjacent to the North Pacific means there are no iceberg-producing regions. There are only a few valley glaciers in Alaska, and these calve only a small number of icebergs per year.

Icebergs in the southern oceans, though they too are land ice, are more regular in shape than those of the Arctic. Easily the most common shape is long, flat, table-like, or tabular icebergs (Fig. 3–49). Around the continent of Antarctica is a margin of ice. It is formed on the continent but extends out into the sea (Fig. 3–50). The shelf moves northward about 100 m a year, breaking off and forming icebergs. This shelf ice calves in much larger pieces, forming the large, flat icebergs of the southern Atlantic, Indian, and Pacific oceans. The

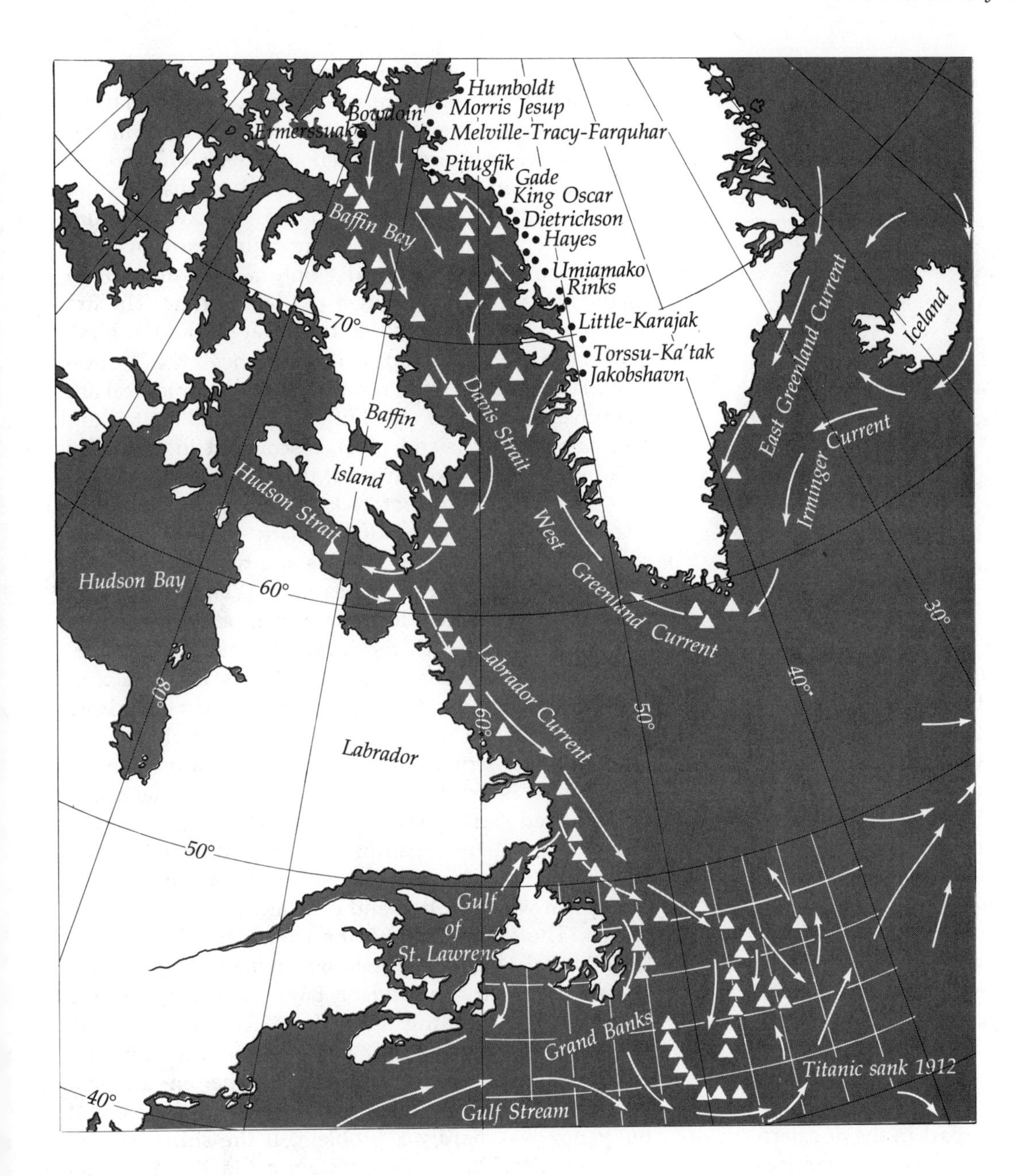

3–44 *The drift of icebergs from their source into the North Atlantic. The map also shows the principal ice-producing glaciers of western Greenland. Of the many thousands of bergs that break off these glaciers each year, about 390 ultimately drift south of Newfoundland (48°N.) Information from the United States Navy Hydrographic Office and the United States Coast Guard.*

largest tabular berg thus far recorded was 350 km long by 100 km wide and stood out of the water over 33 m. A berg of this size is easily the exception rather than the rule, though some bergs 100 m high have been seen. The tabular icebergs are larger on the average than North Atlantic bergs. The dimensions of the shelf ice vary with the season (Fig. 3–51). In winter months some of the tabular bergs may be trapped in the pack ice until spring, and the melting of this ice allows the icebergs to float free.

Until recently, little has been said in novels and stories about the occurrence of icebergs in the southern oceans. The reason for this, of course, is that most sea traffic has been in the northern oceans. Only lately has there been large-scale sea traffic in the southern oceans. Supertankers, too big to fit through canals,

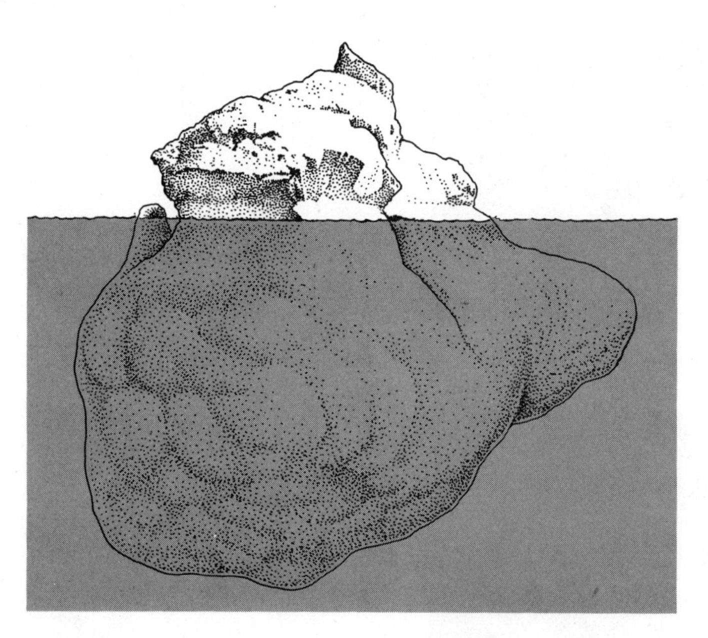

3–45 An iceberg above and below the waterline.

regularly ply the southern ocean in moving from the Middle East to Europe and the United States around the Cape of Good Hope. Antarctic tabular bergs are not an uncommon sight off Cape Horn and the Falkland Islands, and they have been sighted well into the South Atlantic, even as far north as 30°S. With the increased transport of oil by sea and presumably the continued closure of the Suez Canal, tanker and supertanker traffic in the world's southern oceans will increase. The ice is less dangerous in the southern hemisphere because weather conditions are more consistent than in the north. Radar is used to detect ice, but this method is not infallible and visual sighting is still the best method of dealing with icebergs.

Ice Islands. The only counterpart in the northern hemisphere to the southern hemisphere tabular iceberg is the ice island. Throughout the pack ice of the Arctic Ocean are occasional homogeneous flat sheets or patches of land ice. Once located these are easily distinguished from the pack ice, although location is somewhat of a problem considering the weather for much of the year at the North Pole. The existence of these ice islands has been known since 1946, when one was first identified on August 14 by a U.S. Army Air Force reconnaissance aircraft because it showed up on radar as clearly different from the surrounding polar ice. Ice islands are often much thicker than the polar cap ice. They are of land origin, forming apparently when pieces of shelf ice break off northern Greenland or Ellesmere Island and probably other northern islands. The ice islands drift clockwise around the Arctic Ocean with the prevailing currents and, unlike ice floes, which exist only a year or two, ice islands exist for much longer periods of time. The first one that was discovered, T-1 (for Target, Radar, First), has been followed sporadically since 1946. These islands have been used for manned meteorological and geophysical stations by both the United States and Russia. Fletcher Island (T-3), discovered in July 1950, is 8 by 16 km in size, drifts at the rate of about 4 km a day with a maximum of 14 km per day, and has been manned by the United States many times since 1952, usually for a number of months at a time.

Ice, like the water it comes from, is an unusual substance. It is extremely hard. If the temperature remains constantly below 0°C, its durability makes ice an excellent building material. Air Force and warning stations in Greenland and the northern parts of Alaska constantly use it as a construction material. Anyone who has ever struck ice with an ice pick knows, however, that it is brittle. Perhaps the most unusual use of ice occurred during World War II when W. Pyre suggested manufacturing ships from ice. To overcome the brittle quality of ice, about 2 percent fiber content from newsprint was included and constantly mixed in as the ice set. The fibers through the ice added toughness to the material. A pilot program was actually set up by the British Admiralty in Hudson Bay, Canada, to test the feasibility of building ships from ice. Several small vessels were constructed, and the ice was set into molds around cooling equipment. The engine and quarters of the crew were insulated from the ice and buoyancy was hardly a problem. If the ship were hit by a torpedo, the cooling system could be turned on higher at that particular location, and water could be sprayed onto the fractured area to rebuild the hull. All tests indicated not only that this type of vessel was feasible but that its lifespan, even in waters toward the equator, was appreciable and that the vessel could pass from the northern to the southern ocean with no apparent loss in hull. By the time much of the technology was perfected to produce these vessels, American liberty ships were being produced at an extremely

3–46 An iceberg propelled by a subsurface current moving through surface ice. (Official United States Navy Photograph.)

3–47 Icebergs assume fantastic shapes after they calve from the mother glacier and start the long, slow journey through Baffin Bay and Davis Strait toward the North Atlantic shipping lanes. (Official United States Coast Guard Photograph.)

3–48 A North Atlantic iceberg, often more irregular and "pinnacled." As the left of the photo indicates, the exposed segment is generally little indication of subsurface extension. (Courtesy of the National Oceanic and Atmospheric Administration.)

3–49 Dwarfing the 269-foot U.S. Coast Guard icebreaker Westwind, *this tabular iceberg measures approximately three-fourths of a mile long, one-half mile wide, and shows 60 feet above the water. The berg extends 560 feet below the surface. (Official United States Coast Guard Photograph.)*

rapid rate. Had this not been the case, the British quite possibly would have built a number of ice ships to fight in World War II.

The suggestion to tow southern ocean tabular bergs to large cities with possible water shortages, such as Los Angeles, and to use the melted berg as a new source of pure, fresh water, is an interesting example of inductive thought. But this is not yet practical. There are numerous other more dependable and cheaper methods, though less grandiose and fanciful. It is more reasonable to reuse water already present. Rather than dumping 100 million gallons of primarily treated sewage (most of which is water) into the ocean daily, as is done by San Diego, Los Angeles, and San Francisco, most of the water could easily be reclaimed from the sewage. The thought of drinking treated sewer water may not seem appetizing; yet the water, once treated, would be pure and in no way worse than the water most of Los Angeles and San Diego now drinks. Actually, it would probably be better.

What would happen if all the polar ice melted? The amount of frozen water at the poles is tremendous. The Antarctic, for example, contains about 6.5 billion mi^3 of ice. If all the ice at both poles melted, sea level would rise by perhaps 100 m or more and cover 75 to 80 percent of the earth rather than the present 70.8 percent. This degree of melting is rather unlikely, yet an increase in mean sea level of 30 m would inundate most of the Midwest and most of the Atlantic seaboard. Some geologic data indicate that approximately 5000 to 6000 years ago the ocean level was 3 to 4 m higher than its present level, and that this rise occurred over only a few centuries. Possibly this was the Great Flood referred to in the Old Testament.

With the exception of substances such as DDT, the polar regions of the world are relatively unscathed by man. The poles act as a tremendous heat sink, serving in large part to moderate and control the earth's weather. A thought of more than passing concern to those interested in changes in the earth's climate is the present disposition of this polar ice. The exploration of Alaska's north slope by U.S. oil companies is of concern not only to the Canadian government and ecologists but to meteorologists. If, for example, an oil well or wells should burst, as they did at Santa Barbara, or if an icebreaking tanker like Humble Oil's S.S. *Manhattan* (or an even larger one) should split open, millions of gallons of black crude oil could be unleashed on the white Arctic Ocean. Like the ashes used to melt snow by the increased absorption of light energy, one such mishap could make a black blob over thousands of square meters of Arctic ice. At this temperature, crude oil would tend to form into unsightly little blobs, and even if it formed a film one oil spill probably could not change the world's weather. But each little change adds to man's total impact on nature. One wonders not only about the possible changes or limitations of caribou migrations created by a tremendous pipeline running south from Alaska's north slope but also about the tremendous ecological damage that

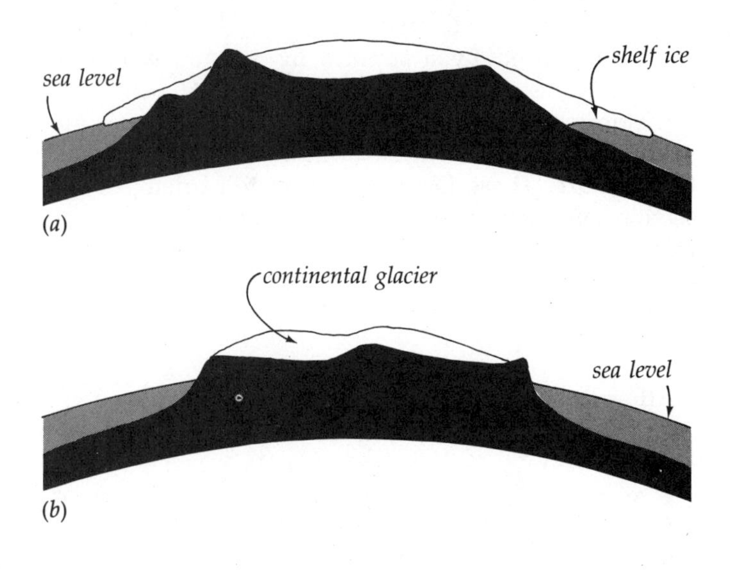

3–50 (a) Cross section of the Antarctic continent. (b) East-west cross section of Greenland.

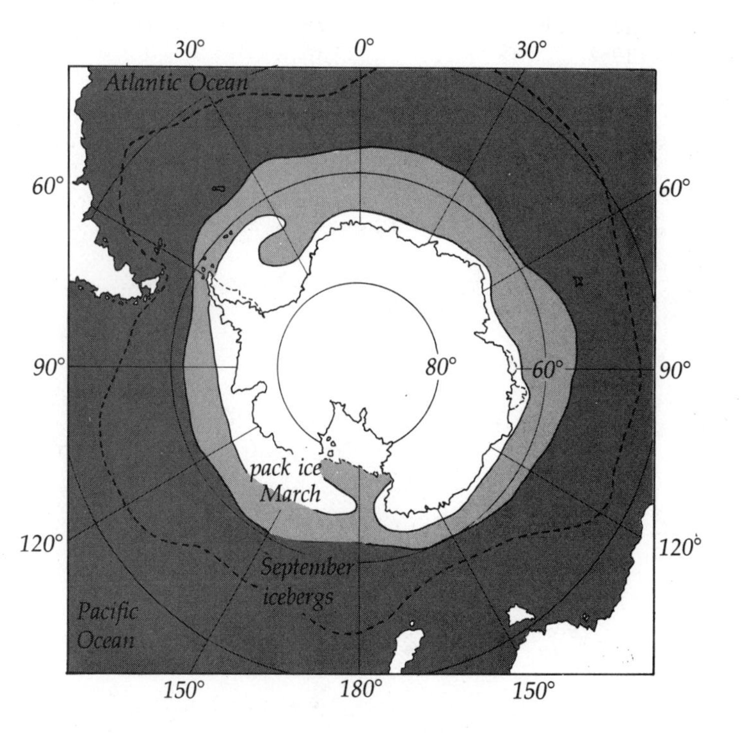

3–51 Average limits of pack ice boundaries for winter (September) and summer (March) in the Antarctic Ocean.

could be caused by a rupture in such an extensive pipe network.

Monitoring the extent of polar ice gives some evidence about the increase or decrease of atmospheric temperature. These factors were treated briefly in the section on light. Present evidence shows that polar ice has been increasing over the last few years. Is the earth cooling?

Further Reading

Adams, K. S., and J. A. Day, *Water: The Mirror of Science.* Garden City, New York: Doubleday-Anchor, 1961.

Albers, V. M., *Underwater Acoustics Handbook.* University Park: Pennsylvania State University Press, 1960.

Groen, P., *The Waters of the Sea.* Princeton, N.J.: Van Nostrand-Reinhold, 1967.

Hardy, Sir Allister, *Great Waters.* New York: Harper & Row, 1967.

Hardy, Sir Allister, *The Open Sea: Its Natural History.* Boston: Houghton-Mifflin, 1965.

Harvey, H. W., *The Chemistry and Fertility of Sea Waters.* Cambridge: Cambridge University Press, 1960.

Kuenen, P. H., *Realms of Water.* New York: Science Editions, 1963.

Martin, Dean F., *Marine Chemistry,* 2 vols. New York: Marcel Dekker, 1968.

Martin, O. L., Jr., "The Titanic—50 Years Later," *Science and the Sea.* Washington, D.C.: U.S. Navy Oceanographic Office, 1966, pp. 35–44.

Menard, H. W., *The Anatomy of an Expedition.* New York: McGraw-Hill, 1969.

Revelle, Roger, "Water." *Scientific American,* September 1963, pp. 93–108.

Riley, J. P., and G. Skirrow, *Chemical Oceanography,* 2 vols. New York: Academic Press, 1965.

Strickland, J. D. H., and T. R. Parsons, "A Practical Handbook of Sea Water Analysis." Ottawa: Fisheries Research Board of Canada, Bulletin No. 167, 1968.

Sverdrup, H. U., Martin W. Johnson, and Richard H. Fleming, *The Oceans: Their Physics, Chemistry, and General Biology.* Englewood Cliffs, N.J.: Prentice-Hall, 1942.

Von Arx, William S., *An Introduction to Physical Oceanography.* Reading, Mass.: Addison-Wesley, 1964.

Wald, George, "Life and Light." *Scientific American,* October 1959, pp. 92–112.

Weyl, P. K., *Oceanography: An Introduction to the Marine Environment.* New York: John Wiley & Sons, 1970.

Williams, Jerome, *Optical Properties of the Sea.* Annapolis, Md.: U.S. Naval Institute, 1970.

Motion of the Waters 4

The wind increased to a howl; the waves dashed their bucklers together; the whole squall roared, forked, and crackled around us like a white fire upon the prairie, in which unconsumed, we were burning; immortal in these jaws of death! . . .

Now, in calm weather, to swim in the open ocean is as easy to the practised swimmer as to ride in a spring-carriage ashore. But the awful lonesomeness is intolerable. The intense concentration of self in the middle of such a heartless immensity, my God! who can tell it? Mark, how when sailors in a dead calm bathe in the open sea—mark how closely they hug their ship and only coast along her sides. . . .

An intense copper calm, like a universal yellow lotus, was more and more unfolding its noiseless measureless leaves upon the sea.

Herman Melville

I. Waves

A. Nomenclature

B. Ideal waves

C. Actual waves

1. *wind*
2. *swell*
3. *deep and shallow waves*
4. *breaking waves*
5. *surfing*
6. *rip currents*
7. *refraction, diffraction, and reflection*

D. Seismic sea waves (tsunamis)

E. Storm surges

F. Seiches

G. Internal waves

II. Tides

A. Terminology

B. Cause

1. *solar tides*
2. *lunar tides*

C. Tidal ranges

1. *tidal currents*
2. *tidal bore*
3. *energy from tides*

III. Oceanic Circulation

A. Surface currents

1. *wind*
2. *the Coriolis "force"*
3. *monsoons*
4. *upwelling*
5. *Eastern and Western Boundary currents*
6. *Equatorial currents*
7. *the Cromwell Current*
8. *divergence and convergence*

B. Deep currents

1. *water mass*
2. *water age and turnover time*

C. Basins and estuaries

D. Waterspouts and whirlpools

Waves

The seashore has always been of particular interest to man, even before he began to sail on the sea. Travel was easier and food always more abundant. Even today most of humanity lives near the ocean. The sea holds a strange fascination for man. Much of this fascination must come from its size and its mystery.

Man depends mainly on his visual sense. He seems to be innately a wave watcher, drawn to the endless train of waves ending as surf on the beaches and rocks.

Nomenclature. Waves come in virtually all sizes, but the same terminology is applicable to all of them (Fig. 4–1).

1. Wavelength is simply the length of one wave, measured from any particular point to the same equivalent point on an adjacent wave; for example, from the top (or crest) of one wave to the next crest, or from the bottom (or trough) of a wave to the next trough.
2. Height is the total vertical distance of a wave from crest to trough.
3. The amplitude is equal to one-half the height—in other words, the distance that the wave moves above or below the normal surface level (the maximum water displacement either up or down).
4. Period is the time it takes for one complete wave to pass a point.
5. Frequency refers to the number of waves passing a point in any given time, such as a minute. A wave that passes in half a minute (period) has a frequency of two waves per minute.
6. The rate of propagation is the speed that the wave travels, measured usually in miles or kilometers per hour.

Wavelength may vary from less than 1.73 cm as in capillary waves to one-half the circumference of the planet. The velocity and length of water waves are determined by their period, except in shallow water. One of a number of possible ways to classify waves is in terms of their period (Table 4–1).

Ideal Waves. The ideal characteristics of waves can be described and explained mathematically. One good way to look at them is with the aid of a rope. Hold one end of a rope 20 feet or more long that is secured at the other end and whip it in an up and down motion. Each up and down cycle produces a wave, which moves along the rope. Another, more valuable experiment is to drop a pebble or stone into a small, still pond or some placid water surface. The waves emanate evenly in concentric circles from the stone's point of impact as the water surface at the point of impact vibrates vertically, much like a drumhead. These waves are about as close to ideal as one can find in the physical world. In the ocean, a typical wave profile might be that shown in Fig. 4–2. Although actual waves do vary from the ideal, they are, of course, essentially the same in character.

Actual Waves. In the same pond place a bottle that will float or, better still, a cork. Drop another stone into the water. The cork bobs up and down but doesn't

Table 4–1 *Ocean Waves Classified by Period*

Period	Classification
less than 0.1 sec	capillary waves
0.1 sec to 1 sec	ultra-gravity waves
1 sec to 30 sec	gravity waves
30 sec to 5 min	infra-gravity waves
5 min to 12 hr	long-period waves
12 hr to 24 hr	tidal waves

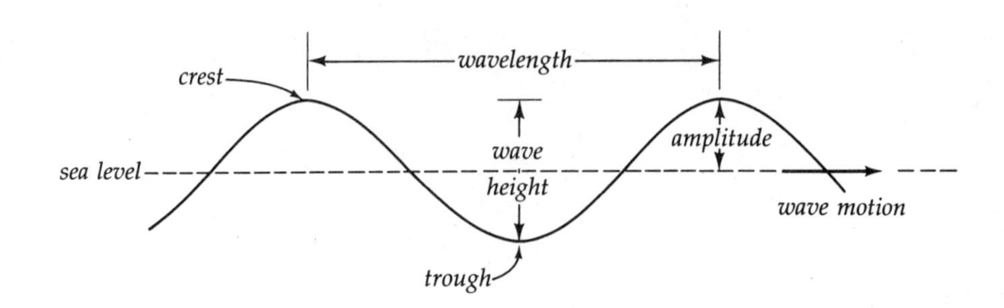

4–1 *Wave terminology (see text).*

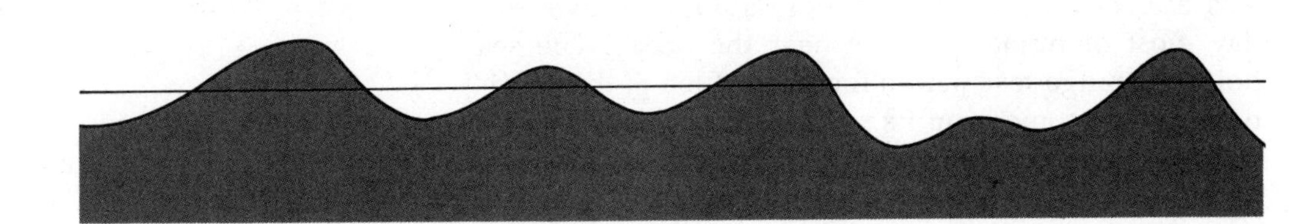

4–2 *A typical wave profile in the ocean.*

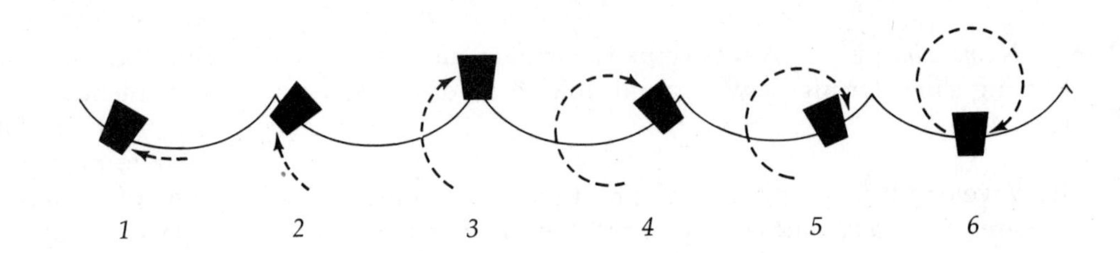

4–3 *The motion of a cork on a wave.*

move with the waves, just as the rope moved only vertically and not horizontally. The cork's motion, however, is not exactly vertical (Fig. 4-3). Like the cork, water particles move in this up and down circular manner and not horizontally with the waves. A wave is simply an energy transfer through the water. Over a period of time the cork might move a little in the direction of wave travel, because some energy is transferred from the wave form to the water and therefore to the cork, causing its movement. But this movement is only slight, unless a wind blows the cork along. If several corks are weighted so that each stays suspended at a slightly different depth, their shape of motion is seen to change with depth. In shallow water (as Fig. 4-4a indicates), the circular motion becomes more flattened (into an ellipse) and ultimately becomes a back and forth motion. If the same experiment were done in deep water, the motion would occur as in Fig. 4-4b. After a few feet there would be no movement. So deep and shallow water waves must be different (see p. 126).

Wind. Waves are produced when energy is imparted to the water surface. In a lake, this may come from a rock or a boat, but in the ocean wind is the primary energy source. Yet, on a calm, perfectly windless day with clear blue skies, waves of different sizes arrive at the shore. Evidently the production of waves by the wind cannot be a phenomenon only at the shoreline. In reality, most waves result from storm winds and are generally formed in the open ocean, often thousands of miles from the beaches they ultimately break on.

The ocean is never motionless. Even on a windless day one can often feel his boat slowly move up and down. But picture a perfectly smooth ocean surface with no waves of any kind and, of course, no wind. As a wind picks up, the sea surface becomes rippled. A breeze of only one-half knot can form these ripples (1 knot = 1 nautical mile per hour; 1 nautical mile = 6080 feet). As these small waves form, they offer a greater surface area to the wind. Increased friction and pressure cause the waves to become longer. The surface then becomes choppy. The waves move in roughly the direction of the wind. These waves are not all the same size. A jumble or cluster of waves in the storm or wind area is referred to as *sea* (Fig. 4–5). As the wind velocity increases, the average wave height also increases, but other factors—the length of time the wind blows and the extent of water, or fetch, over which it blows—also contribute to wave size (Fig. 4–6).

To produce large waves, high velocity winds must move in the same direction over a wide area for a reasonable period of time. A 20 knot wind blowing for at least 10 hours along a minimum distance of 125 km will generate, when fully developed, waves with the average height of 3.3 m for the top 10 percent. A 50 knot wind blowing for three days over a fetch of 2500 km will produce waves 33 m high for the top 10 percent. But this seldom happens in the ocean. Most waves in the sea are less than 3 m high. The old sailors' rule about every seventh wave being a big one is a myth. Large waves at sea are not dangerous to ships if the vessel can move up and over the waves; but if the crest of the wave has broken free, causing "white caps," the water can collide with and damage the ship. Normally a ship at sea during a storm is not an easy place to measure wave height. But the largest wave ever measured was measured under such conditions. During a storm on February 7, 1933, with winds up to 68 knots, Lt. Commander R. P. Whitemarsh of U.S.S. *Ramapo,* a Navy tanker 478 feet in length en route from Manila to San Diego, calculated the height of one wave as 37 m.

While standing watch on the bridge, he saw seas astern at a level above the main crow's nest and that at the moment of observation the horizon was hidden from view by the waves approaching from astern. Mr. Margraff is 5 feet 11¾ inches tall. The ship was not listed and the stern was in the trough of the sea.[1]

[1]Proceedings of the U.S. Naval Institute.

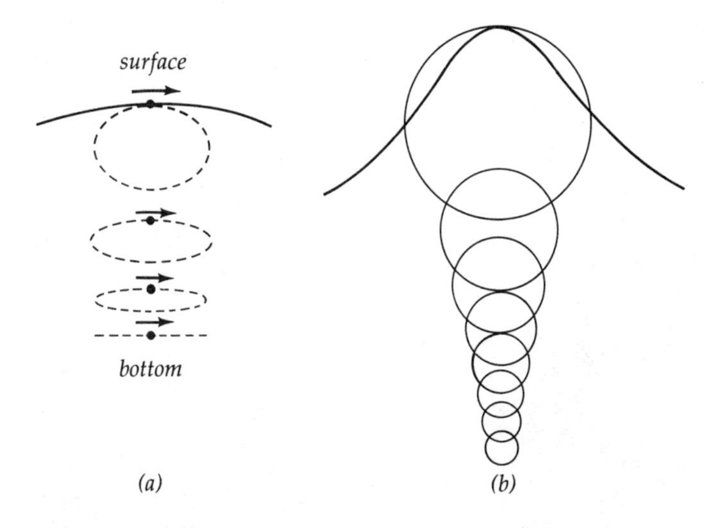

4–4 *Wave orbits: (a) in shallow water; (b) in deep water.*

Swell. As the wind abates or the waves leave the storm system, they change, sort themselves out, and become known as swell (Fig. 4–7). These waves generally move faster than the storm centers that created them, which is the reason that waves from a storm may reach shore well ahead of the storm—if the storm ever gets there.

The shortest waves tend to die out shortly after leaving the storm region because they interfere with one another. The remaining waves, of all sizes, tend to lose height and sort themselves out according to wavelength. A common figure for speed of swell with a period of 10 sec is about 60 km/hr. The longer wavelengths move more rapidly than the shorter and arrive at land sooner. This means that as the newly formed swell moves from a storm region, the wave train spreads out. The farther from a storm region, the longer it takes for the wave train to pass any one point. Waves from a storm generated at 65°S take two weeks to reach the shores of Alaska. At sea the swell may hardly be noticeable as it passes, often creating no more than a slight rocking motion of a boat. At other times, it may appear like a series of slowly rolling hills.

A wave that breaks on the coast of California or Oregon may have been formed in the Antarctic Ocean. It would have traveled across the Pacific a distance

4–5 *Sea. (Courtesy of Jan Hahn, Woods Hole, Mass.)*

of approximately one-half the circumference of the planet in two weeks to finally spend itself on a beach. Waves hitting the California coast may also come from north of Hawaii, while those hitting the New Jersey coast may have come from the South Atlantic or off of Iceland. In this trip across the ocean, once out of the storm area, where the energy loss by waves is the greatest, the swell loses very little energy even in crossing the equator. This energy travels through water as a wave and is expended on beaches.

The energy of the waves, referred to as kinetic energy, is tremendous. About 35,000 horsepower per mile of coastline is generated when a 1.3 m wave strikes the shore. At Tillamook Rock lighthouse in Oregon, a 61 kg rock was thrown through the roof, which was 46 m above the water. On the Scottish coast, a piece of cement weighing 1,350 tons was torn loose and moved by waves; the force of these surf waves has been measured at over 6,000 lb/ft.^{2} Nearly all of a wave's energy is lost on breaking. Because the energy of the wave is released in such a short period of time, the energy density is actually greater than that of the storm that originally made the wave. This energy discharged at the coast shapes the shore and beaches. Quite possibly, some of the energy these waves contain might be harnessed to generate electrical power. In an experiment on the Algerian coast, the waves are funneled along V-shaped cement troughs into a reservoir. The water returning from the reservoir to the sea turns turbines.

Deep and shallow waves. It has already been mentioned that deep and shallow water waves are different. Shallow water waves are the ones that hit the beaches, but deep water waves are the source of the shallow water waves. Ocean surface waves can be divided into these two distinct classes. Because gravity and inertia are the forces that maintain waves with lengths longer than 1.73 cm, they are often called gravity waves. The shape of waves shorter than 1.73 cm is controlled by the surface tension of the water. Shallow water waves, or long waves, occur in water whose depth is less than one-half the wavelength of the wave. Deep water waves, or short waves (even though these may have a longer wavelength), occur where the water depth is greater than one-half the wavelength. The speed of shallow water waves depends only on the depth of the water and, therefore, they travel through a certain body of water with the same speed. Although they move through different regions with different speeds, the speed of deep water waves depends only on the wavelength of the waves. The formulas for the two types of waves are:

$$V_{\text{deep}} = \sqrt{\frac{gL}{2\pi}} \qquad V_{\text{shallow}} = \sqrt{gd}$$

where V = velocity, g = gravity, L = wavelength, and d = water depth.

Because the depth to the floor decreases when a wave moves in from the ocean to the beach, the wave changes from a deep water to a shallow water wave. Here begins perhaps the most exciting event of the ocean. The wave begins to experience the drag of the shallow bottom. If the swell crests are 6 m apart, then the wave will "feel bottom" at the depth of 3 m (one-half of its wavelength). As this occurs, the wave speed decreases causing the wavelengths to decrease. Crests crowd more closely together. Each wave tends to "pile up" or increase in height as the circular path of the particles is squeezed together. The top of the wave moving faster than the bottom causes cresting, or curl-

4–6 *Wind and wave conditions.* (*a*) *Capillary waves, shown close to actual size, superimposed on larger waves. (Courtesy of Jan Hahn, Woods Hole, Mass.)* (*b*) *Wind less than 1 knot.* (*c*) *Wind 4 to 6 knots.* (*d*) *Wind 22 to 27 knots.* (*e*) *Wind 37 to 44 knots. The waves are not fully developed.* (*f*) *Wind 48 to 55 knots. The sea is fetch limited because the sea ice is only 30 nautical miles to windward. (b-f courtesy of Dr. F. Krügler, Hamburg.)*

4–7 (a) Swell (Photo by William J. Wallace); (b) ocean swell "lifting" up and over a treacherous sand bar west of San Francisco known locally as the "Potato Patch" (Courtesy of the United States Coast Guard); (c) a Coast Guard cutter in heavy ocean swell (Courtesy of Jan Hahn, Woods Hole, Mass.); (d) surf (Photo by William J. Wallace).

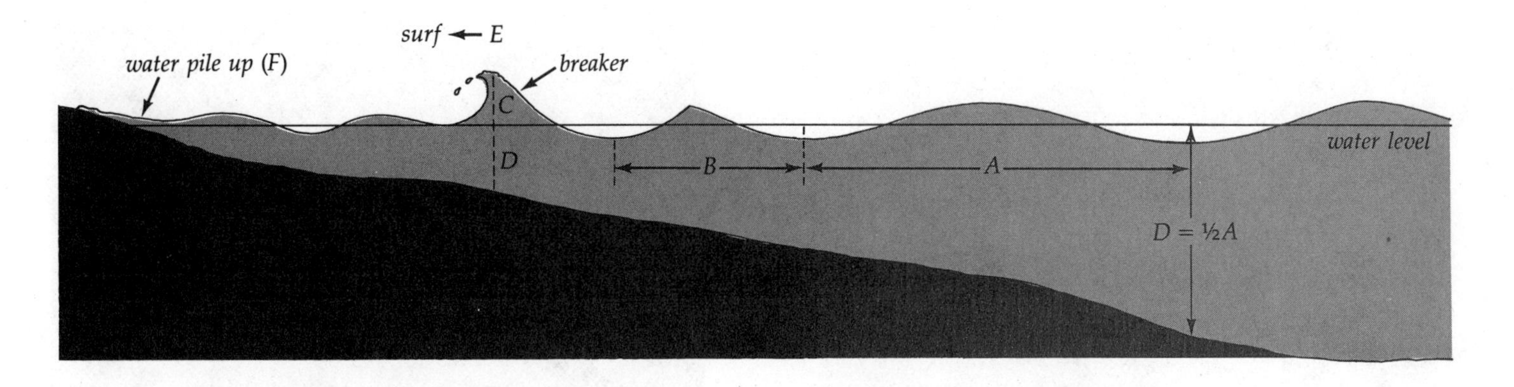

4–8 Wave of length A *in shallow water shortens to length* B *and piles up in height* (*at* E) *when breaking occurs at a ratio of* 3(C) *to* 4(D), *height to water depth. From this point to the shore the wave has become surf, with foam and water pileup on the beach* (F).

ing over at the top. Traveling further still, the crest leans too far forward, and the wave topples and breaks into foam and surf (Fig. 4–8). The velocity of the crest for short periods of time just before breaking may be twice that of the wave.

Breaking waves. To determine when a wave will break, this simple rule can be applied: a wave will break when the depth of the water is four-thirds its own height. A wave 3 m high will break in a depth of 4 m, or a 6 m wave will break in 8 m of water. If one knows the approximate slope and depth of a beach at various distances from shore, he can estimate the height of a wave there. One can walk down the beach toward the water until his eyes align the crest of the breakers with the horizon. The height of the wave equals the vertical distance from eye level to the lowest level of the retreating water from the last wave.

Breaking waves, or breakers, can be divided into spillers and plungers (Fig. 4–9). Spillers roll in evenly for some time (Fig. 4–10). The more impressive plungers (or dumpers, as they are sometimes called by surfers) break over a short distance. They literally pound the beach, with an accompanying roar and splash of flying water and foam. The spillers usually give the longest ride. Various things, such as strong winds or currents, determine whether a swell becomes a spiller or a plunger, but the primary factor is the nature of the bottom. The greater the bottom slope, the more abruptly the wave will slow down and break. The composition of the bottom must also be considered, though generally its effect is not as pronounced as the slope. A smooth, steep underwater beach slope will create the best plungers. Spillers are most apt to be produced on gently sloping bottoms, possibly with small rocky irregularities. Because weather conditions vary and especially because the nature of a beach and its bottom are affected by the continual bombardment of the waves and their release of large amounts of energy, the type of breakers on a particular beach will not necessarily be constant. The very shallow and gentle slope of the coral reef off Waikiki, coupled with a small tidal range, means that the swell forms some of the world's best spillers.

In the numerous Allied amphibious invasions during World War II, it was desirable to know the nature of the beach slope before sending thousands of men against the hostile region. As always, the more man knows about a region, or his environment in general, the less hostile and more compatible it becomes from both sides—his and the environment's. It was necessary to know how close landing craft could come to the beach to dispatch men and supplies. This was discovered by aerial photography of the beaches and the waters preceding them. From the decrease in wavelength as the waves approached the shore, depths were determined and a profile of the beach slope was drawn.

4–9 Breakers: (a) plungers; (b) spillers.

4–10 A spilling breaker. (Photo by William J. Wallace.)

Surfing. When a canoe or a porpoise takes some of the energy out of a wave to be propelled down the forward surface, it is surfing. Porpoises (like submarines, hopefully) are neutrally buoyant, and apparently have learned to position themselves on or under the bow wake of ships and ride the wave indefinitely. One can do this with a small boat on the wave of a passing ship, by attaining and holding a position on the forward face of the first large wave in its wake.

The forces involved in surfing with either a board or a body are given in Fig. 4–11. In effect, the surfer is slipping down the constantly formed front "hill" of the wave. Gravity is the motive force. If the surfer is positioned correctly and moves with the wave and in its direction, he moves at the wave-crest speed. If he is moving sideways across the forward face of the wave, he may move quite a bit faster than the wave itself is moving. The idea in surfing is to get the board moving and the weight balanced just as the wave passes, so that the downhill force on the board is greater than the resistance of the water and the wave's energy can propel the board.

The surfer, or any other observer of the waves, is often puzzled by the variance in breakers. At a given beach over a short period of time, one can experience a group of perhaps a dozen moderate or low waves, then a few high ones, and then a period with relatively few waves. This effect is caused primarily by the arrival at the same time of more than one set of swell from separate storms at sea. Larger waves are formed when waves of separate trains merge and reinforce one another. Smaller waves occur as a result of the opposite effect, when waves are out of phase and interfere with one another.

Rip currents. Associated with waves at the edge of the sea or surf is the notion of undertow. This name is deceiving. There is no real undertow to draw or suck swimmers under. What often occurs, however, is the phenomenon termed *rip currents.* As mentioned in Chapter 2. the wave action at the beach interface often sets up a longshore current. Rip currents can be

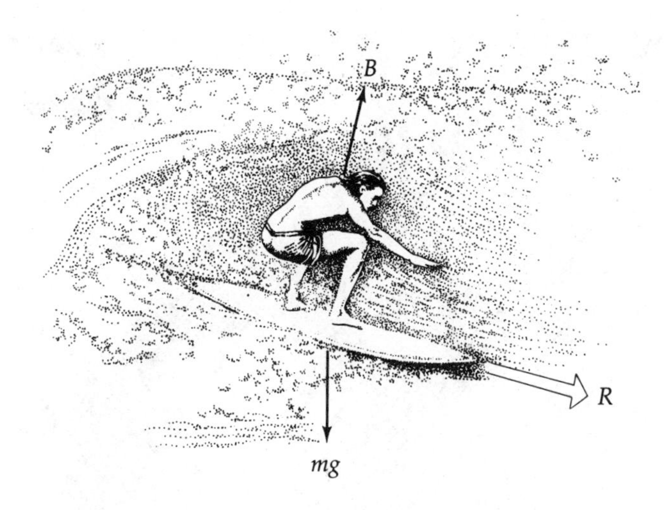

4–11 *The forces involved in surfing. The upward bouyancy of the board (B) and the downward force of weight (mg) yield the resultant vector R.*

4–12 *The position of regularly spaced rip currents on a beach is indicated by concentrations of tracer just outside the breaking waves. Tracer (light areas) is being removed from the surf zone by rip currents. (Courtesy of Scripps Institution of Oceanography, University of California, San Diego.)*

set up whenever these longshore currents funnel or flow back to the sea through a narrow opening formed by storms or, more commonly, by the erosion of a channel or gap across the lowest part of an offshore sand bar. The rip currents may move at speeds of 3 km per hour (Fig. 4–12). They are localized, and their position on a beach can change with a change in wave conditions. As one might guess, the higher the waves the stronger the rip currents. If a swimmer is caught in a rip current and carried out from the shore, he should not swim against it. Even a current of 3 km/hr is exhausting. At a rate of 50 m/min (3 km/hr), the swimmer could not change position in the current. It is simpler to swim along the beach sideways in the current and, because rip currents are narrow, one will soon be out of it. Or if the swimmer relaxes, he will be carried out until the current soon slackens and he can swim to shore at a different point along the beach. Many beaches have sand bars as temporary resting places, perhaps only waist deep in places expected to be well over one's head. Foaming water that is lighter in color is generally more shallow.

Recall that the motion of the water particles of a wave is circular. If a swimmer near shore goes through a wave rather than over it, he will experience a short slight downward pull, part of this circular motion. This is only of small duration and not dangerous in itself. If close enough to shore, he may be thrown landward somewhat ungracefully and deposited not always lightly on the beach.

In the example of a cork floating on a wave, whenever the cork moves even slightly, some of the water must also have moved in the direction of the waves. This very slow transference of water along the direction of wave motion is called mass transport. This transport of water by swell waves is negligible in the open ocean, but inshore it obviously must have some effect. In shallow areas, water tends to pile up on the shore, raising the average water level. There must also be some corresponding seaward transport, as the water build-up cannot continue. This occurs below the surface. Mass transport seaward, especially during

4–13 *Wave refraction.* (*Courtesy of Walter H. Munk.*)

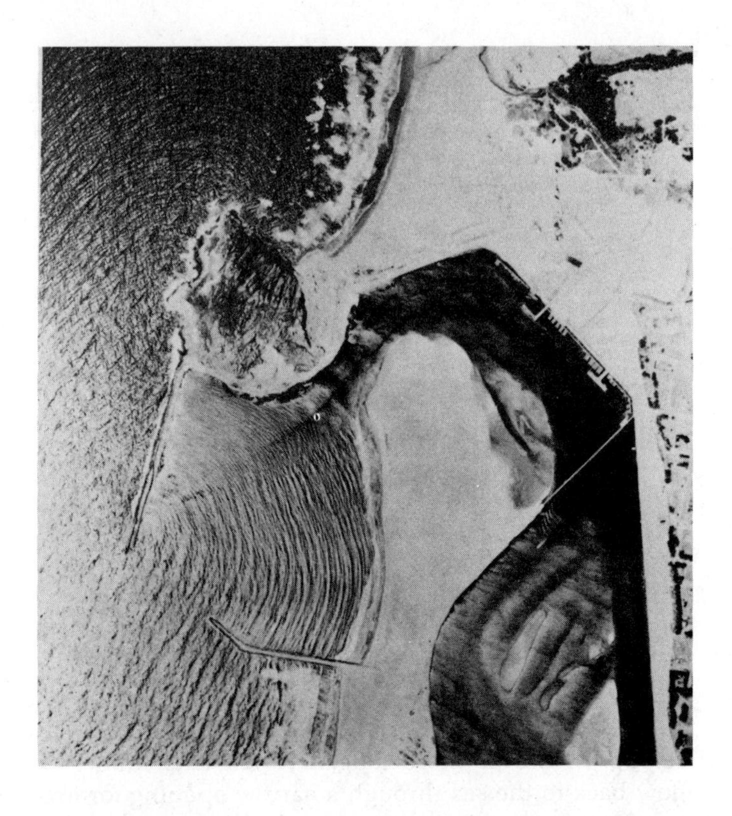

4–14 *Diffraction of waves by a breakwater gap at Morro Bay, California.* (*From* Oceanographical Engineering *by Robert L. Wiegel.* © *1964. Reprinted by permission of Prentice-Hall, Inc., Englewood Cliffs, New Jersey.*)

storms with high waves and onshore winds, helps to account for the extensive eroding of beaches under these conditions.

Refraction, diffraction, and reflection. When waves approach the shore, they may undergo other phenomena before breaking. Among these are refraction, diffraction, and reflection.

When deep water waves enter shallow regions at an angle, they are bent or refracted as they "feel bottom." This is caused by the reduction of wavelength with a decrease in depth. The part of the wave front in shallow water will move more slowly than the part in deep water. The net effect is that wave fronts approaching the shore at an angle will bend so as to more nearly fit the shore. This bending almost makes the wave strike parallel to the shore—but not quite. Because the direction from which waves come and the bottom topography can vary considerably from place to place, numerous possible patterns result (Fig. 4–13).

The diffraction of light can easily be viewed by holding a phonograph record perpendicular to the light. The rainbow effect is the scattering or diffraction of the light. The diffraction of water waves is not strictly the same phenomenon, but definite similarities do, of course, exist. Diffraction accounts for the presence of waves in sheltered regions, such as harbors, that might not be expected to possess them (Fig. 4–14). Like light, water waves can be diffracted into shadow zones behind steep-sided obstacles. Such diffracted waves in a harbor or bay or inside a breakwater will be smaller and often have a grating or choppy pattern. Diffraction patterns of choppy sea have been observed from the air extending for miles downwind from Pacific atolls, which have an almost vertical drop in depth around them. Early Polynesian mariners, perhaps over 1000 years ago, may have used these choppy regions to help navigate because an atoll, being so low in the water, cannot be seen from a canoe or outrigger more than ten miles away.

Waves do not necessarily break. If a regular wave train strikes a vertical or nearly vertical sea wall or cliff that sits in deep water, the waves may be reflected back with very little energy loss. This means, then, that reflected waves exert little force on the vertical wall (Fig. 4–15). This phenomenon is called clapotis (*le clapotis* in French refers to a standing wave), and occurs if the wave strikes perpendicularly. If the waves strike the wall at an angle other than 90°, then waves of equal and opposite angle will be formed off the wall. The two waves may interfere with one another so that crests and troughs will coincide and cancel each other, or at times they may reinforce one another.

4–15 *Wave reflection.* (*Photo by William J. Wallace.*)

Seismic Sea Waves. Tidal waves (tsunamis) have nothing to do with the tides. To emphasize this point, American oceanologists use the Japanese word *tsunami* to designate this phenomenon. Tsunami means "large waves in harbors," which still doesn't describe the phenomenon accurately but eases confusion. Tsunamis are called seismic waves because they originate in some sudden rapid movement of the earth's crust (Fig. 4–16). Most commonly this would be an earthquake and the associated faulting (vertical displacement of ocean floor), an underwater landslide or avalanche, long period earthquake waves, or resonance in submarine trenches that sets the adjacent water in motion. To initiate a tsunami, the surface of the water above a landslide may only have to fall a few feet.

Tsunamis are very long waves. Their wavelength may be 240 km. They travel at considerable speed. Because the wavelength of a tsunami is so great, even considering the great depth of the Pacific Ocean (4,600 m average), it is a shallow water wave, so its velocity is 760 km/hr, almost the speed of a jet plane. A particular seismic action may produce a packet of three or four waves, about fifteen minutes apart. The first is not necessarily the most severe. At sea these waves may only be a foot or two above the surface, so a person on a ship at sea may be completely unaware of a tsunami passing beneath him. As the waves approach the shore, however, they are slowed down and water piles up. They may rise 20 m on flat low-lying shores. At the head of V-shaped inlets, this figure may be over 30 m, a veritable wall of water and a pulverizing force (Figs. 4–17 and 4–18).

On April 1, 1946, a tsunami struck Scotch Cap, Alaska, completely obliterating a lighthouse that had been a concrete two-story building 15 m above sea level (Fig. 4–19). Before the arrival of a tsunami at any shore region, the water moves some distance out to sea for a time. Often a harbor may almost appear to dry up for a brief span between waves. During the tsunami of April 1, the fishing fleet at Half Moon Bay, Alaska, was suddenly deposited on the sandy bottom of the bay where normally it floated at anchor even at the lowest tide. When the next wave passed, boats were relocated on an asphalt road 4 m above sea level.

The violent, catastrophic earthquake that shook Chile (approximately 39°S 74°W) on May 22, 1960, caused by a major marine underwater fault, generated one of the greatest tsunamis of modern times. In a region within 800 km of the epicenter (where there also was a volcanic eruption) hundreds of landslides occurred and 4,000 people died. The tsunami caused much loss of property and life across the Pacific. Along the United States coast, Los Angeles and San Diego harbors experienced $1 million in damage to piers and craft. Nine thousand miles (14,500 km) and 23–24 hours from their origin, the waves that reached Honshu, Japan, were 4.6 m high, killed 180 people, and caused

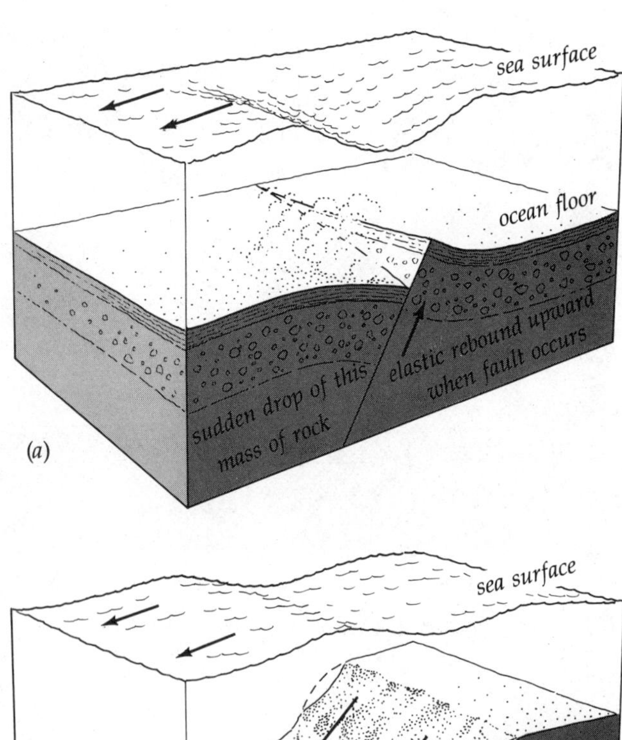

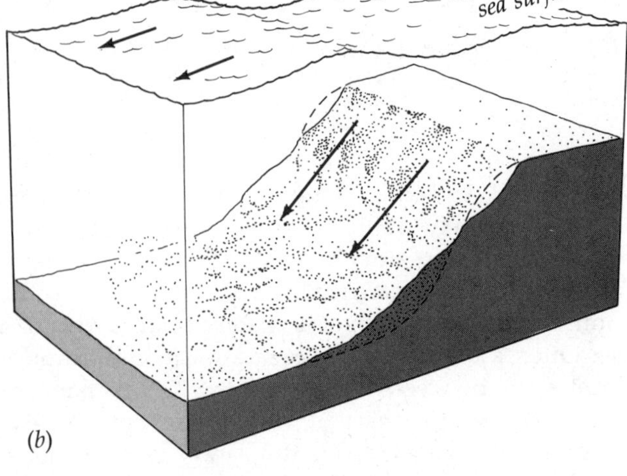

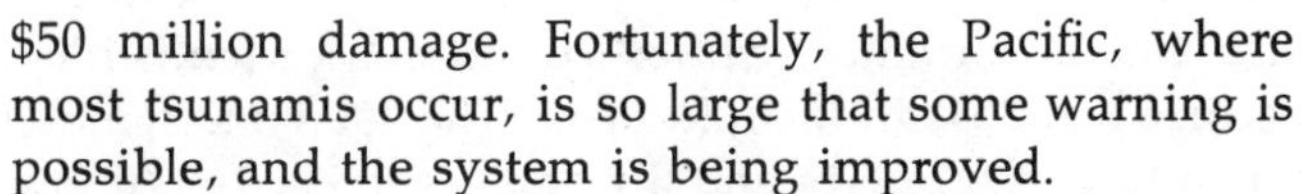
4–16 Some causes of tsunamis. (a) Underwater faulting from an earthquake; (b) an underwater landslide.

4–17 Tsunamis. (Upper photo by Francis P. Shepard.)

$50 million damage. Fortunately, the Pacific, where most tsunamis occur, is so large that some warning is possible, and the system is being improved.

Tsunamis may also be created by large explosions. The classic and often cited example of a natural explosion was that of the volcano Krakatoa on the island of Krakatoa on August 27, 1883. This natural explosion, one of the largest known, blew away about four cubic miles of the island. The sound of the explosion was heard 4,800 km away. The dust it sent into the air was so thick that in some places where it settled ships could not make their way through it, and sunsets all around the world were colored for several years by the fine dust in the sky. Some distance away, a warship was moved almost 3.2 km inland to a height of 9.1 m above sea level.

4–18 Tsunami damage caused by the Good Friday earthquake in Alaska, March 27, 1964. (a) Boats beached by the tsunami at Kodiak. (Official United States Navy Photograph.) (b) Spruce trees 0.7 m in diameter at elevations between 29 and 34 m above lower low water that were broken by a landslide-generated wave. (c) Seward waterfront, showing damage from destructive waves and submarine sliding. (d) Whittier waterfront, showing damage and extent of wave run-up (dark area washed clear of snow near shore). (b, c, d: photos by George Plafker. Courtesy of the Geological Survey, United States Department of the Interior.)

(*a*)

(*b*)

4–19 (a) Scotch Cap lighthouse, Unimak Island of the Aleutians, taken before the tsunami of April 1, 1946. The foundation of the lighthouse is at an elevation of 14 m, the light 30 m above mean low water, the upper radio mast 33 m, and the upper plateau 36 m. (b) The same place after the tsunami. Note that the lighthouse and the radio masts have gone, the slopes are heavily washed almost to the plateau level, and debris has been deposited on the plateau. Note also the exposure of stratification on the slopes, revealed by removal of the overlying sediment. (Courtesy of the United States Coast Guard.)

Who knows? Perhaps the violent explosion of Santorini volcano off Greece (near the present islands of Thera, Therasia, and Aspronsis) in 1400 B.C. created a tsunami that briefly turned the Bardawil Peninsula east of Port Said into a land bridge. This may have enabled Moses to lead the Jews out of Egypt to freedom, and the arrival of the tsunami shortly thereafter drowned pharaoh's horsemen.[2]

An unnatural explosion, namely, a thermonuclear bomb, may also generate tsunamis. There was some concern that the first large thermonuclear explosion detonated at Eniwetok atoll in the Marshall Islands might cause waves similar to Krakatoa's. Geologically, Eniwetok is similar to the island of Krakatoa and the calculated energy of the bomb similar to that of the famous eruption. A tsunami did occur when the bomb went off, but it was only a small one, because the bomb crater did not broach the outer edge of the atoll reef and come in direct contact with the sea.

Storm Surges. Another spontaneous phenomenon, the storm surge, is often disastrous to areas adjacent to the sea. Its effects may be similar to a tsunami's, but the phenomenon itself is quite different. A tsunami often appears in an area experiencing good weather, and if advance radio warning is not given, there would be only a few minutes warning at best. Storm surges occur during bad weather and result from the combination of a number of factors. First, there are the tides themselves. Second, large storms, especially hurricanes, can raise the level of the sea surface. These large storms are characterized by regions of low atmospheric pressure, which actually causes a bulge in the sea surface. Coupled with high winds blowing in one direction for 10 to 12 hours or more, these forces can pull water up on the shore. The result is a large amount of water arriving at the shore causing considerable flooding and surging water, which may last one or two tidal cycles (Figs. 4–20 and 4–21).

The gale that swept down from the North Sea across England and piled waters on the Dutch coast

---

[2]See John Lear, "The Volcano That Shaped the Western World." *Saturday Review of Literature,* Nov. 5, 1966, pp. 57–66.

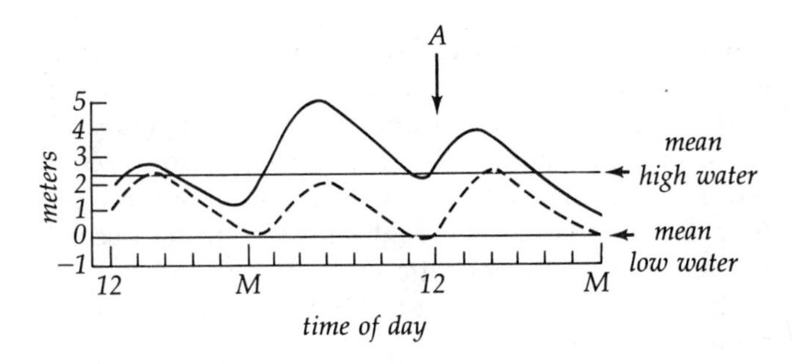

4–21 A water level diagram for a storm surge. Note that the heavy upper line, which indicates the water level during the surge, at its low level (point A, for example) is higher than the mean high water, even though the coming and going of the water over this 36-hour time period corresponds to normal tidal timing sequence (the dotted line).

4–20 U.S. Coast Guard cutter General Greene *aground after a severe North Atlantic storm. (Courtesy of the United States Coast Guard, First Coast Guard District, Boston, Mass.)*

on February 1, 1953, was such a storm surge. Almost 1800 people drowned as 800,000 acres of this dike-rimmed, low-lying country, previously ocean, were flooded. It has been calculated that the probability of all these factors combining at a particular time and place to form such a superstorm is one in 400 years. The Galveston flood of 1900 popularized in American folk songs was actually a storm surge. A hurricane with winds of 193 km/hr pushed the normal 0.7 m tides up to 5 m, along with storm waves an additional 8 m high. Over 5,000 people died as the city was virtually destroyed. The New England coast received a severe battering from Hurricane Carol in August of 1954. The damage exceeded $40 million.

A storm surge that occurred on October 31, 1960, near the mouth of the Ganges River in India picked up an American freighter and set it down over a mile inland. The cost in lives and property from these destructive tsunamis and storm surges is given here to emphasize their danger.

Unfortunately, although warning systems to report the advances of these phenomena are rapidly being improved, their potential danger is on the increase. This does not result from any increase in the frequency or intensity of the phenomena but simply because the world's population is growing especially rapidly in regions near the sea. On November 12, 1970, a massive bulge of water 6 m high, with the large cyclone (the East Asian equivalent of a hurricane) that caused it, struck and buried the densely populated coast of East Pakistan. Over 500,000 people died in the worst natural catastrophe since the Yellow River flood of 1887 in China.

Seiches. Virtually all the waves previously discussed move or progress in a direction—and, as such, are classified as progressive waves. Some other waves, like the waves set up in a vibrating guitar string, are called standing. One type of standing (or stationary) wave, the seiche, will be discussed here. It has long been known that the water level in some lakes rises and falls regularly with a period much shorter than the tide—often in the vicinity of about an hour or more. Long waves are set up in the water as it sloshes back and forth. Perhaps the best way to visualize these waves is to fill an oblong container like a bathtub with 6 to 10 inches of water, raise one end, and then set it down. The water moves back and forth reflecting off each end. In this simple case, the center of the water, which doesn't move up and down, is the node. The wave is called mononodal (Fig. 4–22a). Two, three, or more nodes may exist in an oscillating water basin. In a mononodal seiche, high and low water occur at the same time on opposite ends. In a dinodal seiche, high water and low water occur on each end at the same time (Fig. 4–22b).

Seiching is caused by a variety of factors. Strong winds may temporarily drive the water toward one end of a basin. A difference in barometric pressure between the ends of a basin could cause the surface imbalance that begins the oscillation. In bays or marginal seas that have open access to the ocean, tides often cause seiching. But in these open bays the primary cause is long period wave trains. A tsunami may also cause seiches in these locations, although this cause is rare because seiching is a much more common occurrence than tsunamis.

Many bays and harbors around the world, even those with exceptionally irregular shapes, seem to oscillate with remarkable regularity. Because the wave height of the oscillation is so low and the length so great, seiches generally go unnoticed by people. While seiches seldom cause problems, in some harbors difficulties do occur. Perhaps some may have noticed all the boats in a harbor area jointly straining against mooring lines even though no wind or waves were present at that time. This is a common occurrence in Los Angeles harbor, where seiching is a problem. Often for no apparent reason when the water surface is calm and windless, heavy ships strain against their mooring lines, sometimes snapping them and damaging pilings, piers, and other ships. The period of these oscillations is 3, 6, and 12 minutes, which corresponds to the natural periods of the basin.

Internal Waves. An old nautical term still used by Norwegian fishermen is translated as "lying in dead water." This refers to small vessels, either oar-, motor-, or sail-powered, getting "stuck" in the water. In other words, it is as if the craft were being held in place. Making any headway is difficult and the boat doesn't respond well to the helm. This is especially a problem for a sailing craft, which has the constant tendency to fall away to leeward. Hours might elapse before the craft frees itself. Sometimes the boat breaks

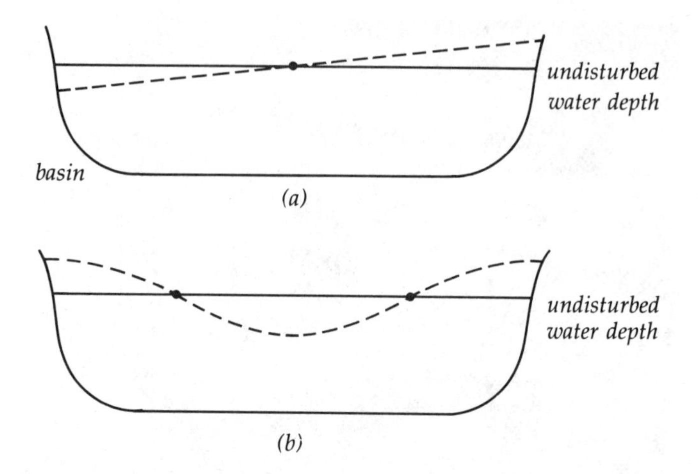

4–22 Seiches: (a) mononodal; (b) dinodal.

4–23 A boat "lying in dead water." The energy of the propeller (or given to a sailboat by the wind) generates internal waves in the lower, more dense water layer.

free for no apparent reason or perhaps the wind may have picked up. While this phenomenon occurs elsewhere, it is especially prevalent along the Scandinavian coasts, where the tides are not too active and much fresh water runs off from the land. Layering of fresh and salt water is quite common here, and the two are slow to mix. It has been shown that this "sticking" phenomenon is caused by internal waves.

All waves so far discussed occur on the surface of the ocean—or phrased differently—at the sea–air interface. There is an obvious density difference between these two fluids. Recall from Chapter 3 the density structure of the ocean. If the water is well mixed, there

4–24 Ocean slicks. (Courtesy of Dr. E. C. LaFond.)

is no layering. But seawater is seldom that mixed, and the layers of seawater have different but increasing densities. As on the surface, waves can be generated along any of these subsurface density discontinuities. The waves move along the interface of two waters with different densities on the surface of the denser, just as they do on the surface of the ocean. The surface for an internal wave need not be flat with respect to the ocean surface. Internal waves probably move most often as smooth undulations, traveling at about one-eighth the speed of surface waves. They have an average height of 4.1 m (but may be well over 30 m), a period of 5 to 8 minutes, and a wavelength of 0.6 to 0.9 km.

These waves may be caused by tidal currents or tidal forces, by impulse pressure of surface winds for short periods of time, or by the propeller of a slow-moving ship. This helps to explain "dead water." In stratified water the propeller or bow (if not motor-powered) may put so much energy into the lower water layer in the production of internal waves that none is left for propulsion (Fig. 4–23). Powerful, modern, motor-driven vessels do, however, find this less of a problem.

Long parallel bands of smooth, slowly moving water called slicks (which sometimes can be seen from the air) may be related to the presence of internal waves in a layer just below a thin surface layer containing plankton or silt (Fig. 4–24).

Submarines submerged at a region of a density discontinuity are known on occasion to rise and fall with internal waves. In such situations the sub may experience difficulty maintaining a specific depth. Possibly, submarine tragedies such as the loss of U.S.S. *Thresher* might be related to internal waves. Very much remains to be learned about internal waves and the ocean bottom as well. The unfortunate loss of the nuclear-powered *Thresher* with 129 men in April 1963 and the subsequent search graphically pointed out how little man really knows about the deep ocean. The *Thresher* went down in only 2,600 m of water in a region where the ocean bed is relatively flat, yet after months of searching with the best equipment available only a few pieces of debris had been located, whereas the main pressure hull was never found. Thus far, less than 5 percent of the ocean bottom has been adequately mapped.

The subject of waves is a complex one. For example, the wave pattern generated as wake by a passing ship is comprised of many individual waves. Different parts of the same ship generate different waves. In addition to all the waves covered above, there may well be waves that oceanologists do not yet know about.

Tides

Unlike waves, which are highly variable, the tides relentlessly come and go every day. Tides, like waves, have always been as fascinating for man to watch as they have been difficult for him to explain. For centuries before the time of Christ, man has known that the moon was somehow related to tides. The Arabian philosophers of the fifth to seventh centuries A.D. attributed the tides to the moon's heat, which at regular intervals heated the ocean more effectively and caused it to expand, resulting in tides. Before steam engines, those whose living depended on the sea felt the dominance of the tides. Early sailing stories always mention sailing with the tide. Today, if one lives in a region with a large tidal range, nearshore pleasure boating is largely at the tide's whim.

Tides are much more significant and extensive than most people believe. Waves affect only the ocean's surface, but tides move the entire ocean (Fig. 4–25). Not only the oceans have tides; ponds, lakes, and even the land comprising the continents and the atmosphere experience this phenomenon. The tides of the Great Lakes measure only a few inches. The continents are not as fluid as water; they rise and fall about 0.2 m during an equivalent 3.3 m oceanic tide. The gaseous fluid envelope called the atmosphere, being the most elastic, bulges a distance of many miles. Like the ocean, people are mostly water. Like lakes and ponds, they, too, have extremely small tides. Corresponding to the rise and fall of the tides, a person's weight is increased and decreased by a few grams.

4–25 *These free-falling instrument capsules, which record deep sea tidal measurements and return to the surface on command, were developed by the Institute of Geophysics and Planetary Physics of the University of California. They are capable of detecting and recording a water-level change of a few millimeters in a water column 8,000 m high.* (Courtesy of Scripps Institution of Oceanography, University of California, San Diego.)

Terminology. The discussion of several familiar terms will be helpful. The words *ebb* and *flood,* of course, refer to the fall and rise of the tide. The sig-

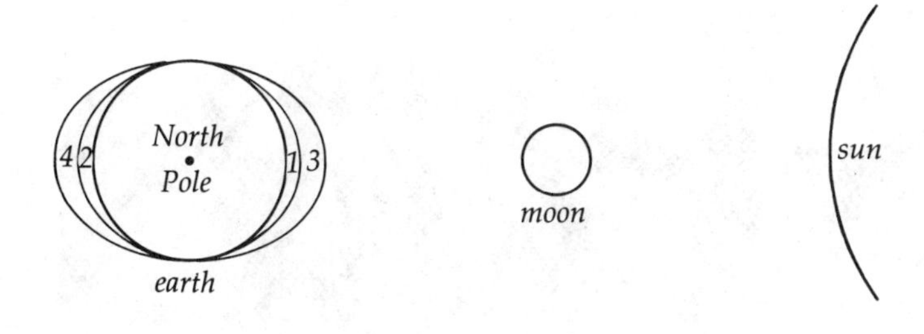

4–26 *Tidal forces. The solar tides are 1 (gravitational) and 2 (centrifugal). The lunar tides are 3 (gravitational) and 4 (centrifugal).*

nificance of high tide and low tide is obvious as are low water, high water, and mean (average) sea level (MSL) or mean tide. This can get complicated. Mean of lower low waters (MLLW), which appears on tide tables for the U.S. Pacific coast, refers to the average height of the lower of the two low tides per day. In European charts LLWS stands for low-low-water spring, which is the lowest spring tide (see p. 142) per month averaged over several years. The depth most charts use as reference level is MLW (mean low water). A region with a diurnal tide has one high and one low tide in a 24 hr 50 min period; semidiurnal refers to the occurrence of two high tides and two low tides in each 24 hr 50 min.

Cause. It is convenient for explanatory purposes to break the tide into separate components. Furthermore, several initial assumptions will help. Assume, for the time being, that earth is entirely water-covered —with no land sticking through the surface. Assume also that the ocean is equally deep everywhere and that the land beneath it is smooth. Lastly, assume that the earth does not rotate on its axis.

Tides are caused by gravitational attraction of the sun and moon on the earth and especially on its water (or hydrosphere). Unlike winds and earthquakes, which are terrestrial, tide-generating forces are astronomical in origin. Tides can be referred to, then, as solar and lunar.

Solar tides. Solar tides are caused by the gravitational attraction all mass has for all other mass according to the theory of universal gravitation proposed by Sir Isaac Newton (1642–1727) in 1687. Newton, who also first described wave motion mathematically, explained that tides were caused by the gravitational pull of the sun (and moon) on earth and its water, producing bulges (Fig. 4–26). The water, being fluid, is pulled somewhat toward the sun causing a bulge (*1*, in the figure). But another bulge (2) equal and opposite to *1* exists. Recall that the earth, in moving around the sun, is held in its orbit by a balance of two forces—gravitation, which would pull them together, and centrifugal force, which would hurl them apart. If one were whirling a ball on a string above his head, the force that would make the ball fly off if the string broke or was released is centrifugal. The revolution of the earth about the center of the earth–sun system generates centrifugal force (the center of mass of the system, for all practical purposes, is the center of the sun). Because this force must equal the gravitational or the earth would not stay in orbit, there must be an equal force in the opposite direction; thus, the bulge on the opposite side (2) is equal in size.

Lunar tides. Although the moon is much smaller than the sun, it is so much closer that the tides it causes, called lunar tides, are greater than solar tides. On the side closest to the moon is a bulge (*3*) of water, and on the opposite side a bulge of the same size (*4*). The bulge between the earth and moon (*3*) is caused by gravitational attraction of the moon to earth and especially its water. The bulge caused by centrifugal force (*4*) is generated by the revolution of the earth–moon system about the earth–moon system's center of mass.

It is fairly easy to visualize a centrifugal force generated on the moon by its motion around earth, but not so easy to imagine the earth generating a centrifugal force of its own. The problem here is the fact that one normally considers the moon to be circling the earth. In reality, they both revolve as a system about a point that is the center of mass of the earth–moon pair. It is easily seen that a rolling ball revolves about its center. But if a wrench is thrown, it too revolves about its center—not its geometric center, but

4–27 *A thrown wrench rotates around its center of mass.*

its center of mass, the point at which all its mass appears to be concentrated (Fig. 4–27). Thus, a hoop can revolve about a center of mass that isn't even in the matter of the hoop.

The earth and moon have mass. Unlike the sun–earth system, in which the sun is incredibly massive compared to the earth, the masses of earth and moon are not too different (the earth is about six times as massive as the moon). As a dumbbell if dropped will revolve not around either ball but around a point *X*, its center of mass, the earth–moon system will revolve around its center of mass, which is not the center of the earth. Because the earth has more mass, the center of the system is actually 1700 km below the earth's surface (its geometric center is 6,373 km). The bulge (*4*, in Fig. 4–26) is caused, then, by the centrifugal force generated by rotating about this point. If bulge *4* were not equal and opposite to *3*, then the force could not be balanced and would change the earth–moon position. The forces and bulges produced by them are balanced. Bulges *3* and *4* are larger than bulges *1* and *2*. The pairs of bulges occur independently but, because the relative position of the moon and sun varies, so do the sizes and positions of the bulges. In Fig. 4–28, bulges *1* and *3* (and *2* and *4*) occur together when the moon is at *A* or *B*, giving the highest and lowest tides, called spring tides. These have nothing to do with that season but occur every 14 days at the new or full moon. When moon and sun are at right angles to each other, intermediate (lower high and higher low) tides, called neap tides, result.

If the earth revolved on its axis once a year, then the solar tide would always be in the same place. But if earth did not revolve at all on its axis, the solar tides would occur once annually at any particular place, and lunar tides would progress around the earth giving the spring tides, which would occur exactly every 14 days. This may not seem too different than what actually does occur, but remember that there would be no daily tides. Daily tides occur because the earth does rotate on its axis. As the moon is also moving in the same counterclockwise direction with respect to the earth, it takes a bit longer for a point on the earth's surface to return to the same exact relative position with the moon. The timespan for one complete rotation of the earth relative to the moon is 24 hr 50 min (one lunar day). This point on the earth experiences two low tides and two high tides in a lunar day.

The moon doesn't rotate about the earth in the exact same plane as the earth but is said to have a declination or elevation above earth's equatorial plane (Fig. 4–29). Imagine a stake fixed into the earth and extending well above water surface no matter what the depths of water bulges are. As this stake moves with the rotating earth, it experiences water of different depths because it moves through different parts of the two bulges. Tides for a point on earth are expected to have this diurnal inequality (Fig. 4–30).

The land of course, sticks out of the water. Tidal bulges attempt to stay aligned with the moon and sun, but the continental land masses interfere. When a continent is directly aligned with the moon, there is no oceanic bulge (although there is a terrestrial tide). The water level at the shores of the continent is higher. When the continent moves away from alignment and the ocean moves into alignment with the moon, the bulge re-forms and continental edges experience low tide as the water moves to form the bulge. The net effect is a continual oscillation of the water that may be considered a wave—a very long wave, with a wave-

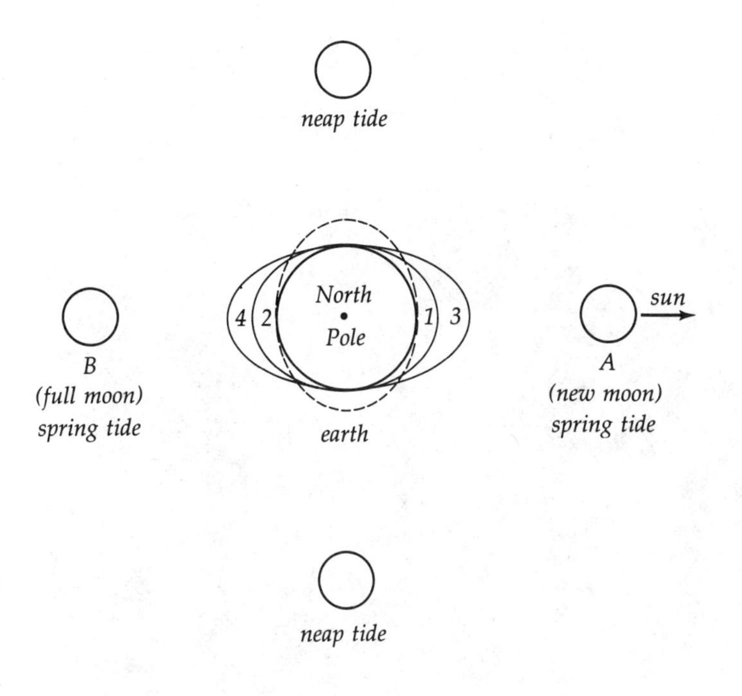

4–28 *Spring and neap tides (see text).*

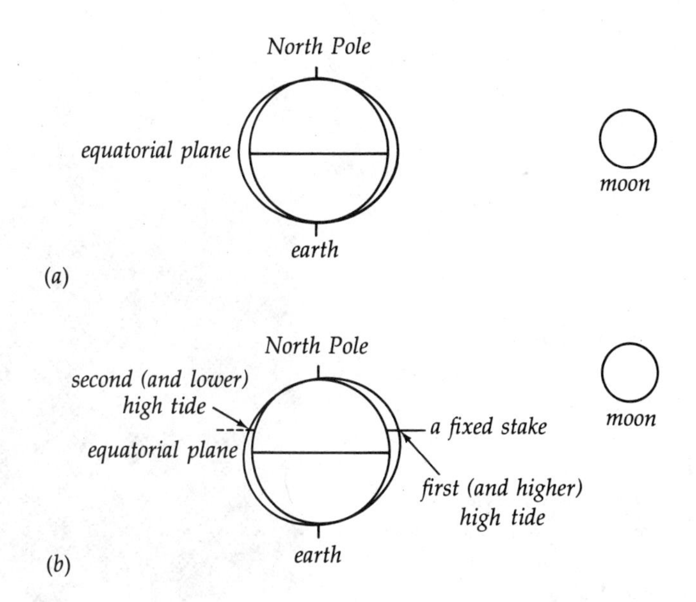

4–29 *If the moon rotated about the earth in the same plane as the earth, the position of the tides would be as shown in a. Because of the moon's declination, the actual position of the tides is as shown in b (see text).*

length one-half the planet's circumference and a period of 44,700 sec (12 hr 25 min). The crests and troughs of the waves are high tide and low tide. These tidal bulges may be considered shallow water waves whose speed (about 200 m/sec on the average in the deep open ocean) depends on the water depth, but whose period depends on the periodic motion of the sun and moon. As these bulges can hardly travel at the speed the sun and moon move across the sky, there is a time lag between the sun's and moon's positions overhead and the tide at that place.

Tidal Ranges. Based on the above discussion, one might expect that all shorelines would experience two high tides and two low tides a day, one high tide being somewhat higher (except at the equator, where both are the same; see Fig. 4–29b). If high tide occurred at 10 A.M., the next expected high tide would be at 10:25 P.M., and the following high tide the next morning at 10:50 A.M. These generalizations are unfortunately too simple. Each ocean basin is shaped differently and varies considerably in dimensions and depth. As a consequence, tides vary from place to place from what would ideally be expected. The tides around the Atlantic occur twice a day (two high and two low) and have the same range, but the regions around the edges of the Pacific have tides of unequal height. The charts in Fig. 4–31 for tides at seven U.S. cities help to illustrate the tidal variances. Note that San Diego, Seattle, Boston, Honolulu, and Galveston have two high tides and two low tides per day, but Pensacola has an entirely different profile. The harbor of Portland, Maine, is only 160 km from Boston, but notice the difference in tide heights. Some areas are almost tideless, such as the Mediterranean, the Gulf of Mexico, and the Baltic Sea. Islands away from the shore and nearer the center of the tidal basins (bulge) have small tides. Nantucket (off Massachusetts) and Tahiti have tides seldom more than one-third of a meter in range. The tidal variances at the ends of the Panama Canal, separated only by 80 km, illustrate the differences that can occur. The tides at Colon on the Caribbean end of the canal have a range of about 0.3 m and are diurnal; but at Balboa, on the Pacific side of the canal (which is actually further east than the Caribbean end), the average range of the tides is 4.3 m and is semidiurnal. Fig. 4–32 shows the variances along the coast of the United States.

(a)

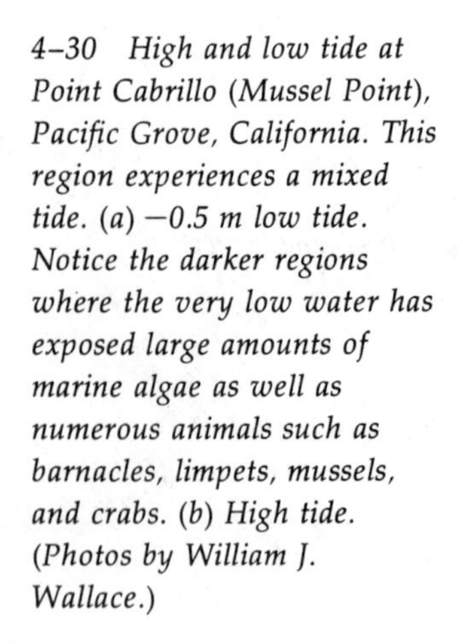

4–30 High and low tide at Point Cabrillo (Mussel Point), Pacific Grove, California. This region experiences a mixed tide. (a) −0.5 m low tide. Notice the darker regions where the very low water has exposed large amounts of marine algae as well as numerous animals such as barnacles, limpets, mussels, and crabs. (b) High tide. (Photos by William J. Wallace.)

Tidal currents. Near land the horizontal and vertical flow of water (especially as viewed against a pier or sea wall) caused by tides becomes apparent. The effects of local winds or weather, sometimes called meteorological tides, cause tidal anomalies that at times can be more pronounced than a tide itself. The storm surges discussed previously illustrate this. Changing tides can set up currents, especially in restricted areas (Fig. 4–33). The current under the Golden Gate Bridge in San Francisco during a change in tides is 6 knots, about the same as the flow of the East River in New York and a value that is not uncommon in the world's straits and channels. This tidal current may reach a velocity of 10 knots in some places, such as the Straits of Georgia (between mainland British Columbia and Vancouver Island). A ship crossing the English Channel toward the North Sea as the tide comes in will have the benefit of as much as 6 hours of favorable current, and if it reaches the Straits of Dover just at high tide, it can continue with the outgoing tidal current to the North Sea.

Tidal bore. Perhaps the most striking effect of local geological conditions on the tide is known as the tidal bore. This occurs particularly during spring tides in estuary or estuarine-like regions where the basin is long and funnel-shaped and depth diminishes regularly. As the tide enters, it is slowed by the constriction. Additional water constantly entering continually catches up with the initial water. The result is a foaming, churning wall of water moving up the bay. Most of the world's rivers do not have these bores. They are most common in Asiatic rivers; but there are bores in French rivers, such as the Seine, the Sirande, and the Orne, and in the Severn and Trent rivers of England. Easily the most famous in North America is the bore of the Petitcodiac River at the head of the Bay of Fundy, which is between two of Canada's maritime provinces, New Brunswick and Nova Scotia (Fig. 4–34). Here, twice a day, a 1.2 m wall of water surges up the bay and fills all of its rivulets. The tidal range in Fundy is the highest in the world, exceeding at times 15 m (Fig. 4–35). Approximately 3,680 billion cubic feet of water

(*b*)

come into the bay with each tide. This is equal to all of the water consumed in three months by all people in the United States. The bore of the Amazon, the world's greatest river, moves up that river for over 480 km at a speed of 12 knots. Its roar can be heard for 24 km. The most famous of all tidal bores is that of the Tsientang Kiang River estuary in northern China, which may attain heights of 7.6 m. When this waterfall-like wall of water rumbles up the river, all boating stops. Natives either remove or securely tie up their craft. Once it passes, one can ride upstream with the following current.

The forces that affect the tides are much more complicated than pictured here. For example, the moon is not at all times equidistant from the earth. The difference between apogee and perigee is about 24,000 km. About twice a year, perigee coincides with spring tide conditions and produces tides that are the highest of the year—by about 20 percent. Or in some areas tides may be reinforced by local resonance—a standing oscillation in the water—to such an extent that only one high tide and low tide occur per day. There are many other facets of the tides.

Energy from tides. Harnassing the tremendous energy contained in the tides to generate electric power has been suggested a number of times. To accomplish this, several conditions would be necessary. A large basin would open to sea with a narrow entrance and with a tidal range of over 3 m. A dam containing two-way turbines would be built across the mouth of the bay to generate electricity as the tide comes and goes. This method of obtaining power has certain advantages. First, one wouldn't expect to run out of tides. The environment is not polluted—except visually by the dam and by the disturbance of regional sediment balance—whereas most electric power now produced is done by burning large amounts of coal and oil, the fumes from which are released into the atmosphere. Nor would there be the problem of radioactive waste disposal, as is the case with nuclear power plants.

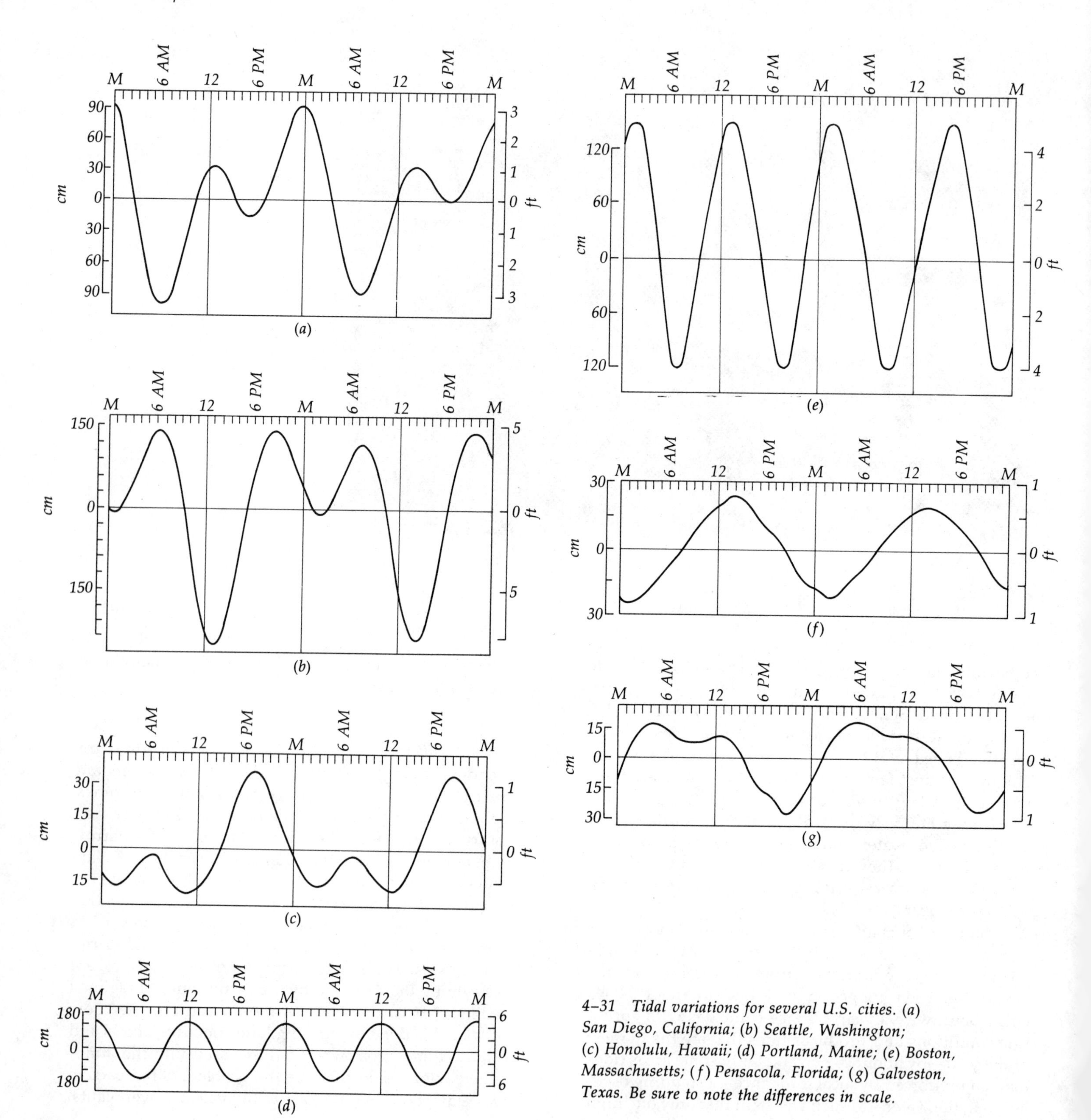

4–31 Tidal variations for several U.S. cities. (a) San Diego, California; (b) Seattle, Washington; (c) Honolulu, Hawaii; (d) Portland, Maine; (e) Boston, Massachusetts; (f) Pensacola, Florida; (g) Galveston, Texas. Be sure to note the differences in scale.

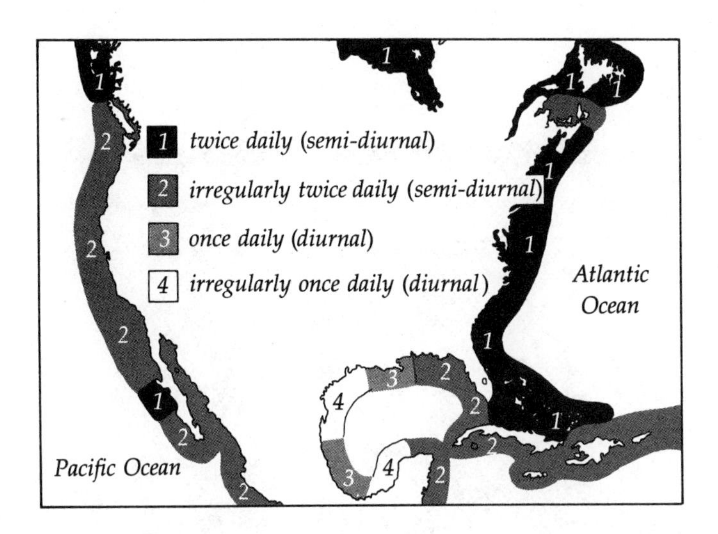

4–32 *Tides along the United States coast.*

4–33 *Tidal currents and resulting rough water (called the "Chopping Block") at junction of the Ipswich River, Parker River, and Ipswich Bay, Massachusetts. (Photo by William J. Wallace.)*

Further, these sea dams may not have the silting problems of regular dams.

As far back as 200 years ago, the tide was used to run grist mills in New England, but so far only a few modern attempts have been made to use the tides for power. One of these, in limited operation since 1960, is located in the Rance estuary of Brittany, France. If the plant goes into full-scale operation, it could supply the electrical power requirements of an industrialized city of about 75,000 people. This amount of power is roughly equivalent to that produced by burning one-half million tons of coal annually.

But there are a number of problems. While places such as Passamaquoddy Bay in northern Maine and Cook Inlet in Alaska would be ideal for such plants, there are few areas where tidal range is sufficient. If all these regions were used, they could supply only about 10 percent of the world's total electrical power requirement. Such dams are also extremely expensive —more so than conventional or nuclear power stations —costing now about $250,000 per linear foot of dam. At present the total cost for this manner of power production would not be economically competitive with other methods.

It has been estimated that the friction between the tides and the ocean bottom as the water moves back and forth is slowing the earth's rotation on its axis by about one second every 120,000 years. A recent calculation indicated that total use of the tide (2 billion kw/yr) would slow this rotation by about 24 hours in 2,000 years. Talk about meddling with one's environment. But there is no reason why some of these tidal power stations cannot and should not be built.

Oceanic Circulation

Surface Currents. The existence of surface currents in the ocean has long been known by mariners, though these currents are difficult to see. Early sailors probably first became aware of currents by the difference in a course attempted and the one actually made. The commanders of the sail-driven Spanish galleons of the fifteenth and sixteenth centuries not only knew which of the prevailing winds in the Atlantic to take advantage of in their passages, but also of certain currents that existed in the sea. Ponce de Leon, remembered for his search for the fountain of youth, described the Florida Current in 1513. The first great American scientist (and diplomat), Benjamin Franklin (1706–1790), as Deputy Postmaster General in 1770, had a chart of the Gulf Stream drawn to improve the speed

4–34 Tidal bore in Petitcodiac River, Bay of Fundy. (Courtesy of Canadian National Railways, Atlantic Region. National Copyright Reserved. Made in Canada.)

of mail delivery between Europe and the colonies. He had noted that merchant ships and whalers often took as much as two weeks less than the mail packets to cross the Atlantic to New England. In his numerous trips to Europe, Franklin, himself a great experimenter, made numerous temperature measurements and water color observations as he crossed the Atlantic to help plot the course of this current.

Unlike waves, which at sea cause no water transport, the water of the ocean currents is actually moving and being mixed. The currents described by the Spanish in the sixteenth century have essentially the same pattern today. It may seem odd that these currents undergo little or no change in course, whereas the path of a river, though bounded on both sides by land, may wander considerably over the years. But ocean surface currents, namely those in the first several hundred meters of depth, are caused by the wind. In 1686 the great English natural philosopher Edmund Halley (1656–1742), after whom the comet is named, first suggested this relationship. If the pattern of the seas' surface circulation is fixed, then the cause of these currents, the wind, must also have fixed patterns. Although this may seem unlikely judging from one's own experiences with the changeability of wind, such is actually the case.

Wind. A chart of the ocean's surface circulation is shown in Fig. 4–36. Note that in the northern hemisphere oceans (including the Arctic Ocean), the currents form a large, circular, clockwise cell or gyre. In the southern hemisphere, the direction of the gyres is reversed. Under ideal conditions, this clockwise motion of water in the northern hemisphere can be seen even when the plug is pulled in a sink full of water, but other variables, such as the currents set up by the water that filled the sink, often make this an unreliable demonstration.

All weather, more correctly climate (weather averaged over a long period of time such as a year), results from the heating action of sunlight. The heat from sunlight causes regions of high and low pressure

(a)

(b)

(c)

4–35 Tides in the Bay of Fundy. (a) The distance from the berth to the top of the dock is about 9 m. Note the difference between the dock to deck height in (b) high tide and (c) low tide. This change in water level represents only part of the total tidal prism in many parts of the Bay of Fundy. (Courtesy of the Reid Studio, Windsor, Nova Scotia, and Minas Basin Pulp and Power Company Limited, Hantsport, Nova Scotia.)

and all the winds and storms. Because the relative position of the sun and earth does not change much from year to year, the heating, and therefore the climate and winds, are similar from year to year. There are always local variances, but on the average the general weather pattern of an area from season to season is predictable.

The horizontal wind pattern at sea level for the planet is shown in Fig. 4–37. The doldrums (largely between 0 and 5° N and S, but much further north and south in some areas) is the equatorial belt of variable winds. Here there are no prevailing winds, and about one-third of the time no winds at all, although small violent thunderstorms and squalls are not uncommon. The northeast and southeast trade winds (roughly 5° to 30° N and S) are the most persistent of all winds both in steadiness of blowing and in direction (coming from only one-sixteenth of the total possible angles of the compass). The region between 30° and 40° N, the horse latitudes, is a subtropical high pressure area with light and variable winds and calms as often as one-fourth of the time. In the days of sail, ships commonly were becalmed in this region and in the doldrums. The westerlies, between 35° and 60° N and S, is the belt where winds blow predominantly from the west, but may blow from any direction. The areas of doldrums contain light and variable winds and warm, moist air that causes frequent showers. The horse latitudes are marked by low rainfall and sunny skies most of the time; the westerlies, however, are frequently cloudy, with highly changeable weather and much precipitation. Although storms in this region are not as intense as those of the trades, they are more common and cover a wider area.

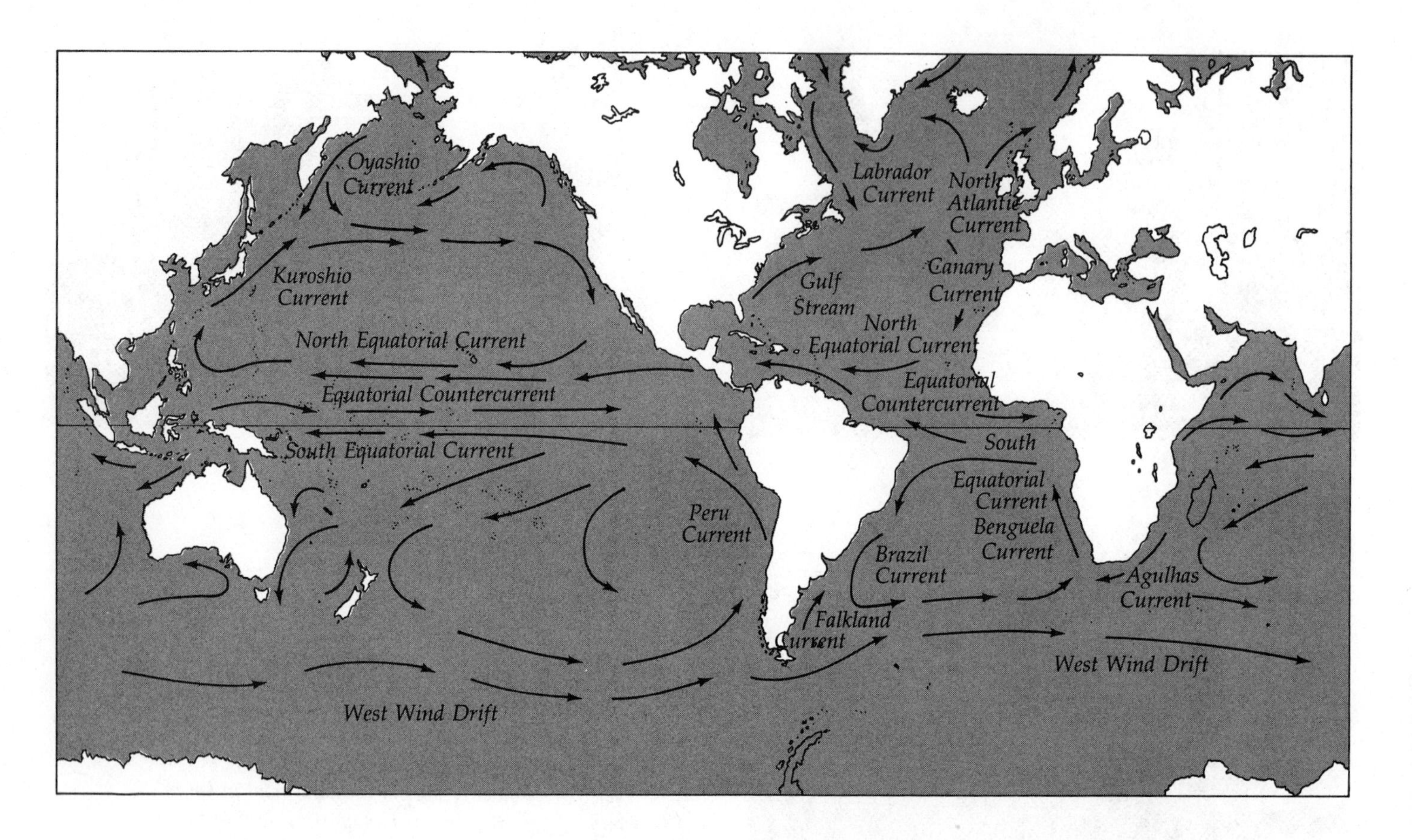

4–36 Oceanic surface circulation.

If the continents did not exist, the pattern of water currents would look somewhat similar to that for the winds, although as a sailor knows, water is transported to the right of the wind in the northern hemisphere. A wind in the northern hemisphere moving due north moves water roughly to the northeast, at a speed about 2 percent that of the wind. A north wind, by the way, comes from the north, but a north current moves to the north. The terminology that has arisen over the years leaves something to be desired. But the continents, oceanic islands, and ridges block these hypothetical ocean currents, changing their expected pattern. Other effects also occur.

The Coriolis "force". As the earth spins on its axis, it may be thought of as tending to move out from under the water of the ocean. Because the earth spins counterclockwise, or to the east, as viewed from outside, the water of the oceans tends to pile up slightly on their western edges. Another factor must also be considered. When a rocket or a spacecraft reenters the atmosphere anticipating contact or splashdown at a particular point, the fact that the earth is moving under the craft must be taken into consideration. A long-range cannon of a battleship like U.S.S. *Missouri* firing a shell (with a conventional range of 27 km) at a target that distance away will miss unless the rotation of the earth is allowed for. The target, being fixed to the earth, has moved while the shell is traveling. In reality, no force changes the course of the projectile (the target has simply moved away from the shell), but it appears to an observer on the earth's surface as if the shell had been deflected. This apparent force is called the "Coriolis effect" after the nineteenth-century French mathematician and physicist Gaspard Gustave de Coriolis, who first described it. This effect can alter existing motion but cannot cause acceleration. You feel this effect when an automobile moving in a straight line at constant

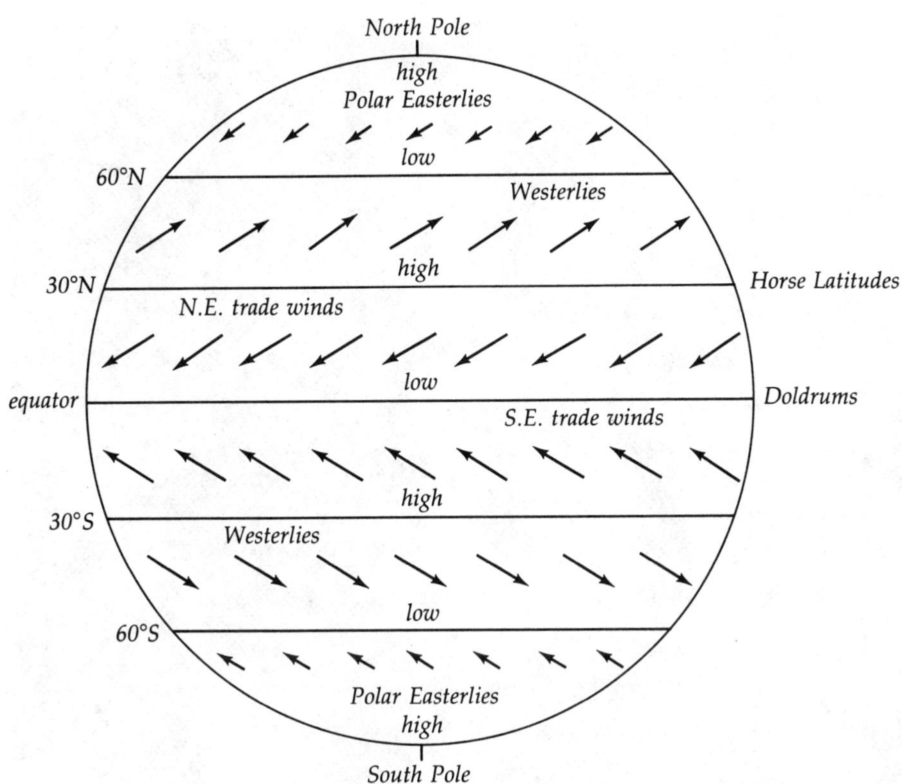

4–37 *General circulation of winds and pressure patterns at sea level.*

speed takes a sharp curve. If the curve is to the right, a person experiences a force to the left. In reality no actual force is applied against the person. His body is trying to continue to move in a straight line while the car is moving to the right, so he feels a push to the left (see Appendix VIII).

Suppose a person takes a penknife and cuts a straight groove across a phonograph record from center to edge and then performs the same cut when the record is spinning on a turntable. In the second trial, the cut made is not a straight line but a curved one. The record has moved as the knife cut across it and a curved mark resulted. So, too, the air as winds and the water as currents experience a deflection or curved path. The combination of all these factors creates clockwise gyres of the oceans in the northern hemisphere and counterclockwise ones in the southern hemisphere. The circular motion of the winds is easily noted in hurricanes and tornadoes. Hurricanes are very large swirling masses of moist air forming over the water in warm regions, but away from the equator so that they begin the spinning motion. They occur in the North Atlantic, the Gulf of Mexico, the Caribbean, and the southeastern part of the North Pacific. The same type of storm occurring in the western North Pacific is called a typhoon, but because of the greater expanse of warm water in the western Pacific, they often are larger and more intense than hurricanes. Tornadoes are much smaller funnel-shaped cyclonic storms. Although they contain the most violent winds on earth, their velocity over the earth's surface is slow, and their lifespan is short.

The heating of the earth's surface varies with the time of year, depending on the exact angle of the earth's attitude with respect to the sun's rays. Because this relative angle changes through the year, the weather pattern, specifically the areas of high and low atmospheric pressure (or highs and lows), must change. But

4–38 A weather satellite picture. Baja California and the southwestern United States are visible in the upper right (outlined by dots) as is New Zealand in the lower left. The large clear areas are high pressure regions—good weather. (Courtesy National Oceanic and Atmospheric Administration.)

as the heating cycle each year is reasonably constant, a series of seasonally permanent highs and lows are set up (Fig. 4–38). These are not to be confused with the wandering local highs and lows that cause everyday weather, although they too are associated with the big seasonal highs and lows. With the seasonal shifting of these areas of pressure (Fig. 4–39), the oceanic surface currents do shift somewhat. On the western edges of the continents this shift often results in a dramatic change in weather between summer and winter.

Monsoons. In Fig. 4–39b, note the immense low pressure region in the North Pacific during January. This is replaced by a high, more to the east, in summer. These two systems determine the overall yearly weather of much of the west coast of the United States. In winter, the counterclockwise winds of the low, laden with moisture from contact with water warmed by the Kuroshio Current, blow in and over the Pacific Northwest, Canada, and Alaska. This warm, moist air gives the Pacific Northwest, west of the mountains, its cloudy, rainy, but temperate winters. This rainy season is correctly termed a *monsoon*, often associated incorrectly only with tropical regions. As the summer comes with the formation of the high, the winds are moving in a different direction—off the land. Now, they are dry and warm giving the area its dry, warm summers. A monsoon, then, is really a seasonally reversing wind caused by seasonal variation in atmospheric pressure. This pattern need not be exactly the same as that of the U.S. west coast, where the winter monsoon effects are felt as far south as San Diego. The summers of all of Southeast Asia and India, for example, have wet onshore winds and the resultant rainy season.

Upwelling. Another effect associated with this reversal of winds is upwelling. Off Oregon the winds of winter drive the waters onto the coast, making surface level higher than normal, although only by 3 cm. In summer, the winds are moving down the coast, and the surface water near the coast is moved at an angle to the right of wind direction, or offshore. Deep water moves up to replace this outward surface transfer (Fig. 4–40). Upwelling is important because it brings nutrient-rich waters to the surface and results in remarkable biological productivity. Without vertical transfer of water by this and other means, the surface waters would soon be stripped of nutrients except, of course, near the world's many sewage outfalls. This mixing also means that the water is actually colder off the coast of Oregon in summer than in winter. Only when the wind stops for a few days and upwelling ceases for that time does the surface water warm up enough for pleasant swimming, but this usually isn't too often.

The northward flow of the cold, nutrient-rich Humboldt Current along the west coast of South America, coupled with upwelling (on shoreward side of the current) caused by southerly winds, furnishes a supply of nitrates and phosphates that support one of the world's most abundant populations of marine life. The recently mushroomed Peruvian fishing industry depends on the current, as do millions of fish-eating birds, whose droppings of guano are the basis of the huge Peruvian fertilizer industry. Periodically (roughly every 7 years), the winds drop off in intensity, warm equatorial waters move in displacing the cold waters of the Humboldt Current, upwelling ceases, and the entire picture changes. This phenomenon has become known as *El Niño* (the child) referring to the Christ child because the phenomenon occurs during the Christmas season. The warm water and its rapid depletion of nutrients causes massive plankton and fish kills. Tons of dead fish pile up on the beaches, and the oxygen content of the water is quickly used up causing gases such as the foul-smelling hydrogen sulfide to be produced. This may cause the paint on ships to blacken and produces other discoloring effects, which give rise to the phenomenon's other name, Callao Painter. With the vanishing of the anchovies, thousands of birds die of starvation, birds already pressed by the invasion of man and DDT. There have been occurrences of *El Niño* in 1911, 1932, 1939, 1941, 1951, 1957–58, and 1965, with particularly severe consequences in 1891, 1925, and 1953.

Because the rotation of the earth piles up the waters (though only a few centimeters) on the western edges of the oceans, different types of ocean currents are produced. These surface currents may be divided arbitrarily for ease of study, like the synthetic divisions of science itself, into Eastern Boundary, Western Boundary, and Equatorial.

Eastern and Western Boundary currents. Eastern Boundary currents are, of course, on the east side of the oceans, and the water runs toward the equator, especially at low latitudes. The speed of the water is slow, about 10 cm/sec or one-fifth knot, and these currents are not deep, usually only about 500 m, but they are quite wide, often greater than 1,000 km. The rate of water transport would be about 10×10^6 m^3/sec (10 sverdrups—sv—named after the famous Norwegian oceanologist H. U. Sverdrup; one sverdrup = 1×10^6 m^3/sec). Currents on the western side of the oceans, the Western Boundary currents, move in a general northerly direction, are narrow in width (only a few tens of kilometers), are deep (1,000 m or better), and have a high speed of several knots. As a consequence, they have a large transport, 25 to 75 sv. Man has been aware

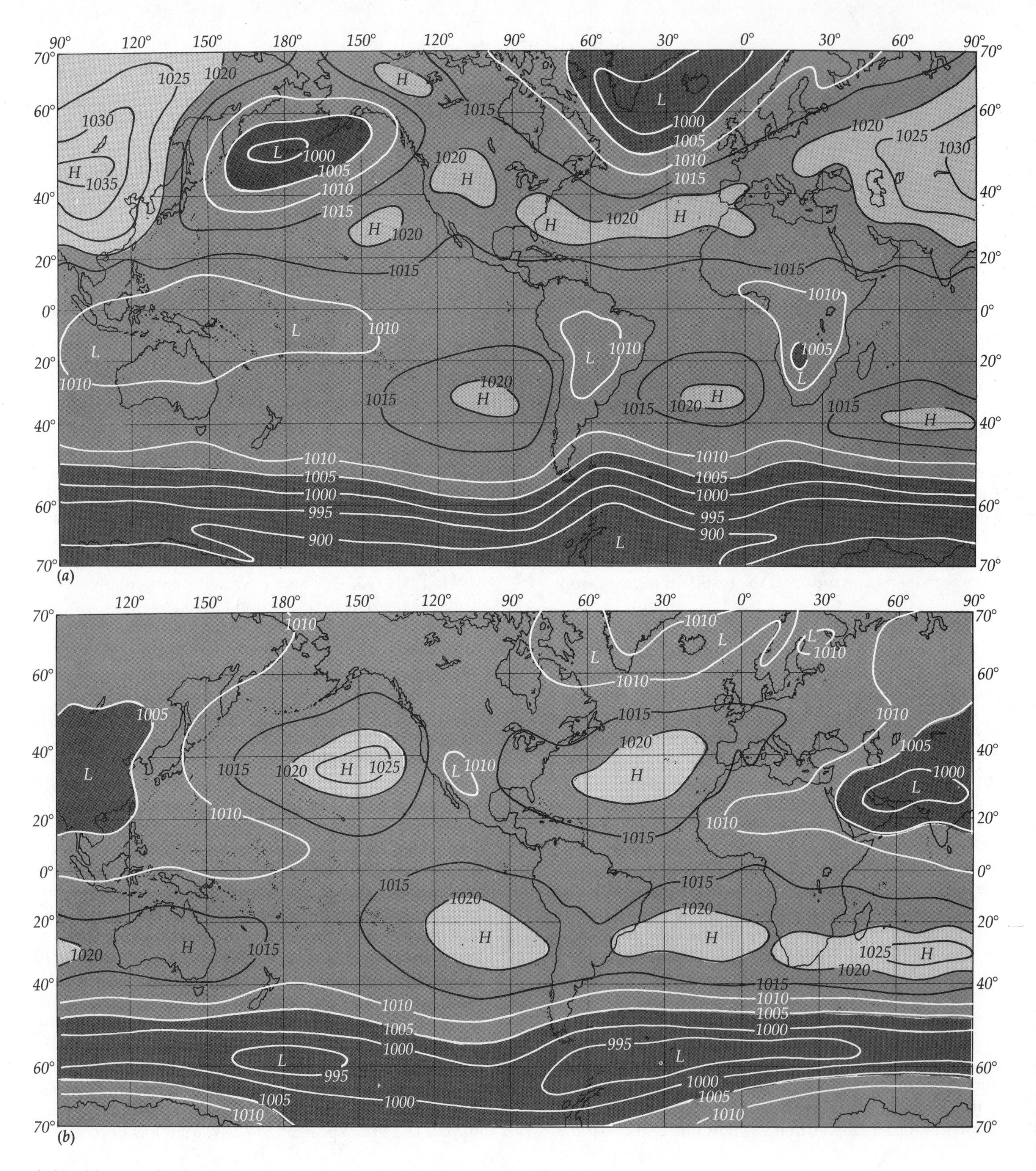

4–39 *Mean sea level pressure (a) in January (millibars); (b) in July (millibars).*

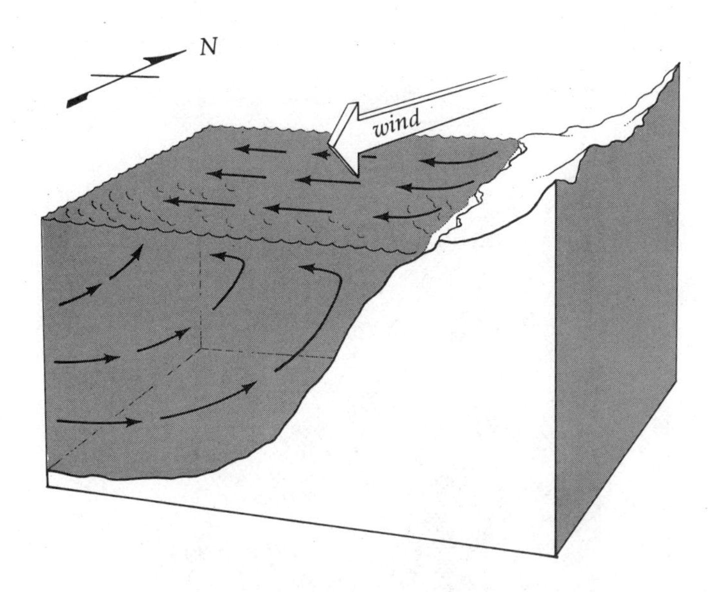

4–40 Upwelling, as along the coast of Oregon during the summer.

of these Western currents the longest because they have a more marked effect on navigation.

The Gulf Stream system is a Western Boundary Current. Composed of the Florida Current, the Gulf Stream, and the North Atlantic Current, this system has been and is still the most studied of all ocean currents. Its name, a misnomer, stems from the earlier belief that its source was the Gulf of Mexico. It actually results from the merging of the North and South Equatorial currents, which pile up water in the Gulf. This water escapes between Florida and Cuba as the Florida Current. The amount of water flowing in the Gulf Stream is tremendous. At Miami more than 4 billion tons of water pass per minute in a stream which is, at this point, 800 km wide and over 300 m deep, a volume of flow more than 100 times that of the Mississippi. Off most of the United States, the Gulf Stream is generally clearly defined, especially by temperature. Merchant vessels using temperature sensing devices can save time and money by staying in the current when traveling up the coast. By the time the Gulf Stream reaches the Grand Banks, the current is more vague and has numerous eddies or branches (filaments) associated with the main flow. Off New Brunswick these filaments may be as large as one degree of latitude and are common out to approximately 60° W. It is best to remember that the Gulf Stream is just part of the giant oceanic swirl that is the total North Atlantic subtropical gyre system.

Inside the great circle that is the North Atlantic gyre is a region fabled in maritime legend, a place where the world's eels go to spawn, a place having almost no surface currents and almost stagnant water. This is the once-dreaded Sargasso Sea. The area between the Bahamas and the Azores abounds with patches of several species of floating seaweed—sometimes aligned in rows—but seldom longer than 2 to 3 m in diameter. Sailors as far back as Columbus feared that the seaweed of the area would ensnare their ships or that, since the presence of seaweed is normally associated with shallow water, they might go aground. The small flotation air bladders common to seaweed reminded early Portuguese sailors of a small grape called the *sargaco* indigenous to their country, hence the name of the region. There is no truth to these legends of the strange Sargasso. Even the floating brown seaweed is not present in great abundance. For the warm, blue Sargasso, mixing little with deeper nutrient-rich waters, is more devoid of the microscopic plant life on which all other sea life depends than any other region of the sea. It is truly the oceanic equivalent of a bleak, barren desert.

Equatorial Currents. The third type of current is the Equatorial. These are well established and quite permanent. They have low velocities (0.5 to a little over 1 knot), are fairly shallow (to about 500 m), but are wide so that the water moved is appreciable. The North Equatorial Current of the Pacific has a flow of 30 sv, and the South Equatorial more than 30 sv. There is some seasonal variation to these currents, the entire set moving slightly north in summer.

The Cromwell Current. In 1951, personnel of a U.S. Fish and Wildlife research ship who were testing a tuna fishing technique in the westerly flowing waters of the Central Pacific's South Equatorial Current noted with surprise that the deeply paid-out fishing gear drifted east instead of following the prevailing western surface current. A study by Townsend Cromwell of the Fish and Wildlife Service the next year showed the existence of a subsurface current below the equator, running from west to east. This current, named the Cromwell Current after its discoverer, though well below the surface, is still shallow and is associated with surface currents. Its depth varies between 50 and 300 m and it has speeds up to 3 knots with 1.5 knots a probable average. In its travel from New Guinea to Ecuador, well over 13,000 km, the Cromwell Current is closer to the surface in the eastern Pacific than at its western end. The total water transport is approximately one-half that of the Gulf Stream. One result of the International Geophysical Year more significant than the naming of Bagel Seamount (one homesick oceanographer missed his morning bagels) was the discovery of a current below the Gulf Stream flowing in the opposite direction at depths of 2,000 to 3,000 m, much deeper than the Cromwell. A great deal of specific data concerning these currents remains to be known. For example, why is there no core water of a specific temperature and salinity range in the Cromwell Current as is the case in many currents? And where does all this water go? Perhaps a partial explanation for the motive power of these currents is the Ekman spiral. In 1902 the Scandinavian physicist V. Walfrid Ekman published a mathematical model that now bears his name (Fig. 4–41). As wind moves steadily over the surface, the water moves at a 45° angle to the air. With increasing depth through successive layers of water, the deflection is more and more to the right until at some depth, about 100 m for average winds, the current has a speed of about 4 percent of that at the surface and is now 180° or opposite that of the surface. This phenomenon has been observed in the hydrosphere as well as being produced in the laboratory. Aside from the tendency for sea ice to move at some angle right of the wind, little evidence existed until recently that the Ekman spiral does exist in the open ocean (Fig. 4–42). There are too many difficulties with this theory now to use the Ekman spiral as the entire explanation for subsurface currents.

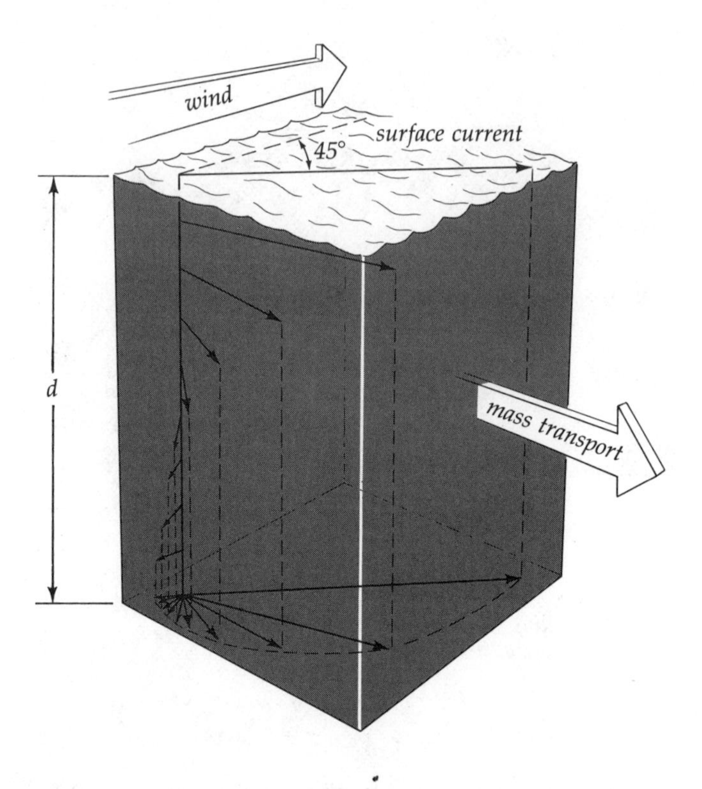

4–41 *The Ekman spiral (northern hemisphere).*

Divergence and convergence. Upwelling is a specific instance of a more general phenomenon called divergence, which means waters moving away from one another. Divergence makes it necessary for deep water to move up and take the place of the vacating water. In the case of upwelling, water moves away from the coast and not other water. In some regions of the ocean the opposite occurs; waters meet one another. Because they cannot go up, they sink—the water is forced downward. The best known and probably the most important sinking region is the Antarctic convergence (Fig. 4–43). All around the Antarctic continent between latitudes of 50° and 60° a convergence is formed where cold waters from the Antarctic and warm waters from mid-latitudes meet and sink. When traversing the area in a north–south direction, one notices that the temperature drops abruptly by 5° C over a short linear distance as the convergence is passed. The Antarctic convergence is not too important to surface currents but is very significant in deep ocean circulation.

It is often asked just how far a message has ever traveled by bottle. Woods Hole Oceanographic Institution has been releasing between 10 and 20 thousand clear, eight-ounce beverage bottles per year. These contain some sand for ballast to keep them upright and a self-addressed postcard. The return rate runs about 10 to 11 percent. These drift bottles have often traveled as far as Europe, and many have made a complete circuit of the North Atlantic gyre—a distance of 8,000 to 10,000 km. The longest trip thus far was made by a bottle released in 1962 at Perth, in southwestern

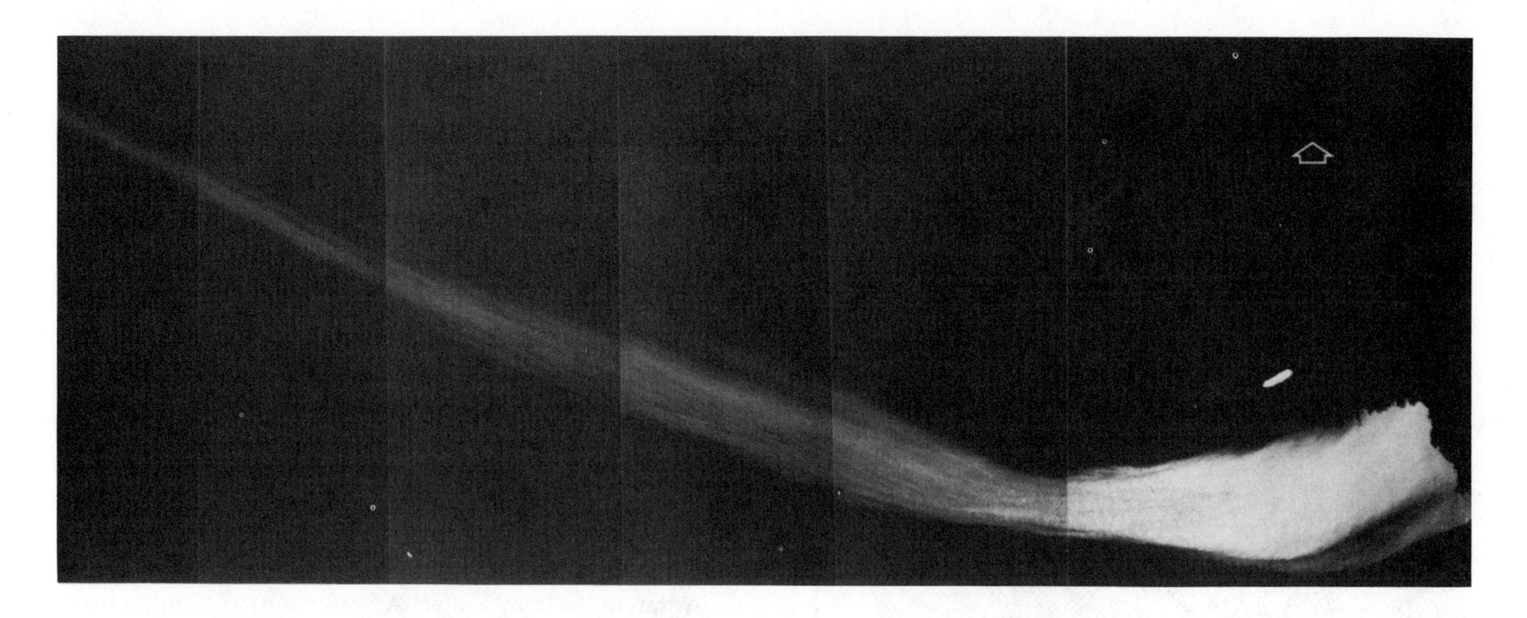

4–42 An Ekman spiral in the ocean. The arrow points north—the direction of mass transport. The wind is northwest. (Courtesy of Richard Linfield.)

Australia, and found near Miami five years later—a distance of 26,000 km. From the current chart (see Fig. 4–36), what is the most probable route of travel?

Deep Currents. The ocean is never still. In photographs taken as deep as 7 miles, the bottom shows ripple marks, indicating that even there the water is in motion. Deep water movement is quite different from that of the surface. The deep waters, for example, move much more slowly. The sun's heat generates the winds, which are the prime cause of surface currents, but it also plays a more direct role in the movement of the surface waters. The waters warmed at the equator expand and create a hill only a few inches high. This small slope accounts for some net transport of surface waters in moving "downhill" from the equatorial regions toward the poles. On cooling in the far northern and southern areas, this high-salinity water increases in density, sinks, and slowly moves along the bottom back to the equator. Ideally, then, one would expect the pattern of currents to be that in Fig. 4–44. But as usual the picture is more complex; the moving earth, the Coriolis effect, and the fact that the equator isn't the barrier once thought all change this idealized pattern.

Where surface winds are the primary cause of surface currents, temperature and salinity govern the deep currents. In other words, the motion of deep currents, which is also vertical rather than purely horizontal as in surface currents, is density dependent. Because temperature and salinity determine density, these deep currents are called thermohaline.

Water mass. A fluid with homogeneous properties is referred to as a mass. A body of water that has a characteristic temperature and salinity range is a water mass. Although these two factors determine density, different water masses can have the same density but different temperature and salinity values. Thus, identification and tracing of water masses may be done through the determination of these two parameters. Such methods as ship drift and drift bottles obviously cannot be used in deep waters, and their velocity is too slow to be measured with the free floating and fixed current or flow meters now available. Chemical methods are also useful in tracing these currents. These are based on the assumption that living organisms present in the water mass as it leaves the surface change the

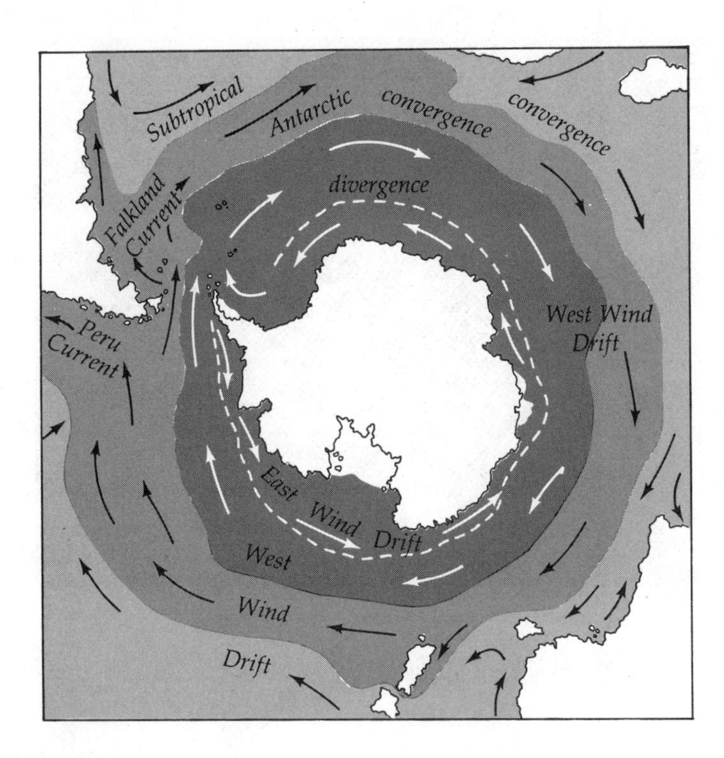

4–43 *Surface circulation and convergence around the Antarctic continent.*

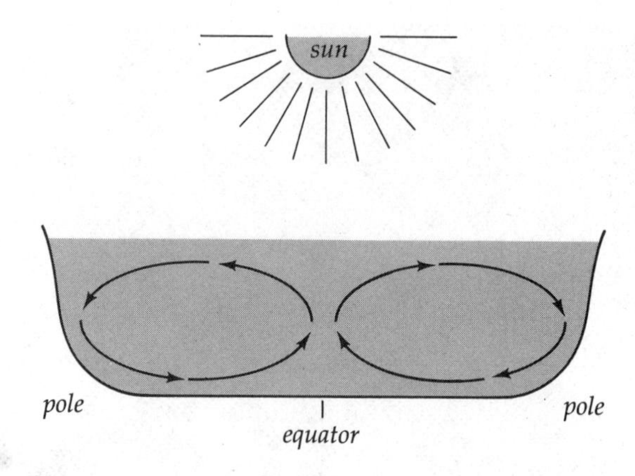

4–44 *Theoretical deep ocean circulation.*

water composition. For example, as the water moves away from the surface, the amount of organic matter decreases and the amount of inorganic material increases. Radioactive dating, using carbon 14, is also useful.

Water masses may be divided into five types: surface (to approximately 100 m deep), central (to the bottom of the main thermocline), intermediate (from the bottom central waters to approximately 3,000 m), deep, and bottom (both of which fill the remainder of the lower ocean). Technically, the surface water is not a water mass because the range of temperature and salinity is too great to classify it as a specific mass. A temperature–salinity (T-S) diagram is a convenient way to look at water masses (Fig. 4–45). At particular points at the ocean surface, specific weather conditions combine with parameters of the water to produce water of an exact temperature and salinity. This would be a unique point on a T-S diagram. Few water types exist for long in an unmixed state, the major exception being Antarctic Bottom Water. A water with a temperature–salinity range is a portion of a curve on a T-S diagram. Water masses are combinations of two or more water types, as most ocean water is the result of mixing. A straight line on a T-S diagram represents the mixing of two water types. The mixing may be lateral, along surfaces of equal density; or vertical, across regions of varying density as, for example, when a cold dense water sinks to the bottom. The mixing of two or more water masses with the same density will produce a new mass of greater density. This mixing and sinking process is termed *caballing*. If a parcel of water moves along a density surface as a unit body, the movement due to gravity is known as advection, and the spreading of a dye stain in water due to the random molecular motion of the water particles and the dye is known as diffusion. In the ocean these occur together, but the water bulk is moved by advection.

Because the deep waters cannot change their salinity except by mixing, they receive their basic characteristics only at the surface. The deep waters are expected to be formed at high latitudes, whereas those closer to the surface are formed near the equator.

Cross sections of the Atlantic, Pacific, and Indian oceans are shown in Fig. 4–46. Notice that one water is common to all three oceans and leaves its fingerprints on these oceans. This water mass, because of its low temperature and high salinity, has a greater density than all others. This is the Antarctic Bottom Water (AABW). It flows in an easterly direction around Antarctica as a result of the surface current, the West Wind

Drift, which itself flows eastward uninterrupted around the continent. The AABW flows well into the Atlantic, to a latitude about equivalent to New York, whereas the entire bottom water of the Indian Ocean is AABW.

The section marked Mediterranean on the profile represents the intrusion of warm and saline waters into the North Atlantic. In the same way, very saline Red Sea water enters the Indian Ocean. The effect of the Mediterranean water has been traced 2,500 km from Gibraltar. This Mediterranean water enters the Atlantic below the surface water, which is moving east into the Mediterranean. During World War II German and Italian submarines attempted to use the two-way currents at the Gibraltar interface to sneak past the British blockade by silently coasting through in the currents with their diesel engines shut off. This was another example of man's continued efforts to use his environment for cover. The Strait of Gilbraltar is only 29 km wide and 320 m deep. It is rumored that some subs were successful in this passage.

Basin-like landlocked seas such as the Mediterranean and Black seas, while similar in structure, can vary appreciably in current profile. The larger Mediterranean has characteristic water masses and vertical circulation as well as such a large flow of Atlantic water that the entire sea is renewed every 75 years. But the Black Sea has only a shallow narrow outlet to the Mediterranean, the Bosporus, through which small amounts of water enter. The net effect is a lack of vertical circulation, due to the great vertical stability. The turnover time for the Black Sea is 2,500 years and, although the surface water is aerated and contains much life, below 65 m the waters are stagnant, the conditions anerobic (suited only for bacteria), and the sea floor is covered with the sulfide-colored black mud.

The Pacific Ocean doesn't have access to the cold northern waters as does the Atlantic. As a consequence the extensive thermohaline convective activity of the North Atlantic is absent in the North Pacific. This affects the entire ocean. There are no really dense water masses in the North Pacific; the boundaries of its central and intermediate waters are not clearly definable, and the deep water of the Pacific is, in general, much more sluggish than that of the Atlantic. North Atlantic Deep Water (NADW) is produced in one region off Greenland (Fig. 4–47).

The boundary regions for these water masses in the cross-sectional diagrams do not, of course, mean that

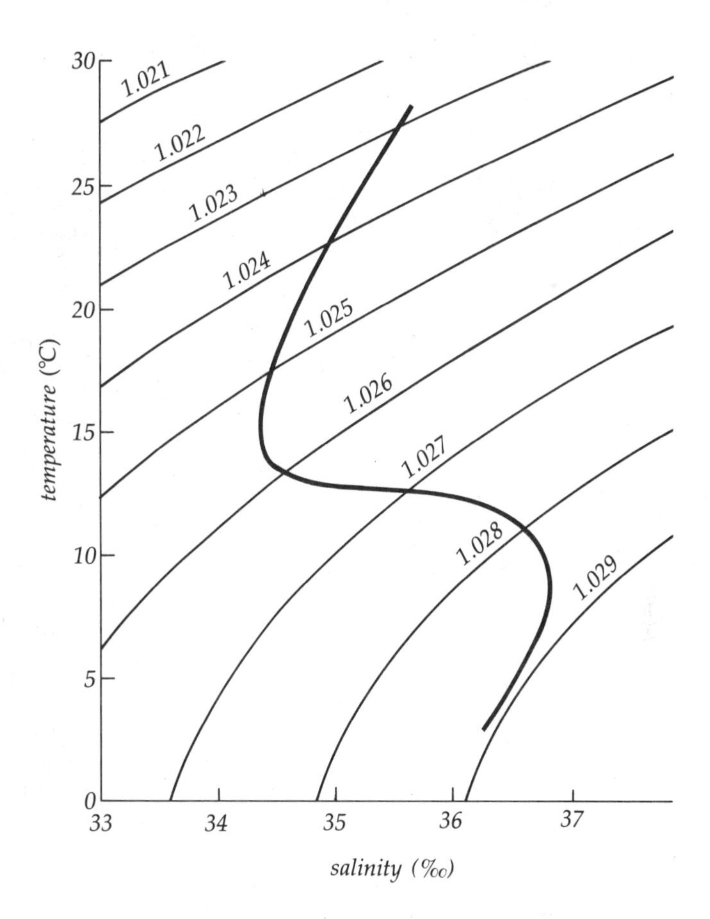

4–45 *A typical T-S diagram, density curves added.*

these masses stop there. Mixing with other waters occurs along the boundaries, and the masses constantly lose their identities and are no longer traceable. It should be emphasized that these currents, although formed at the surface, are not maintained by mixing at depth. But the boundaries indicate, subsurface mixing tends to wipe them out. Cross-sectional diagrams also fail to give an adequate picture of the spread of deep waters with respect to the entire ocean bottom. Fig. 4–48 gives some idea of the deep oceans' lateral circulation.

Water age and turnover time. All these oceans are constantly renewing themselves at all depths, but the rate at depth is slow. Water age refers to the length of time since a water parcel (part of a deep mass) left the surface. Values vary for these maximum

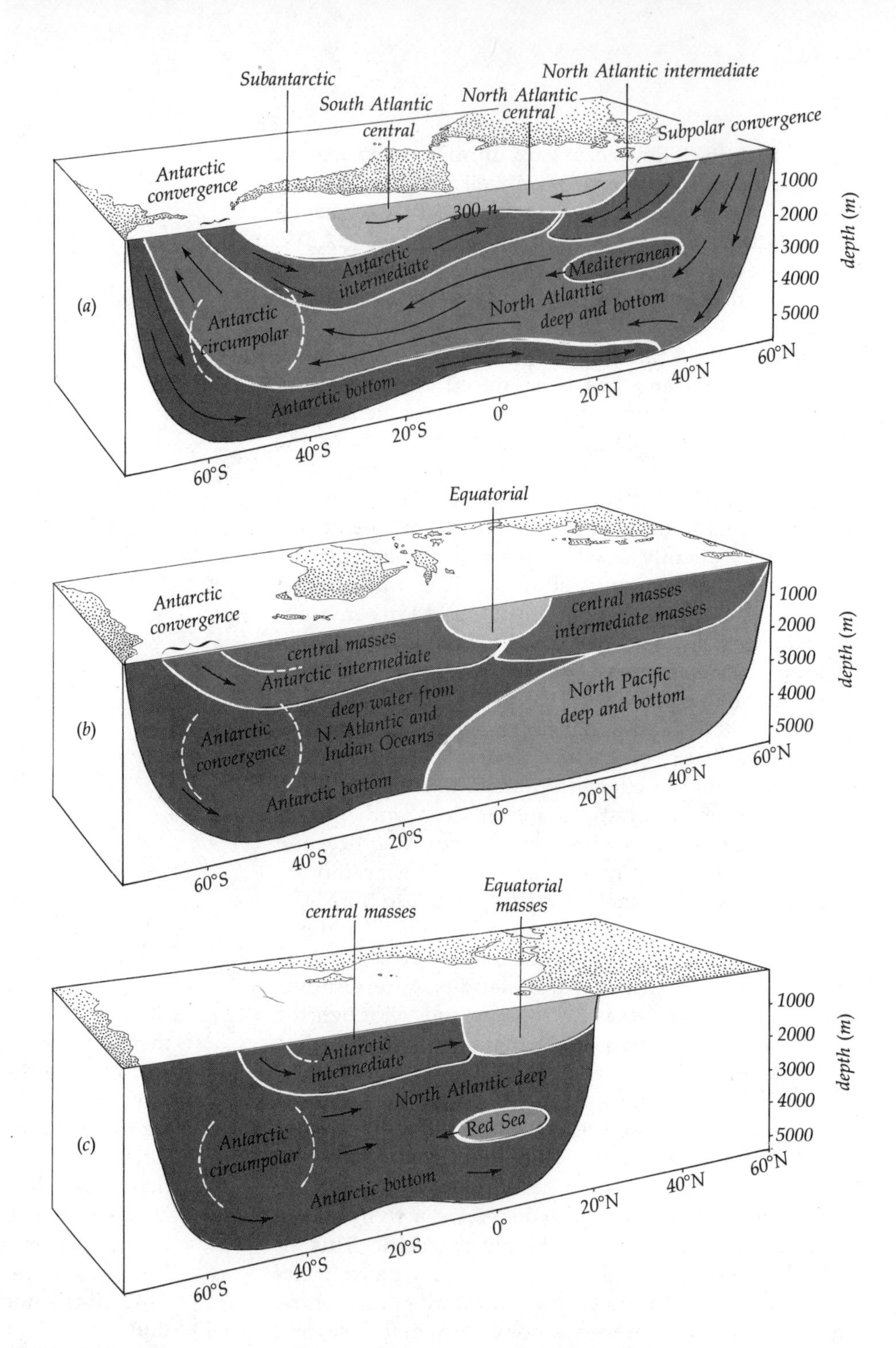

4–46 Subsurface circulation. North–south cross sections of (a) the Atlantic; (b) the Pacific; (c) the Indian Ocean.

water ages but 750 years seems a reasonable estimate for AABW in the Atlantic compared to 1,500 years for the same water in the Pacific. It is obvious that deep currents move much more slowly than surface currents; a particular water molecule may take less than a year to circle the Atlantic at the surface, but included with its fellows as NADW perhaps over 750 years. Table 4–2 summarizes the most important parameters of the world's water masses.

The Antarctic continent is a most unusual place. In many ways it is the antithesis of the Arctic. Antarctica has an area of 5.5 million mi^2 and is covered with the world's largest continental glacier, a mass of ice so great (29 million km^3) that it depresses parts of the continent to below sea level. Yet because of the great amount of ice, the average height of Antarctica is greater than any other continent. Fossil evidence indicates that in the past, the Antarctic contained tropical forests and animals. This implies that the poles might have shifted, or more possibly in light of recent world-wide data that the continents have drifted. The Antarctic even possesses a number of extinct volcanoes as well as a currently active one, Mt. Erebus.

The Antarctic region has an important and profound effect on the rest of the world. It acts as a tremendous heat sink and, with the Arctic, serves to balance the heat budget (the heat received from sun and the heat lost to space annually are believed to be equal; this is termed the *heat budget*) of the earth. The effects of the Antarctic on the weather, felt as far north as 50° S, largely power the wind patterns of the mid- to far-southern hemisphere.

Present-day life in the Antarctic is primarily marine. There are only a small number of species there, such as the Adélie and emperor penguins, skua gulls, crabeater, leopard, and Weddell seals, and whales, from the massive blue to the killer. All of these animals are directly or closely dependent on a small, reddish, shrimplike crustacean, *Euphausia superba,* or krill. The krill in turn feeds on small plants or plankton such as the diatom. The amount of plankton in the water is largely a function of the nutrient content. Unlike the waters in most of the world's oceans, there is no shortage of nutrients in Antarctic waters. The waters around this continent, more correctly called the Antarctic or Southern Ocean rather than an extension of the Pacific, Atlantic, and Indian oceans, provide a constant supply of the essential nitrate, phosphate, and silicate in such a degree that even in late summer they are never exhausted. The amount of light, especially in the winter, is possibly the only limiting factor for phytoplankton productivity in the Southern Ocean. This extremely rich region of life is caused by the constant large-scale overturn of deep nutrient-rich waters of the Atlantic, Pacific, and Indian oceans.

As Fig. 4–36 indicates (although not clearly due to the extreme distortion of the projection near the poles), there is a constant eastward flow of surface waters with some diversions around the Antarctic. This is called the Antarctic Circumpolar Water (AACP).

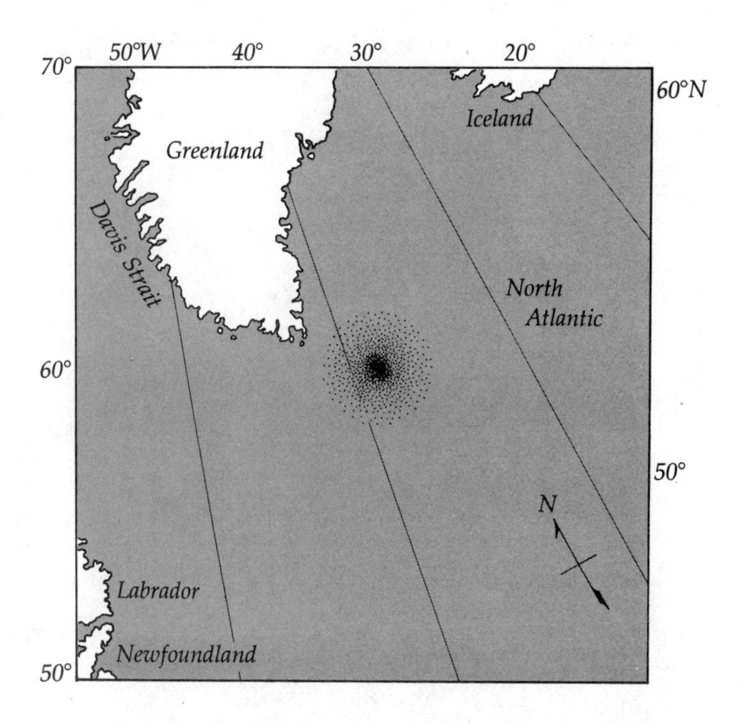

4–47 The region in which North Atlantic Deep Water is produced.

Only recently has the very slow turnover time of the oceans become a source of concern to scientists. Presently, the earth abounds with the pesticide DDT, which is now found concentrated in the fat of animals ranging from reindeer and polar bears to penguins and people (who in the United States, due to their DDT content, now couldn't be classified as USDA choice). Each year tons of DDT and other halogenated (chlorinated) hydrocarbons and their residues find their way primarily by rivers to the oceans. Even if the use of DDT were completely stopped today, it could take 20 years before the total maximum reaches the sea. Not only do scientists not know the possible effects of this chemical invasion of the sea, but because of the slow renewal time of the deep waters, these chemicals will remain in the water for a long time. The catastrophe, if it reaches those proportions, would continue for many years. The ocean is already returning the effects of DDT to man, whether he likes it or not, in the form of fish now often inedible—like the mackerel.

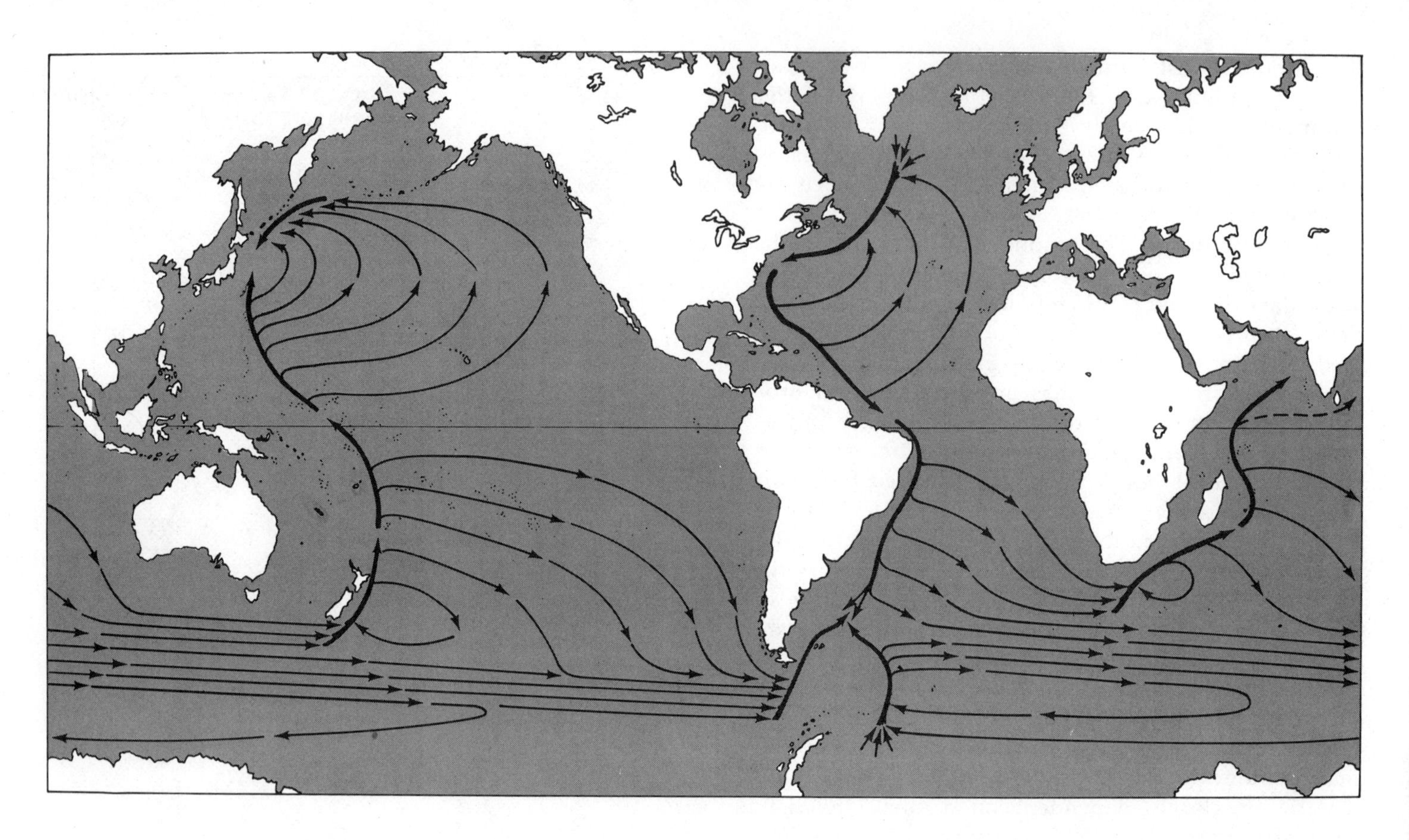

4–48 *Deep circulation below 2000 m. (After Stommel.)*

Basins and Estuaries. Between the ocean and the continents are numerous indentations in the land filled with waters not simply fresh, not totally salt. The vertical structure of the waters in these basins and estuaries is often a complex picture of current and tide-driven fresh and salt water layering.

Basins commonly have a barrier called a sill near their entrance, which may or may not hinder circulation (Fig. 4–49). If the sill is very deep, as in the Caribbean, the flow in and out moves freely and the waters are usually well mixed. The surface waters of a basin in which evaporation exceeds the fresh water input, a combination of precipitation and river runoff, are usually well mixed because the surface waters are constantly becoming more saline and sinking, as in the Mediterranean. In basins like the Baltic and Black seas, and in the high latitudes, fresh water input exceeds evaporation, keeping the surface waters fairly fresh. There the fresh water flows out over the denser waters with little mixing.

Many fjords fit into this last category. Often the bottom salt waters have so little circulation that they become stagnant, as the oxygen content of the water is used up by the organisms therein. When the oxygen is gone, organisms may use other substances for oxidation. One product of these anaerobic processes is hydrogen sulfide (H_2S). In the past, after a vigorous storm, the white houses of a Norwegian fishing village bordering a fjord sometimes turned black. The pigment in the white paint was a lead compound (hydrated lead hydroxy-carbonate). With the churning and mixing that accompanies a violent storm, some of the hydro-

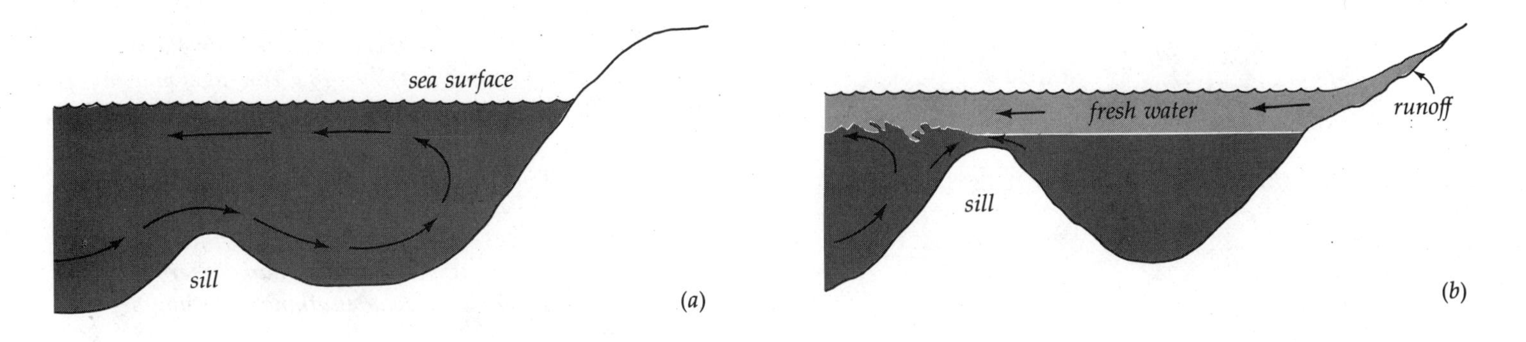

4–49 *(a) A basin with a deep sill, such as the Caribbean. The waters are usually well mixed. (b) A basin with a high sill like the Baltic Sea or the Black Sea. A large amount of fresh water runoff, exceeding evaporation, results in deep waters with little circulation.*

gen sulfide was released from the stagnant water and when it came in contact with the houses and barns formed the black lead sulfide.

Estuaries may be considered the biological factories of the world. They are nutrient traps that support and propagate tremendous numbers of animal species. In many ways, aside from the fact that estuaries contain more life than any other places on the planet, estuaries are unusual. The salt water of the tides has access to these estuaries, and the result is a salt water wedge or prism under the river runoff. Diagrammatically, the estuary would appear as in Fig. 4–50. In reality there is always mixing at the interface of the waters, and the actual structure of the estuary may take many forms depending on the degree of river versus tide flow. The same estuary can vary appreciably with the season. In spring, when most rivers have their greatest fresh water discharge, the salt wedge of the sea water may not proceed as far up a river; there probably will be less mixing and the fresh water at the surface may be noticeable even well out to sea. In the late summer when fresh water runoff is low, the same

Table 4–2 Characteristics of Several Water Masses

Water mass	Salinity (‰)	Temperature (°C)	Density (g/ml)
Antarctic bottom water	34.66	−0.4	1.02786
Antarctic circumpolar water	34.7	0–2	1.02775–1.02789
Antarctic intermediate water	34.2–34.4	3–7	1.02682–1.02743
Arctic intermediate water	34.8–34.9	3–4	1.02768–1.02783
Mediterranean water	36.5	8–17	1.02592–1.02690
North Atlantic central water	35.1–36.7	8–19	1.02630–1.02737
North Atlantic deep and bottom water	34.9	2.5–3.1	1.02781–1.02788
South Atlantic central water	34.5–36.0	6–18	1.02606–1.02719
North Pacific central water	34.2–34.9	10–18	1.02521–1.02634
North Pacific intermediate water	34.0–34.5	4–10	1.02619–1.02741
Red Sea water	35.5	9	1.02750

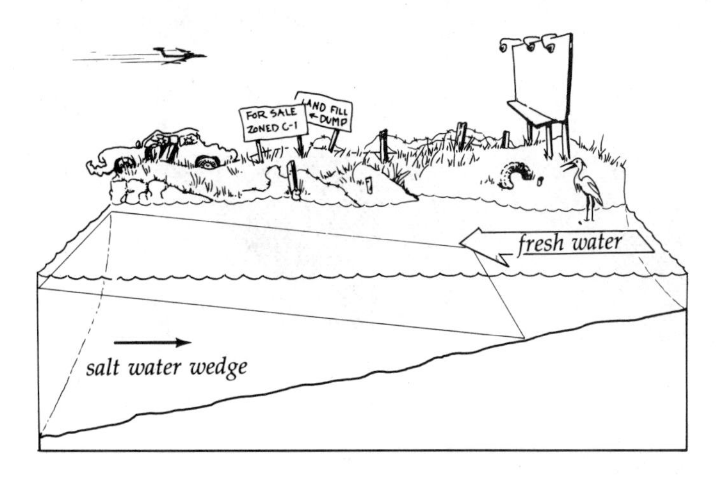

4–50 *An idealized salt wedge estuary.*

estuary may be salty and fairly well mixed for some distance upstream because the tidal wedge has easier access.

Basins generally are much larger than estuaries and differ not only in physical structure, but in the causes of their circulation. The primary factors that determine the estuary's circulation structure are the amount of river discharge, the structure of the estuary, and the tidal velocity. The basin's circulation structure, on the other hand, is determined primarily by the presence of a sill and the competition between evaporation and fresh water input.

Waterspouts and Whirlpools. There are other references, primarily in nonscientific literature, to apparent types of water movement in the ocean. Most of these are not actually different forms of motion but are specific examples of motion previously mentioned. A couple of examples are the waterspout and the whirlpool.

The waterspout, mentioned by ancient to modern mariners, is a deceiving term, because it isn't a spout of water. It is a tornado, formed exactly like those on land, but over water. As the tornado's funnel on land picks up dirt and debris, so the waterspout picks up some water, but it is mostly a rapidly swirling mass of air—short-lived and not very dangerous (Fig. 4-51).

Whirlpools are a fascination to man. In May 1841 Edgar Allan Poe wrote "A Descent into the Maelström."

"The island in the distance," resumed the old man, "is called by the Norwegians Vurrgh. The one midway is Moskoe. That a mile to the northward is Ambaaren. Yonder are Islesen, Hotholm, Keildhelm Suarven, and Buckholm. Further off—between Moskoe and Vurrgh—are Otterholm, Flimen, Sandflesen and Stockholm. These are the true names of the places—but why it has been thought necessary to name them at all, is more than either you or I can understand. Do you hear anything? Do you see any change in the water?"

We had now been about ten minutes upon the top of Helseggen, to which we had ascended from the interior of Lofoden so that we had caught no glimpse of the sea until it had burst upon us from the summit. As the old man spoke, I became aware of a loud and gradually increasing sound, like the moaning of a vast herd of buffaloes upon an American prairie; and at the same moment I perceived that what seamen term the chopping character of the ocean beneath us was rapidly changing into a current which set to the eastward. Even while I gazed, this current acquired a monstrous velocity. Each moment added to its speed—to its headlong impetuosity. In five minutes the whole sea, as far as Vurrgh, was lashed into ungovernable fury; but it was between Moskoe and the coast that the main uproar held its sway. Here the vast bed of the waters, seamed and scarred into a thousand conflicting channels, burst suddenly into frenzied convulsion—heaving, boiling, hissing—gyrating in gigantic and innumerable vortices, and all whirling and plunging on to the eastward with a rapidity which water never elsewhere assumes, except in precipitous descents. In a few minutes more there came over the scene another radical alteration. The general surface grew somewhat more smooth, and the whirlpools, one by one, disappeared, while prodigious streaks of foam became apparent where none had been seen before. These streaks, at length, spreading out to a great distance, and entering into combination, took unto themselves the gyratory motion of the subsided vortices, and seemed to form the germ of another more vast. Suddenly—very suddenly—this assumed a distinct and definite existence, in a circle of more than a mile in diameter. The edge of the whirl was represented by a broad belt of gleaming spray; but no particle of this slipped into the mouth of the terrific funnel, whose interior, as far as the eye could fathom it, was a smooth, shining and jet-black wall of water, inclined to the horizon at an angle of some forty-five degrees, speeding dizzily round and round with

4–51 A waterspout. (Courtesy of the National Oceanic and Atmospheric Administration.)

a swaying and sweltering motion, and sending forth to the winds an appalling voice, half shriek, half roar, such as not even the mighty cataract of Niagara ever lifts up in its agony to Heaven.

The mountain trembled to its very base, and the rock rocked. I threw myself upon my face, and clung to the scant herbage in an excess of nervous agitation.

"This," said I at length, to the old man—"this can be nothing else than the great whirlpool of the maelström."

"So it is sometimes termed," said he. "We Norwegians call it the Moskoeström, from the island of Moskoe in the midway."

The ordinary account of this vortex had by no means prepared me for what I saw. That of Jonas Ramus, which is perhaps the most circumstantial of any, cannot impart the faintest conception either of the magnificence or of the horror of the scene—or of the wild bewildering sense of the novel which confounds the beholder. I am not sure from what point of view the writer in question surveyed it, nor at what time; but it could neither have been from the summit of Helseggen, nor during a storm. There are some passages of this description, nevertheless, which may be quoted for their details, although their effect is exceedingly feeble in conveying an impression of the spectacle.

"Between Lofoden and Moskoe," he says, "the depth of the water is between thirty-six and forty fathoms; but on the other side, toward Ver (Vurrgh), this depth decreases so as not to afford a convenient passage for a vessel, without the risk of splitting on the rocks, which happens even in the calmest weather. When it is flood, the stream runs up the country between Lofoden and Moskoe with a boisterous rapidity; but the roar of its impetuous ebb to the sea is scarce equalled by the loudest and most dreadful cataracts; the noise being heard several leagues off, and the vortices or pits are of such an extent and depth that if a ship comes within its attraction, it is inevitably absorbed and carried down to the bottom, and there beat to pieces against the rocks; and when the water relaxes the fragments thereof are thrown up again. But these intervals of tranquillity are only at the turn of the ebb and flood, and in calm weather, and last but a quarter of an hour, its violence gradually returning. When the stream is most boisterous, and its fury heightened by a storm, it is dangerous to come within a Norway mile of it. Boats,

4–52 A small, rather gentle whirlpool (as evidenced by the ducks present) in New Brunswick, Canada, caused by currents generated when the tide changes. (Courtesy of the Southern New Brunswick Tourist Bureau.)

yachts, and ships have been carried away by not guarding against it before they were carried within its reach. It likewise happens frequently that whales come too near the stream, and are overpowered by its violence; and then it is impossible to describe their howlings and bellowings in their fruitless struggles to disengage themselves. A bear once, attempting to swim from Lofoden to Moskoe, was caught by the stream and borne down, while he roared terribly, so as to be heard on shore. Large stocks of firs and pine trees, after being absorbed by the current, rise again broken and torn to such a degree as if bristles grew upon them. This plainly shows the bottom to consist of craggy rocks, among which they are whirled to and fro. This stream is regulated by the flux and reflux of the sea—it being constantly high and low water every six hours. In the year 1645, early in the morning of Sexagesima Sunday, it raged with such noise and impetuosity that the very stones of the houses on the coast fell to the ground."

Such is the famous Maelström; though the name has been used for large whirlpools elsewhere, it originally applied to the Moskenström, a strong current running between the islands of the Moskenaes and Moskoe in the Lofoten Islands off the west coast of Norway. This whirlpool, the Maelström, though exaggerated above by Poe, is not a specific type of motion, but results from strong tidal currents that change direction frequently. The whirlpool is not always there, but at certain conditions of wind and tide a large swirling area of water results and can be dangerous. The region of greatest violence in these hazardous waters occurs between the island of Moskoe and the coast.

The whirlpool known longest by man is probably the Charybdis (Garofalo or Galofalo), known to the ancients and first described by Homer. This large, dangerous, wind-driven whirlpool exists along the Calabrian coast in the Straits of Messina between Italy and Sicily.

The rotary direction of water flow in a whirlpool is determined by coast lines, bottom topography, and local currents, and is independent of the earth's rotation.

A similar though smaller whirlpool can be seen near Black's Harbour northeast of Passamaquoddy Bay in New Brunswick (Fig. 4–52). The authors have observed this strange spectacular phenomenon, which apparently occurs often enough that tourist information brochures in the region suggest a visit to a park lookout to view it.

All strong offshore currents are potential woe to mariners. These local currents should not be confused with the massive oceanic surface circulation system. One of the most dangerous regions in the North Atlantic is near Sable Island, off Nova Scotia. Here tricky current patterns, coupled with bad weather, strong winds, and an island that moves about a bit, have wrecked many ships. A region of equal peril to shipping is Cape Hatteras, off North Carolina. Severe and sudden storms, large shoals extending well out to sea, shifting sand bars, and currents that both vary in force and reverse direction have earned this region the name "Graveyard of the Atlantic." This title was used as far back as 1790 by Alexander Hamilton, who, as Secretary of the Treasury, had the first lighthouse built on this cape and who founded the Coast Guard.

Further Reading

Bascom, Willard, "Ocean Waves." *Scientific American,* August 1959, pp. 74–84.

Bascom, Willard, *Waves and Beaches.* Garden City, N.Y.: Doubleday-Anchor, 1964.

Clancy, Edward P., *The Tides.* Garden City, N.Y.: Doubleday, 1968.

Chapin, H., and F. G. Walton Smith, *The Ocean River.* New York: Charles Scribner's Sons, 1952.

Defant, Albert, *Physical Oceanography,* 2 vols. New York: Macmillan, 1961.

Hill, M. N., *The Sea: Composition of Sea Water.* New York: Interscience, 1963.

Hill, M. N., *The Sea: Physical Oceanography.* New York: Interscience, 1962.

Knauss, J. A., "The Cromwell Currents." *Scientific American,* April 1961, pp. 105–119.

McDonald, J. E., "The Coriolis Effect." *Scientific American,* May 1952, pp. 72–78.

Munk, Walter, "The Circulation of the Oceans." *Scientific American,* September 1955, pp. 96–108.

Neumann, G., and W. J. Pierson, Jr., *Principles of Physical Oceanography.* Englewood Cliffs, N.J.: Prentice-Hall, 1966.

Pickard, G. L., *Descriptive Physical Oceanography.* New York: Pergamon Press, 1963.

Stewart, R. W., "The Atmosphere and the Oceans." *Scientific American,* September 1969, p. 76–105.

Stommel, H., "The Anatomy of the Atlantic." *Scientific American,* January 1955, pp. 30–35.

Stommel, H., *The Gulf Stream,* 2nd ed. Berkeley: University of California Press, 1965.

Tricker, R. A. R., *Bores, Breakers, Waves, and Wakes.* New York: American Elsevier, 1964.

5 Diversity of Marine Life

All that is told of the sea has a fabulous sound to an inhabitant of the land, and all its products have a certain fabulous quality, as if they belonged to another planet, from seaweed to a sailor's yarn, or a fish story. In this element the animal and vegetable kingdoms meet and are strangely mingled.

Henry David Thoreau

Astern of the boat the repeated call of some bird, a cry discordant and feeble, skipped along over the smooth water and lost itself, before it could reach the other shore, in the breathless silence of the world.

Joseph Conrad

The shell in my hand is deserted. It once housed a whelk, a snake-like creature, and then temporarily, after the death of the first occupant, a little hermit crab, who has run away, leaving his tracks behind him like a delicate vine on the sand . . .

Anne Morrow Lindbergh

I. *Describing and Naming Living Things*

II. *The Interactions of Living Things*

A. *Population interactions*

B. *The food web*

C. *Habitat*

III. *Life Forms in the Sea*

A. *Monera*

B. *Protozoa*

C. *Plants*

1. *Chrysophyta*
2. *Pyrrophyta*
3. *Rhodophyta*
4. *Phaeophyta*
5. *Chlorophyta*
6. *Angiosperms*

D. *Animals*

1. *Porifera*
2. *Coelenterata*
3. *Ctenophora*
4. *Aschelminths*
5. *Platyhelminthes*
6. *Rhynchocoela*
7. *Lophophorates*
8. *Entoprocta*
9. *More Worms: Priapulida, Echiuroida, Sipunculida, Acanthocephala, Chaetognatha*
10. *Annelida*
11. *Echinodermata*
12. *Mollusca*
13. *Arthropoda*
14. *Protochordata*
15. *Vertebrates*

Many fascinating hours can be spent at a tidepool watching the starfish slowly move in search of its prey, limpets and urchins moving about, fish darting through crevices that you can't see, and, with luck, an octopus creeping out to warn all invaders away. One is struck by the great numbers and kinds of living things at the shore, and often wonders about the fantastic creatures that must live deeper in the ocean. Often with much the same awe that he had as a child, the marine biologist uses scientific techniques to answer his questions and to extend his knowledge of many diverse organisms.

Life in the marine environment is characterized by great diversity. Both plants and animals exist in nearly every possible size, shape, color, and habitat. Some groups are more successful than others, as measured by numbers of species and numbers of individuals. Biologists have asked for hundreds of years why there are so many species, and why they have so many sizes, shapes, and ways of life. Many kinds of analysis have been used to try to answer these questions. For example, scientists have examined the fossil record, discovering that even millions of years ago diversity existed. Though the same *kinds* of organisms are not always present, and land forms themselves change, there has always been diversity through time.

Knowing that diversity exists now, and did in the past, scientists tried to make some order of it. They made efforts to categorize the kinds of plants and animals. This was not difficult in a specific locality, for each different kind had a common name. However, over larger areas, there were difficulties. The common name of a flower in one region was not the same name applied to that flower in another region. Therefore, the concept of the type came into being. A specimen was described in Latin, and a Latin name was assigned to it. Usually, characters of the size and shape were used in the description, and other organisms that fit the description could be referred to by the Latin name. It was thought that a species so described could never change, so, once described, it was "known." There was no agreement on the kinds of names to be used or the way that kinds of organisms could be grouped if they were similar but not the same. Then, in 1759, Carolus Linnaeus, a Swedish botanist who was interested in determining how many kinds of organisms there were, proposed a naming system now called binomial nomenclature. Under this system, used today, the scientific name has two parts—the first part designates the genus, the grouping of similar forms, and the second part designates the species, the particular kind of organism.

Linnaeus used the concept of homology, or common origin of morphological structures, to determine what genera and species were, and used shared characters to show how organisms were allied as they were less and less similar. A taxonomic hierarchy was adopted, which has had nearly universal agreement (in style, but not always in placement of organisms). For example, the classification of the surf perch is:

Kingdom: Animalia
Phylum: Chordata
Subphylum: Vertebrata
Class: Osteichthyes
Subclass: Teleostei
Order: Perciformes
Family: Embiotocidae
Genus: *Embiotoca*
Species: *jacksoni*

The kingdom is the most inclusive grouping; almost everything that doesn't photosynthesize (except fungi) is included under Animalia. The phylum level uses broadly unifying shared characters; chordates all have notochords, gill slits, and a nerve tube along the back at some time in their development. The vertebrates all have a backbone; the Osteichthyes are all bony fish, distinguishing them from cartilage skeletoned fish and from nonfish vertebrate forms. The order, family, and genus levels use progressively finer discriminations. All of the above placements require

---

This chapter was contributed by Dr. Marvalee Wake, Department of Instruction in Biology, University of California, Berkeley.

arbitrary decisions based on characters, and scientists do not always agree on the allocation to these ranks. A species, however, can usually be distinguished biologically, for its members can breed with each other, but not with other similar groups, even those placed in the same genus on the basis of morphological characters. This system allows us to categorize organisms, but tells us nothing about the way they live.

Charles Darwin's epoch-making ideas about the continuity and rate of change of living organisms insured that biologists would investigate the dynamism of life on the earth. Once the idea of spontaneous creation was discredited by scientists and evolution acknowledged in the early twentieth century, scientists began to study the *biology* of organisms, rather than just to give them names. They recognized that species of living things and even the earth change or evolve, so that organisms are not fixed or unchanging. The real questions became "How does evolution work?" "How do organisms live together?" "What causes changes in numbers and kinds of organisms?" and many similar questions. Scientists began looking at organisms in their habitats in order to understand how they had adapted to the environment and to study the effects or results of selection. Darwin's contributions gave impetus to the studies of ecology, development, and behavior, and studies of the mechanisms of evolution.

Most current research into problems of what species are and how they live adopts a synthetic approach. The theories and techniques of morphology, ecology, biochemistry, behavior, development, and genetics are all employed to analyze a species and its interaction with other species and with its physical environment. That, really, is what a species "is." In dealing with species diversity, many scientists now look at what controls a population. Such considerations as the kind and amount of food available, light, temperature, and moisture requirements, the medium on or in which the population lives, and many other factors are all analyzed to determine the niche, the sum of the requirements for life, of the population. As the elements of the niche are analyzed, one can begin to understand how some species live together, why many do not live together, and how evolutionary changes affect a species.

The food web (discussed in Chapter 6) is an important ecological concept in analyzing the structure of a community of organisms. Further, specific investigation of what and how organisms obtain nutrients gives information about energy requirements. Most plants harness the sun's energy and convert it to chemical energy. Many animals eat plants (such animals are called herbivores); a number of animals eat other animals (and are called carnivores). In this way, the sun's energy is passed along and the food web is exploited. Some animals use particular components of other forms. Those that feed on dead animals are scavengers; those that feed on the dying parts of algae and the associated bacteria of the ocean bottom are detritus feeders. The habitat of the organisms is also important. The setting in which organisms live influences and is influenced by the physiology, behavior, and evolutionary pathways taken by the organisms. The habitat provides the food, shelter, and physical factors necessary to maintain the life of the organisms. Analysis of the community, that is, all of the populations of various species of plants and animals living together in an area and related by common environmental requirements, provides information on the ways that organisms coexist and distribute the sun's energy.

A study of a specific population (the group of individuals of one species living in an area and capable of reproduction to perpetuate the species) can give detailed information about mechanisms of adaptation and survival. Various populations associate in different ways. If two populations interact, and both benefit, but neither requires the interaction, the association is called mutualism. If both populations benefit but require the interaction for survival, the interaction is called protocooperation. Commensalism is an association in which one population acts as host to another; the host is not affected, but the "guest" or commensal does benefit. Predation is an interaction in which a population of larger animals benefits by eating the smaller, or prey, population, which obviously is adversely affected. In parasitism, a population of small organisms (the parasites) benefits by depleting the host population of larger animals. If two populations each inhibit the other, the association is called competition. Amensalism is the state in which one population is inhibited, but the other is not affected. Finally, if neither of two populations in an area affects the other, the association is called neutralism. "Benefit" means that the growth or survival of a population is increased; an adverse relationship decreases growth or survival.

The following account of the diversity of major groups of marine organisms uses the population interaction, food web, and habitat concepts to show the organisms in their environments and to emphasize both diversity and similarity. After all, the problem of diversity no longer is "How many kinds are there?" or "What is its name?" but "How do all of these kinds of organisms make their livings?" In other chapters, you learn about several habitats in the marine world, so some acquaintance with the organisms who live in that world is in order. Major groups are discussed by phylum (according to their place in the taxonomic hierarchy), and the characteristics that ally them are mentioned. The hierarchy places the least complex phyla at its base and more complex ones toward its peak, so phyla are discussed in a sequence from less to greater complexity of organization and function. Because phyla are assigned using shared characters, they reveal evolutionary relationships as well. Some groups have few kinds of members; some have many. The species in some groups all look very much alike, while the species in other groups have fantastically bizarre modifications. The following account introduces you to that diversity, especially who is related to whom, where they live, and how they make their livings in the marine world.

Kingdom Monera

The bacteria and blue-green algae share morphological and biochemical characteristics that indicate they are closely related to each other, but not to other groups. Their cells do not have discrete internal structures that perform specific functions as do other organisms. They do, however, have the same basic chemistry —DNA is found in "nuclear bodies," the enzymes necessary for metabolism are present, as are sugars, starches, and other basic chemicals. They have cell walls. Most are one-celled or occur in small colonies.

Bacteria, once thought rare in seawater, live in all kinds of marine habitats—wherever there is a suitable environment. Some bacteria manufacture their own food from inorganic materials such as water, carbon dioxide, and nitrates, using the sun as a direct energy source. They do this without giving off oxygen and usually produce a sulfur compound as a by-product, so they can carry on the process where plants requiring oxygen for metabolism cannot. They break down various materials—for example, sulfur bacteria are found in water that contains a lot of decaying material. The bacteria use the hydrogen sulfide that is released through protein decay. Most bacteria, however, obtain organic nutrients from the environment, especially from other organisms. Such bacteria live in water, mud, and debris in abundance and are responsible for much of the decay process. Some organisms feed directly on the bacteria, but of greater significance is the work of decay bacteria in making a food supply available to other organisms. Seaweeds, for instance, are abundant and comprise a great energy store. Few animals feed directly on them, however. As seaweeds break off or die, they sink to the bottom and are attacked by bacteria, which get their own nutrient materials and, in addition, break down the seaweeds to a form more readily usable as an energy source for a multitude of detritus-feeding animals.

Bacteria of many genera and species are found on marine substrates. Because generations are short and living conditions favorable, there may be millions of individuals per cubic centimeter. Shapes vary from round to elongate, and the bacteria may occur individually, in clusters, or in strands (Fig. 5–1).

The blue-green algae show two main evolutionary trends—a one-celled, free-floating form seems to have produced both filamentous and nonfilamentous colonial forms. These algae do not have flagella and have only chlorophyll *a* of the chlorophyll pigments; virtually all forms photosynthesize. Blue-green algae occur in all sorts of habitats, including many marine ones. In shallow coastal areas, they may be attached to rocky or sandy substrates or to larger algae; some occur as algae mats on mud flats or lagoons. The filamentous *Oscillatoria* (Fig. 5–2) is a common marine form. Algae are important in the weathering of rock. Some blue-greens are planktonic and may cause "blooms" that kill many animals either because the algae secrete a toxin or because they use up oxygen in the water. Blue-green algae come in all colors—one species produces the blooms that give the Red Sea its name. Some blue-green algae may contribute to coral reef formation by secreting calcium and magnesium salts necessary for the skeletal formation of the coral animals.

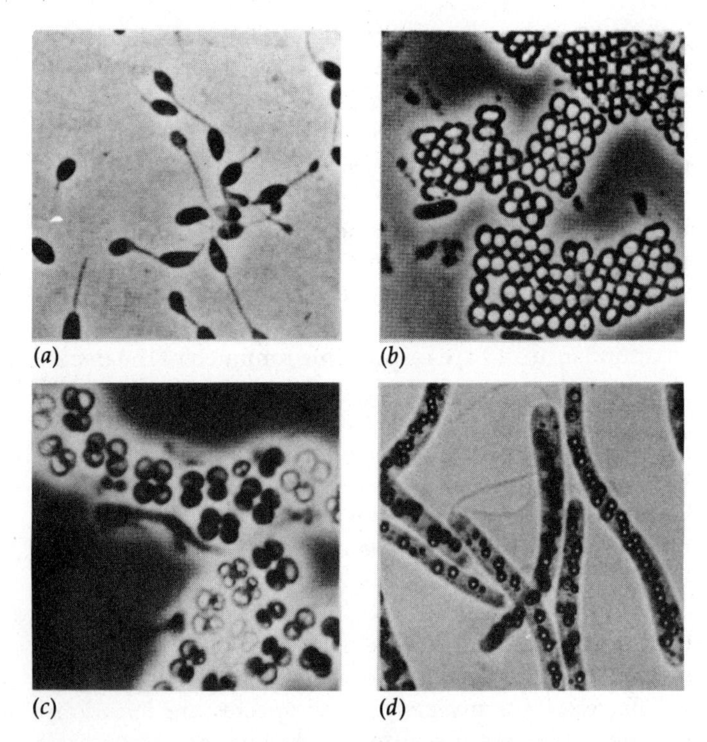

5–1 *Photosynthetic bacteria.* (*a*) Rhodopseudomonas, *a nonsulfur bacteria;* (*b*) Thiopedia, (*c*) Thiocapsa, (*d*) Thiospirillum, *all purple sulfur bacteria. From* Botany: An Ecological Approach *by William A. Jensen and Frank B. Salisbury, Wadsworth Publishing Company, Inc.,* © *1972.*

Protozoa

Protozoa are one-celled organisms having a level of internal organization in which organelles perform specific functions. They have a membrane-bound nucleus and so are considered eucaryotes (true nucleus), as are all organisms except bacteria and blue-green algae, which are procaryotes (pre-nucleus). They have cilia or flagella (or both) or pseudopods for movement and sensing, and vacuoles for food ingestion and excretion; they have mitochondria, ribosomes, and other organelles. Some have chloroplasts and therefore can photosynthesize to produce their own nutrient energy. Other protozoans may be parasites or scavengers that feed on dead organisms, or feed freely on bacteria, diatoms, and other protozoans.

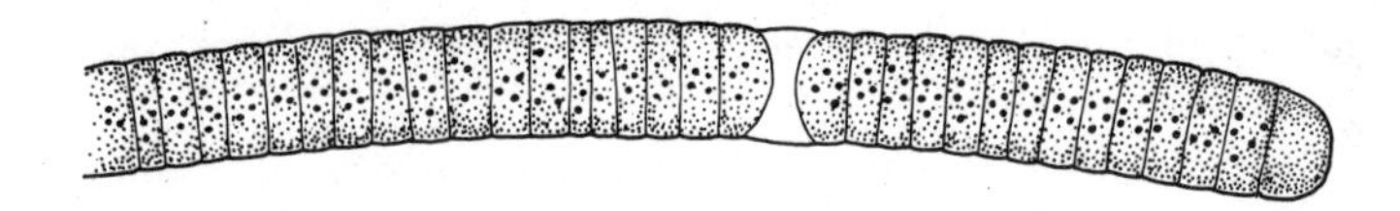

5–2 Oscillatoria, *a blue green alga.*

Protozoans are placed in one of four classes, depending on their organs of movement: Flagellata (with flagella—long whip-like structures), Ciliata (with cilia—numerous short extensions), Rhizopoda (with pseudopods—out-pocketings of the cell membrane into which the cell body flows), Sporozoa (with no organs for movement). Each of these classes has marine representatives, but some groups have such economic importance to life in the sea that they warrant special attention (Fig. 5–3).

Foraminifera, members of the class Rhizopoda, are important because their calcareous shells provide a clue to the age of strata of the earth. As the organisms die, their shells sink to the bottom and are preserved in layers. Great numbers of certain species indicate the age of the layer. Radiolaria, also Rhizopoda, have a silicate shell. These skeletons, too, drop to the ocean bottom, forming "radiolarian ooze," which is useful in determining geologic age. There are thousands of species of tintinnoid ciliates, and any plankton sample from open water will contain several species. Their abundance in plankton provides an important food source for many larger animals.

Chrysophyta

The various algae and the angiosperms are all considered plants. The golden algae and the diatoms, phylum Chrysophyta, are components of plankton. The golden algae, class Chrysophyceae, are predominantly fresh-water, but some are marine. They are usually one-celled, though some are colonial. They are eucaryotes, having a distinct nucleus, chloroplasts, and other membrane-bound organelles. Chlorophyll *a* is their predominant photosynthetic pigment, although the characteristic golden color results from orangish pigments that mask the green. Most of the golden algae move freely, as either forms with pseudopods or forms with

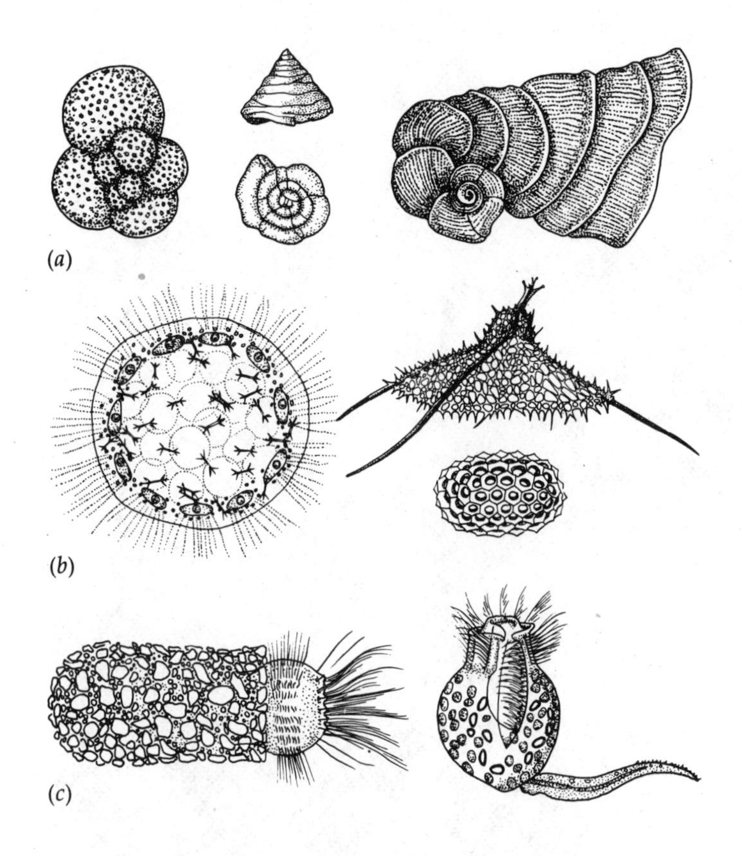

5–3 (*a*) *Foraminifera: (from right)* Globigerina, Tetraxis, Fischerina, Medusetta. (*b*) *Radiolarians:* Spinaerozoum, Dictyophimus, Cenolarus. (*c*) *Tintinnoid ciliates:* Codonella, Tontonia.

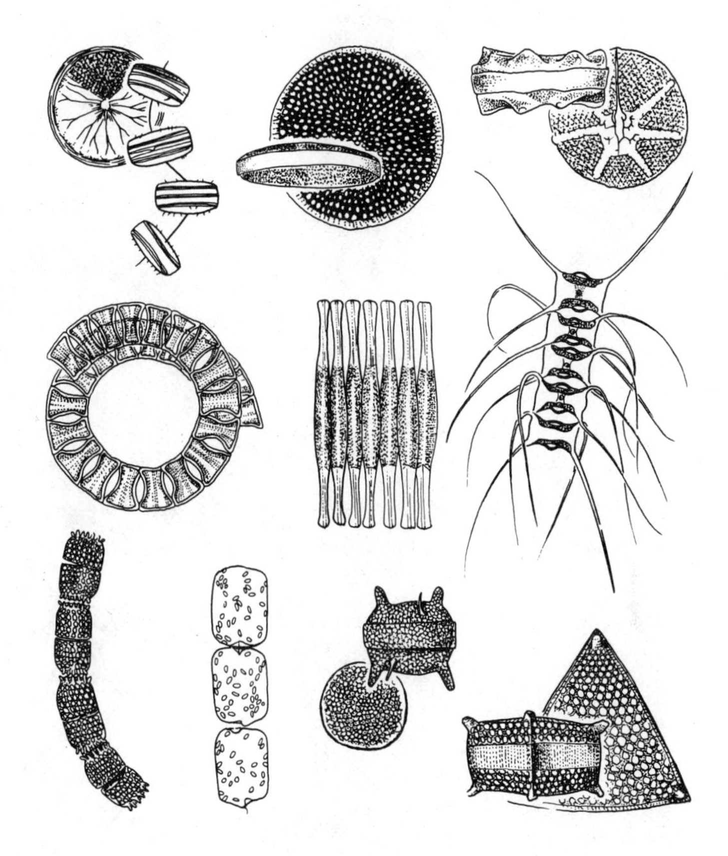

5–4 *Diatoms:* (*from right*) Thalassiosira, Coscinodiscus, Asteromphalus, Eucampia, Fragilaria, Chaetoceras, Stephanopyxis, Rhizosolenia, Cerautulus, Biddulphia.

flagella. As with all photosynthesizing marine organisms, they are primary producers since they harness the sun's energy during photosynthesis and produce oxygen as a by-product. Some forms are planktonic in both warm and cool oceans; others occur in tidepools and salt marshes, often coloring the water golden.

The Bacillariophyceae, or diatoms, are conspicuous in marine environments. Diatoms also have the photosynthetic pigment chlorophyll *a*, and often chlorophyll *c*, plus several other pigments, making them primary producers. Diatoms are complexly organized one-celled plants. The cell walls, composed primarily of silicon, are in two halves. The walls, or "shells," have a great variety of shapes, with patterns of notches, ridges, and depressions (Fig. 5–4). When the animals die, the shells fall to the bottom, often forming extensive deposits. Diatoms occur both in plankton (Fig. 5–5) and on surfaces of substrates, in mud, marshes, and on other algae. They may form extensive blooms in plankton.

A good deal is known about energy use and oxygen production in the ocean because of studies on diatoms. Cool marine waters have two peaks of abundance. In late spring, sunlight and temperature increase; as a consequence, so does production. As essential nutrients are depleted and small animals eat the plankton, production diminishes. By late summer, when nutrients are recycled and consumers fewer, a second bloom occurs. Production again diminishes as temperatures cool and light intensity decreases. Such cycles affect all of the organisms in the marine environment, and knowledge of them is fundamental in determining the total energy available to marine organisms.

35 mm
5 mm

5–5 The open ocean—plankton and nekton.

1 *dolphins,* Delphinus
2 *tropicbirds,* Phaëthon
3 *paper nautilus,* Argonauta
4 *anchovies,* Engraulis
5 *mackerel,* Pneumatophorus, *and sardines,* Sardinops
6 *squid,* Onykia
7 Sargassum
8 *sargassum fish*
9 *pilotfish*
10 *white-tipped shark*
11 *pompano,* Palometa
12 *ocean sunfish,* Mola mola
13 *squids,* Loligo
14 *rabbitfish,* Chimaera
15 *eel larva, leptocephalus*
16 *deep sea fish*
17 *deep sea angler,* Melanocetus
18 *lanternfish,* Diaphus
19 *hatchetfish,* Polyipnus
20 *"widemouth,"* Malacosteus
21 *euphausid shrimp,* Nematoscelis
22 *arrowworm,* Sagitta
23 *amphipod,* Hyperoche
24 *sole larva,* Solea
25 *sunfish larva,* Mola mola
26 *mullet larva,* Mullus
27 *sea butterfly,* Clione
28 *copepods,* Calanus
29 *assorted fish eggs*
30 *stomatopod larva*
31 *hydromedusa,* Hybocodon
32 *hydromedusa,* Bougainvillea
33 *salp (pelagic tunicate),* Dolielum
34 *brittle star larva*
35 *copepod,* Calocalanus
36 *cladoceran,* Podon
37 *foraminifer,* Hastigerina
38 *luminescent dinoflagellates,* Noctiluca
39 *dinoflagellates,* Ceratium
40 *diatom,* Coscinodiscus
41 *diatoms,* Chaetoceras
42 *diatoms,* Cerautulus
43 *diatom,* Fragilaria
44 *diatom,* Melosira
45 *dinoflagellate,* Dinophysis
46 *diatoms,* Biddulphia regia
47 *diatoms,* B. arctica
48 *dinoflagellate,* Gonyaulax
49 *diatom,* Thalassiosira
50 *diatom,* Eucampia
51 *diatom,* B. vesiculosa

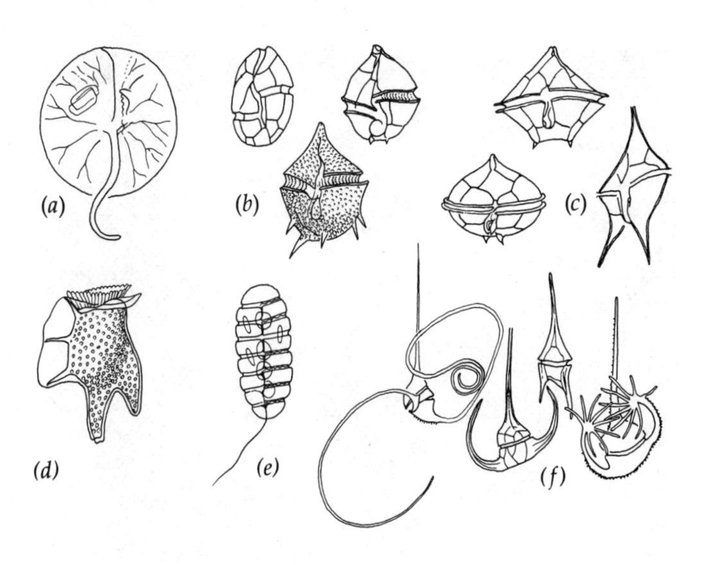

5–6 *Dinoflagellates:* (*a*) Noctiluca; (*b*) Gonyaulax; (*c*) Peridinium; (*d*) Dinophysis; (*e*) Polykrikos (*a colonial species*); (*f*) Ceratium.

Pyrrophyta

The Dinoflagellata are members of the phylum Pyrrophyta. Most are one-celled with two flagella, one in a groove around the body, the other directed backward or forward. The dinoflagellates secrete a cellulose covering and some have chloroplasts. They are an important component of plankton and harness considerable energy; they then may be eaten by other animals (Fig. 5–6). Other dinoflagellates do not photosynthesize, but ingest prey or detritus. Some forms are especially interesting. Members of the genus *Gonyaulax* can occur in numbers so great that the water turns red (the red tide), and another species of that genus is responsible for mussel and clam poisoning. Mussels ingest the species as they feed on plankton, and the toxin of the dinoflagellate poisons people who eat the mussels. Another dinoflagellate, *Noctiluca,* is responsible for luminescence in many marine waters. Huge numbers of this genus occur in the water and "shimmer" with the waves.

5–7 Microcladia, *a lacelike red alga, growing on a blade of* Laminaria *(a brown alga). The foam results from the breakup of marine diatoms. From* Botany: An Ecological Approach *by William A. Jensen and Frank B. Salisbury, Wadsworth Publishing Company, Inc., © 1972.*

5–8 Various red algae on the rocks at Bodega Bay. (Photo by Steve Renick. Courtesy of Bodega Marine Laboratory, University of California, Berkeley.)

Rhodophyta

The Rhodophyta, red algae, are mostly marine and are common along rocky seacoasts. They are relatively small plants, usually less than two feet long. Almost all species are multicelled, including sheets and simple branched filaments as well as complexly branched, lacy forms (Figs. 5–7, 5–8). Many species occur in warm waters of the tropics, where they are small and inconspicuous. In cooler waters, where they are fewer, they often are much larger. Most marine forms die if their cells are exposed to air, for their internal water evaporates. Some genera, such as *Porphyra,* occur in the intertidal zone and are often subject to drying out, but survive because their heavy outer layers reduce water loss. Most grow deeper so that they are not buffeted by waves. Other red algae occur as deep as 150 m, especially in the tropics.

Chlorophyll *a* is the primary photosynthetic pigment, and some species also contain chlorophyll *d.* The red color results from a protein. The actual color of red algae depends on the amount of light available and therefore their depth in the ocean. They are often green, black, or purple when much light is available, but brilliant rose red if they grow in the deeper regions of the oceans. The accessory pigments mask the chlorophyll, but are necessary to absorb more light, which is then transferred to the chlorophyll.

One major group of red algae is found primarily in warm tropical waters. Its members cement together coralline animals and, therefore, are a fundamental component of coral reefs. They maintain reef structure through their cementing action. Red algae contribute to the food web primarily in two ways—the reproductive cells released by red algae are an important component of plankton, and dead algae parts form detritus. Man also uses red algae in several ways. The Japanese have long cultivated it as a food item. Several extracts of red algae have commercial use as sizing, starch, paint binders, and in the production of agar, which is used in scientific media, medicines, and many other situations.

(a)

(b)

5–9 *(a) A large kelp bladder, fronds, and stipe. (b)* Postelsia palmaeformis, *the sea palm, a West Coast brown alga growing in the surf zone. (Photos by Steve Renick. b, Courtesy Bodega Marine Laboratory, University of California, Berkeley.)*

Phaeophyta

The brown algae, Phaeophyta, are large, conspicuous algae that are almost exclusively marine. Some forms are simple branched filaments, others are giant seaweeds that may be more than 60 m long. They grow primarily in the intertidal zone, and are fastened to the substrate. Many forms are differentiated into a basal holdfast region, a stem or stipe, and a broad blade in which much of photosynthesis occurs. Chlorophylls *a* and *c* are used for photosynthesis, and an accessory pigment gives the characteristic brown color.

A few forms are free-floating far from shore. The genus *Sargassum* (Figs. 5–5 and 5–53c) forms many clumps in the western North Atlantic (the so-called Sargasso Sea), and the mat increases in size by vegetative reproduction, or the growth of new plants from parts of older plants without sexual reproduction. Some of the genera of large plants are attached in deeper water and form subtidal "forests." Fig. 5–9a, for example, shows a large kelp whose branches form extensive mats on the surface of the ocean.

The brown algae are very important in the economy of the sea. They are important primary producers, for huge numbers of their reproductive cells are free-swimming planktonic elements. The vegetative plants are less significant as producers, for they occupy the relatively small space available on the continental shelf, whereas plankton occupy much of the ocean's surface. Some animals feed directly on brown algae, and many ingest algal particles as detritus. Their presence is important to animals for other reasons, too—many animals live inside the holdfasts, or fasten onto the blade; others, including fish, seek protection for themselves and their young in the subtidal forests. Man, too, uses brown algae for food, fertilizer, and to obtain the salts of various chemicals.

Chlorophyta

The green algae exhibit fantastic diversity of size, shape, life history, and habitat. Some are one-celled, free-swimming, filamentous, or colonial; others are multicelled, of many shapes. Some have many branches; some have a flattened leaf-like surface; a number have small holdfast organs (Fig. 5–10). The green algae appear to be the evolutionary pathway of development for land plants. Most species live in fresh water, but the several marine species attain larger sizes than most fresh-water forms. The green algae have chlorophylls

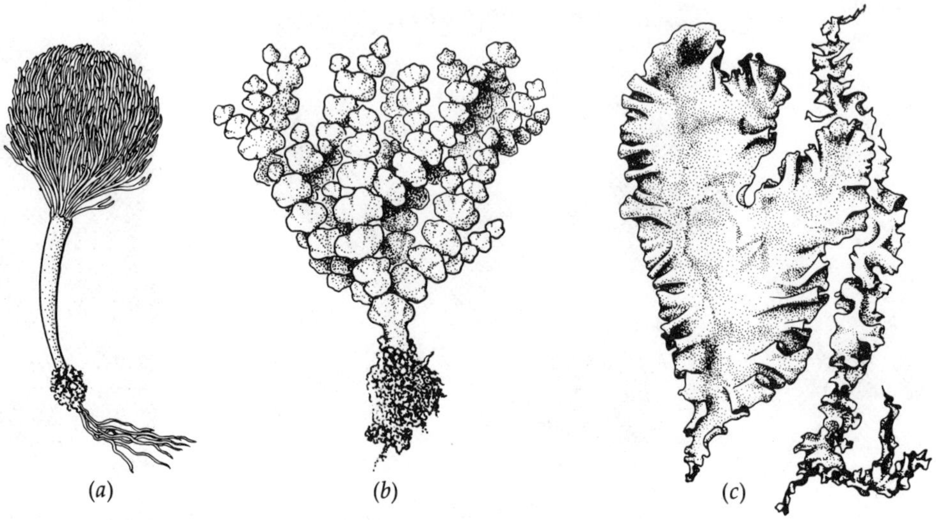

5–10 Green algae. (a) Penicillus; *(b)* Halimeda; *(c) two species of* Ulva.

(a) *(b)* *(c)*

a and *b*, and some other pigments. The marine forms are primary producers, though not a highly significant proportion of the plankton or nekton. Some of the larger marine forms that do attach to rocky substrate form large mats. *Ulva*, the sea lettuce, has a large leaf-like blade that some animals graze on, and on which other green algae attach and grow. Some green algae live as commensals inside the shells of molluscs, and others live inside the cells of some protozoa and other animals. The *Zoochlorella* live within anemones and provide oxygen as a by-product of photosynthesis.

Green algae, especially, as well as reds and browns, are experimenting with reproductive modes. The more advanced of these have alternating generations in which one generation is small, one-celled, and free-swimming (and often a component of plankton), and the other generation is multicelled, much larger, and is fastened to rocks or ocean bottom. An evolutionary trend for the multicelled generation to become larger and live longer is found in both algae and land plants.

Angiosperms

Flowering, seed bearing plants are not often thought of as members of the marine environment, but a few tolerate saline conditions well and live at or near the tide zone. Mangroves (*Rhizophora*) (see Fig. 2–18), perhaps best adapted to the marine environment, are large trees with many butresses that live in shallow warm seas. Their main contribution to the marine way of life is that they provide a habitat for many organisms, and in fact, form "islands." The debris from palms and other vegetation of isolated tropical islands also provides shelter for marine animals. The plants of the coastal strand and of coastal salt marshes may also grow down to the tide line; they provide cover for crabs and other foragers. This vegetation is very low and often is succulent. *Mesembryanthemum* (ice plant) and *Salicornia* (salt plant) are typical of this vegetation, and form extensive mats over the sand (Fig. 5–11). Occasionally lupines, herbs, and sagebrush occur on the strand as well.

Porifera

Members of the Kingdom Animalia are multicelled organisms and include all those to be subsequently discussed. The simplest animals, the sponges, are collections of cells that form large masses, especially in warm, shallow seas (Fig. 5–12). The reasons that organisms became multicelled are not known, but such factors as acquiring ways to increase in size, specialization, and

5–11 Ice plant along the shore. (Photo by Steve Renick.)

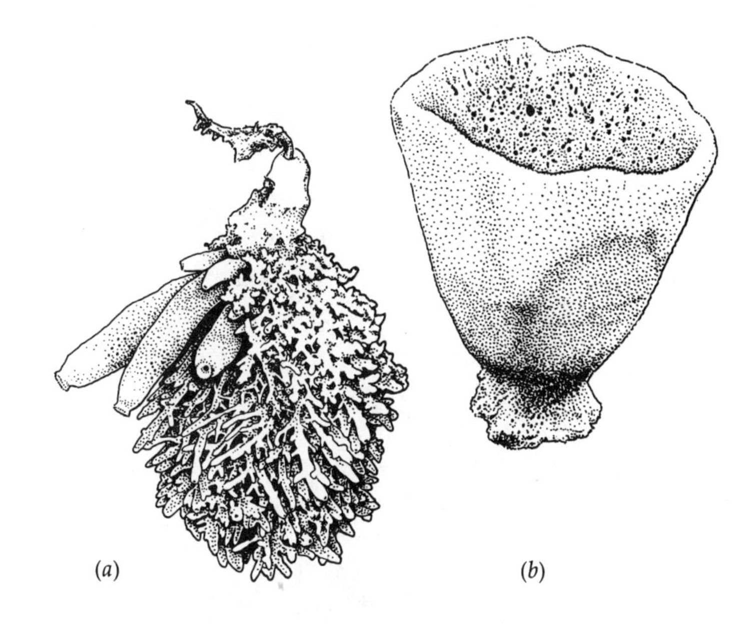

*5–12 Porifera. (a) Calcareous sponges—*Sycon *(on the left) and* Leucosolenia*—which attach to wharf pilings. (b) Skeleton of an elephant's ear sponge, from the Mediterranean.*

efficiency are implied. Single cells must be of limited size, because they must have a great enough surface area to allow the exchange of nutrients and waste material in the cell. The volume of a single cell increases faster than its surface area increases; eventually, there isn't enough surface area to allow adequate exchange for the total cell volume. Multicelled organisms may have arisen (1) as cells increased in size, nuclei divided, and cell membranes grew to partition them and the cell contents, giving rise to a mass of cells, or (2) as the members of colonies of one-celled organisms lost their independence, remained in a group, and divided their functions so that some cells did the feeding, others the reproduction, and so forth, for the entire mass.

A sponge is a mass of cells supported by siliceous spicules or by fibers; some cells of the mass ingest food particles from water flowing through the channels of the sponge and nourish the entire organism, throwing water out the main channel, or "mouth"; other cells secrete the support structures; others produce eggs and sperm for sexual reproduction, and yet others may produce special asexual reproductive units. Budding or branching can also occur. Sponges lack a coordinating system and have other characteristics that suggest they evolved from a different one-celled group from the one that may have given rise to other multicelled organisms.

After the invention of synthetic sponges, some of the commercial value of natural sponges was lost. However, the horny sponges (those without spicules) are still harvested from warm shallow seas to be sold, and calcareous forms are significant because they collect on pilings and may foul boat channels. Sponges are not restricted to warm seas; some occur in the ocean depths, and a few have invaded fresh-water habitats.

Coelenterata

The coelenterates are highly successful and very diverse—anemones, corals, jellyfish including the Portuguese man o' war (Fig. 5–13), and hydras are all members of the phylum. They share a common body plan that shows radial symmetry (concentric structure around a central axis). They have a mouth as the only entrance to the digestive cavity; most have tentacles

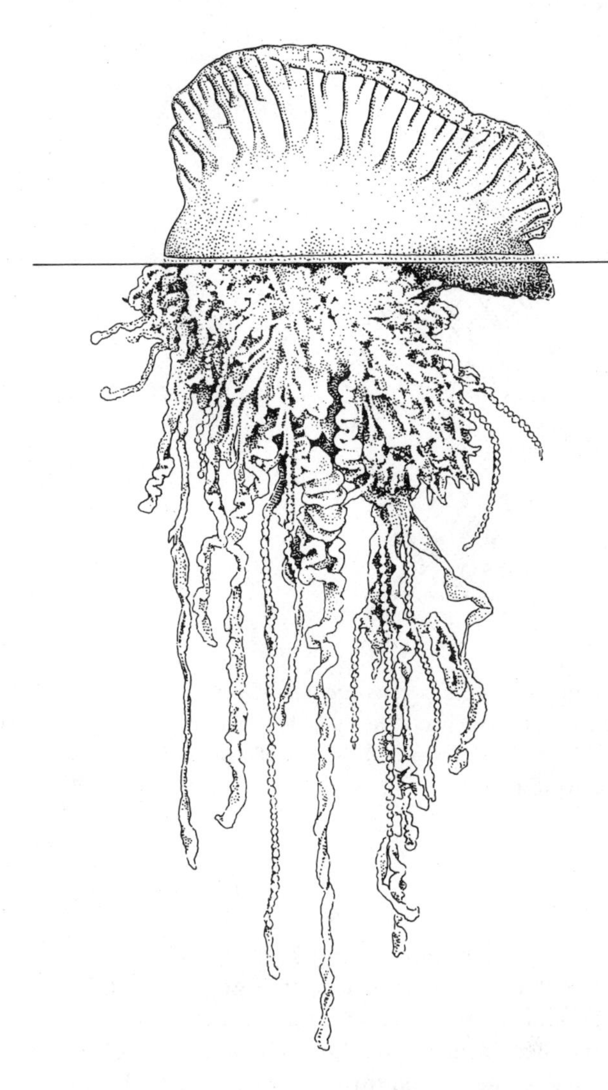

5–13 Portuguese man o' war (Physalia physalia), *a colonial hydrozoan composed of up to a thousand specialized polyps and medusae integrated to act as an individual. Stinging cells on the tentacles contain a powerful nerve poison, which immobolizes prey.*

5–14 A cluster of hydroids (Tubularia).

around the mouth with stinging cells on the tentacles for food gathering and defense. They have a more complex organization than do sponges—the body wall has two distinct layers, or tissues, and in some forms a third layer is evident. With this "tissue level of organization," the cells have become more specialized to include a network of "nerve" cells that gives the animals coordination. Coelenterates may have two different body forms in members of the same species. One form is the polyp, which is attached to the substrate at one end with the free end containing the mouth circled by tentacles, and which may be individual or colonial. The other form is the medusa, which is free-swimming and has marginal tentacles with the mouth projecting from the underside of the bell-shaped body. This dual-form condition is thought to be a balance of form to cope with environmental conditions. The general radial body plan and tissue organization is found in both forms. Reproduction is usually asexual in polyps (producing more polyps or medusae), and sexual in medusae.

There are three major classes of coelenterates, each class having similar tentacles, sex cells, digestive tract, and life cycle. The Hydrozoa include the hydroids (Fig. 5–14) and many of the jellyfish. In many species the polyps are colonial, and division of labor occurs with some polyps being feeding structures, some reproductive, and some serving other functions. Medusae are also usually well developed in this class. The polyp stage is reduced or absent in members of the class

5–15 Sea anemones. (Photo by Steve Renick. Courtesy Steinhart Aquarium, California Academy of Sciences.)

Scyphozoa; medusae are the main life form. The class Anthozoa, conversely, includes only polyps. Sea anemones of many sizes, colors, and forms (Fig. 5–15), sea pens, sea fans, and corals are anthozoans. All have tentacles surrounding their mouths. They are carnivores, as are all members of the phylum, feeding on small animals, especially mollusc and echinoderm larvae. Certain anemones enter interesting symbiotic relationships, for tiny shrimps and fish live among the anemones' tentacles unharmed, and "clean" small parasites from the tentacles (Fig. 5–16). Many other anthozoans secrete a calcareous shell, and some corals have a symbiotic relationship with certain dinoflagellates, the zooxanthellae, which promote calcification and supply oxygen and organic nutrients to the polyp.

Coelenterates have a very long fossil record, beginning in the pre-Cambrian, that traces the diversity of members of the phylum. One group of coelenterates, hydras, are now found in fresh-water habitats, though most species are marine. Medusae are pelagic, and many float in currents on the open ocean. Polyps, because they live on the bottom, are usually found in shallow waters, including tidepools, though they may occur in the depths as well. Corals (Fig. 5–17), particularly, are found in shallow, warm, subtropical or tropical waters, where their skeletons provide the structures for an entire community of plants and animals.

Ctenophora

The ctenophores, or comb jellies (Fig. 5–18), resemble coelenterates in many ways—they have tissue level of organization, no organ systems, radial symmetry, and a mouth as the entrance to the digestive cavity. They look like medusae, but differ in having eight rows of comb-like structures used for movement (from which the phylum and common names are born), a different kind of early development and larva (the pre-adult form), and adhesive organs on their tentacles for prey capture, rather than stinging cells. Most forms have a pair of long tentacles that bear numerous short filaments. Their prey include tiny animals, larvae, and eggs. Comb jellies reproduce sexually, and are bi-

5–16 *A coral reef habitat.*

1 *black-capped petrel,* Pterodroma
2 *sea nettle,* Dactylometra
3 *angelfish*
4 *lobed corals*
5 *sea whips (gorgonian corals)*
6 *triggerfish,* Balistes
7 *sea fans (gorgonian corals)*
8 *tube anemone,* Cerianthus
9 *orange stone coral,* Astroides
10 *bryozoans*
11 *brain coral*
12 *Picasso butterflyfish*
13 *moray eel*
14 *cleanerfish*
15 *tube corals*
16 *snail,* Murex
17 *nudibranch,* Phyllidia
18 *sponges*
19 *colonial tunicate*
20 *giant clam,* Tridacna
21 *purple pseudochromid fish*
22 *cobalt sea star*
23 *soft corals*
24 *barber-shop shrimp (a cleaner shrimp),* Stenopus
25 *sea anemones*
26 *clownfish*
27 *worm tubes*
28 *cowry*
29 *sea fan,* Eunicella

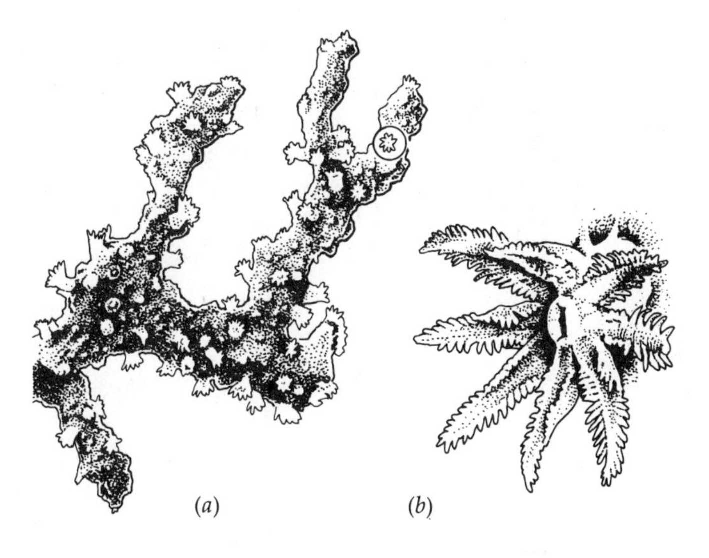

5–17 *(a) A precious, stony coral* (Corallium), *which has a skeleton of red limestone. (b) Closeup of the coral polyp.*

sexual (gonads of both sexes are present in each individual).

All of the comb jellies are marine, and most are pelagic. They are among the most beautiful marine animals—delicate, translucent, and floating near the surface, they are iridescent by day, and bioluminescent by night, giving the currents in which they are found an unusual beauty.

The Aschelminth Groups

The five groups of aschelminths are variously called classes in the phylum Aschelminthes or full phyla. Their taxonomy is unresolved, but the groups share several characteristics. They have a complete digestive tract (with both an anus and a mouth) and are pseudocoelomate (there is a space between the gut and the body wall, which exists from early development, but no true membranous lining of the body wall). They are usually small and cylindrical, have an epidermis which secretes a protective outer layer, as well as nervous, muscular, and excretory systems; all reproduce sexually. The groups that have marine species will be discussed primarily.

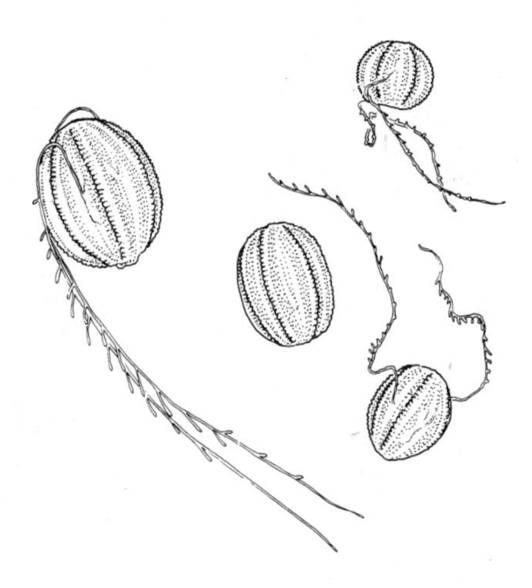

5–18 *Sea gooseberries* (Pleurobranchia), *ctenophores.*

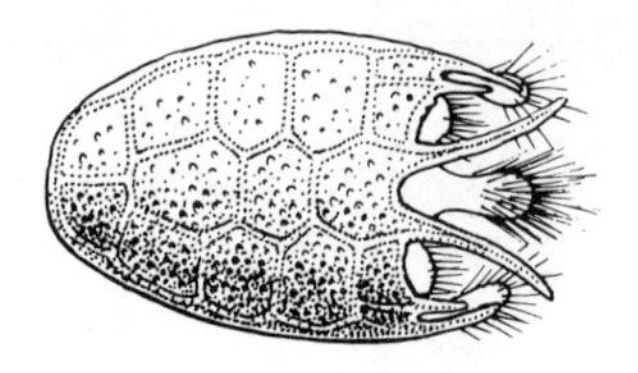

5–19 *A rotifer* (Karatella cochlearis).

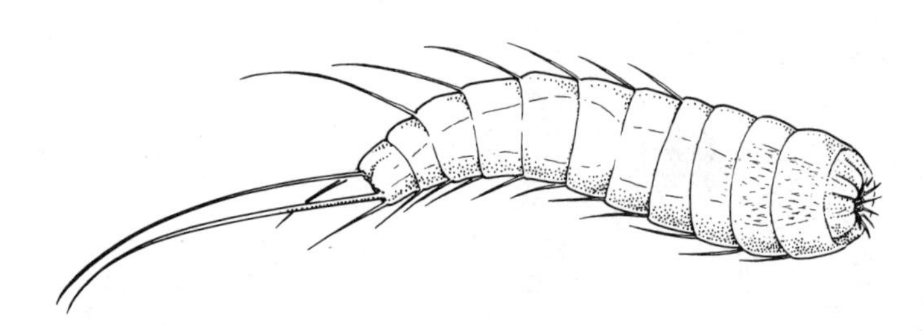

5–20 *A kinorynch* (Echinoderella).

Rotifera live mostly in fresh water, though there are a few marine species (Fig. 5–19). They appear segmented (but are not truly so), have "wheel organs" that create water currents around the mouth to bring in food. Many have the pharynx (or "throat" region) equipped with crushing plates, and most forms have paired adhesive organs on the rear part of the body to grip the substrate.

One of the two orders of the Gastrotricha is marine. These animals are less than 1 mm long, and are flattened on the bottom. They have a "head" region with a mouth and rows of cilia. Some cilia create currents to bring in food, others are sensory, and rows of cilia on the downward side are used for movement. They, too, have a specialized pharynx, paired adhesive organs, and most are bisexual.

The Kinorhyncha are superficially segmented marine dwellers in the muddy bottoms of shallow waters. The body has head, neck, and trunk regions, and the head is equipped with spines (Fig. 5–20). Their existence is similar to that of the marine gastrotrichs.

Nematodes live everywhere! The round worms are unsegmented and often very small, although the marine species may be hundreds of millimeters long and are the most specialized nematodes (Fig. 5–21). Nematodes possess the general aschelminth characters, and aquatic forms often have special anterior sensory organs and lack excretory canals. Some marine nematodes are free-

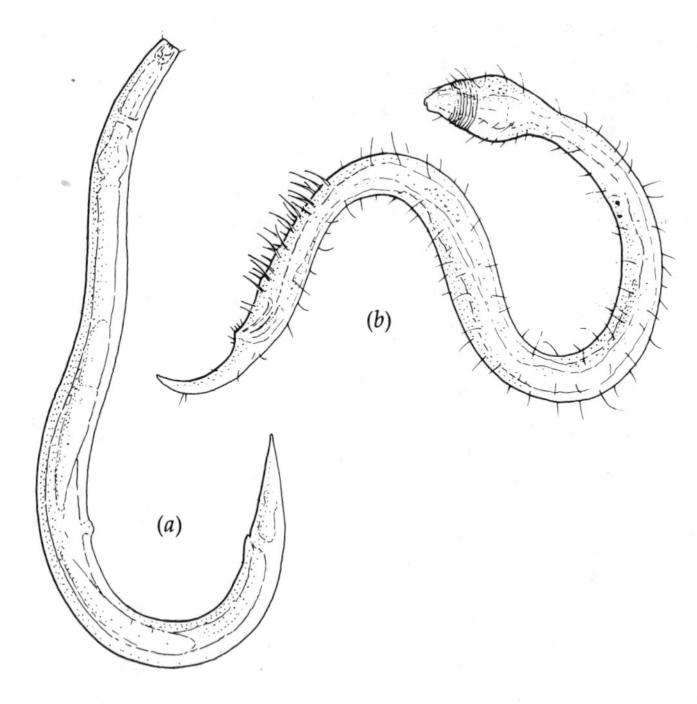

5–21 *Two free-living marine nematodes. (a) A cyatholaimid; (b) a draconematid. The latter moves inchworm fashion using its "stilt bristles."*

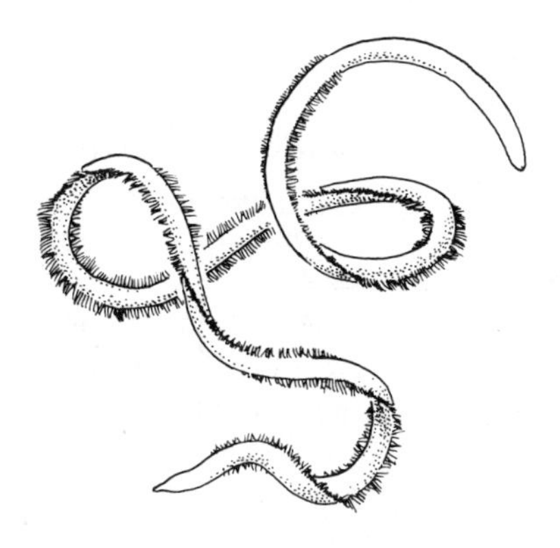

5–22 Nectonema, *an agile pelagic marine nematomorph 50 to 200 mm long, very thin, and translucent. It swims on the surface with an undulating motion. The young are parasitic on crabs; the adults do not feed.*

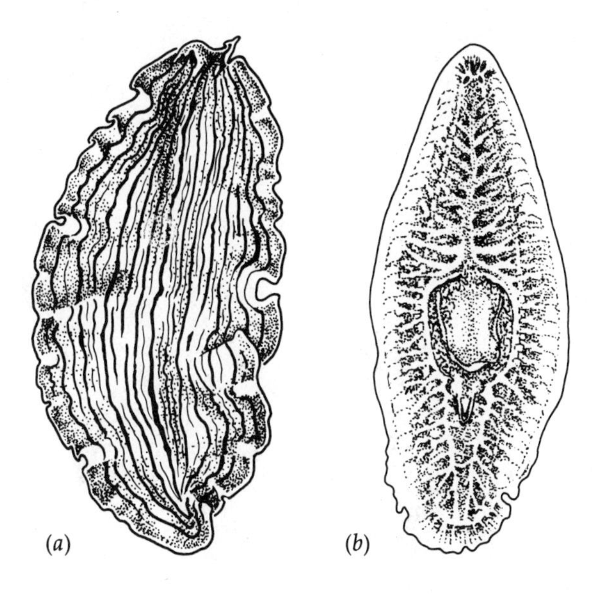

5–23 *Turbellarians:* (*a*) Prostheceraeus, *free-living and tropical;* (*b*) Bdelloura, *commensal on the gills of horseshoe crabs.*

living and feed on detritus. However, many, perhaps most, nematodes are parasites, and the incidence of parasitism of marine vertebrates and larger invertebrates by nematodes is of considerable ecological significance.

The Nematomorpha contain only a single marine genus. Its juvenile stages parasitize crabs, and its adult is a long, slender, planktonic worm with the general aschelminth features (Fig. 5–22). It is distinguished by having two rows of swimming bristles and has only one gonad (organ that produces reproductive cells).

Platyhelminthes

The flatworms are of great evolutionary interest because they "introduce" a number of features that are significant in the adaptation of all higher groups. For instance, flatworms have bilateral symmetry (two halves that are mirror images), three embryonic cell layers, a system of muscles, an excretory system, and a simple central nervous system. *Dugesia* (often called a planarian), a small brownish-black fresh-water turbellarian, is a common subject for regeneration experiments. The nature of the platyhelminth nervous system is such that biologists consider it the "lowest" form useful in memory and learning experiments. The circulatory, respiratory, and skeletal systems are not developed, and the animals are flattened top-to-bottom. Flatworms have a variety of life styles—they may be free-living, crawling about on mucky bottoms or rocks; they may be external parasites, such as the species that lives on the shell of the horseshoe crab; they may be internal parasites that have a complex life cycle in which various stages live on different hosts.

Most members of the class Turbellaria are free-living (Fig. 5–23a). The mouth, located on the underside of the body, takes food into the intestine and also serves as the exit for undigested material. Turbellarians can reproduce both asexually by regeneration (growing a complete individual from a fragment of another individual) and sexually. Most are bisexual, but they crossfertilize through a copulatory mechanism. Most turbellarians are small and transparent, though some are up to 50 cm long, and some are brown, black, or brightly colored in warm marine waters.

The Trematoda (flukes) are a class of internal and external parasites. Marine species live primarily on fish and molluscs. These flatworms are equipped with

5–24 Tubulanus, *a nemertine or ribbon worm. It lives in a mucus tube under stones or in crevices at low tide levels. Several feet long, this species is common on the European coast.*

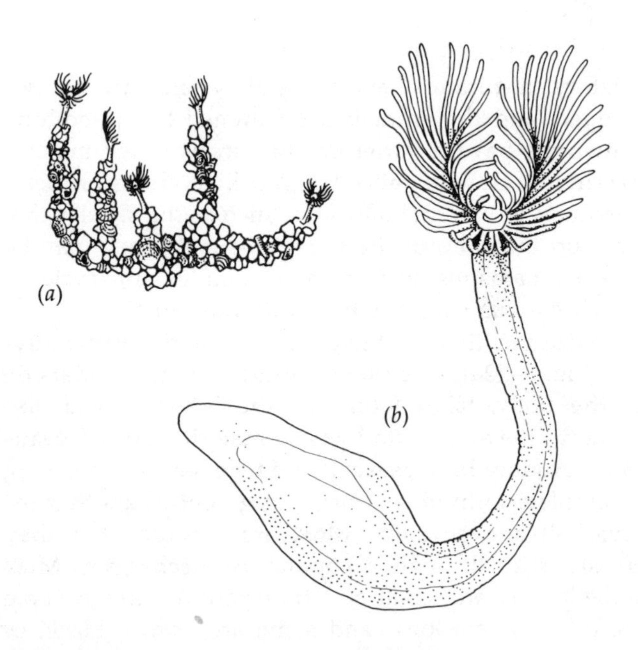

5–25 *Phoronids. (a) A cluster of* Phoronis psammophila *in tubes coated with sand grains and small shells. (b)* P. architecta, *a young adult removed from its tube.*

large adhesive organs or suckers. They ingest the blood, mucus, and tissue of the host.

The class Cestoda includes a small group of non-segmented worms that parasitizes primitive fishes, and the large group of true tapeworms. The adult tapeworm has a head with hooks and suckers for attachment to the host and has many segments—some forms grow to be several feet long. Adults are intestinal parasites of vertebrate animals; larvae may find intermediate hosts among invertebrates and vertebrates. Cestodes are highly specialized parasites having no digestive system; they absorb nutrients from the fluids surrounding them in the body of the host. However, each of the possibly hundreds of segments has a set of both male and female reproductive units. Sexual reproduction may take place in the same segment, between segments of one worm, or between segments of two different worms. Survival is well assured, for by these mechanisms huge numbers of larvae can be produced.

Rhynchocoela

The rhynchocoels, ribbon or nemertine worms, like the flatworms, do not have a coelom (membrane-lined visceral cavity). They do, however, have a complete digestive system and an anterior region that can be projected outward—a proboscis. The proboscis is used for food capture, burrowing, and protection. Because they have a proboscis nemertines have developed more sophisticated ways of dealing with the environment than have flatworms, so they do not need to be parasitic to survive. The nemertines are free-living, usually on muddy or sandy marine bottoms beneath rocks and shells (Fig. 5–24). They may be several feet long and brightly colored—startling the observer who turns over a mass on a coral reef to reveal a nemertine.

The Lophophorate Phyla

The phyla Phoronida, Ectoprocta, and Brachiopoda are coelomate animals that have a lophophore—a fold of the body bearing tentacles that surrounds the mouth but not the anus. The lophophore is used to create water currents that bring food particles to the lophophore to be filtered. All phoronids are marine, and live just below the water's surface in mudflats. They secrete tubes around their bodies, and the tubes may form masses attached to the substrate (Fig. 5–25). Muscular, digestive, nervous, and excretory systems are well developed. Phoronids have a closed circulatory system (one in which blood is completely enclosed in vessels) but no heart. They can reproduce both sexually and asexually.

The ectoprocts, or bryozoans, are widely distributed in marine and fresh water. They are usually colonial and the colony is attached to rocks, vegetation, or debris. Ectoprocts, too, usually secrete a case around their bodies (Fig. 5–26). In contrast to phoronids, ectoprocts have no circulatory and excretory systems, and are bisexual. They may also reproduce asexually by budding. Colonial life is stressed, and members are modified variously for feeding, cleaning, reproduction, and other maintenance tasks. The lophophore is used for filter feeding.

Brachiopods have a very long fossil history. The lampshells were once very abundant, but now few species remain. They are marine, sessile forms that live from the tide level into the deeps. They attach by a stalk to the substrate. They differ from phoronids and ectoprocts in having a shell with two valves (Fig. 5–27) and an open circulatory system (one in which vessels open into body spaces) with a heart. They are related to the other two phyla because they have a lophophore used as a filter-feeding mechanism.

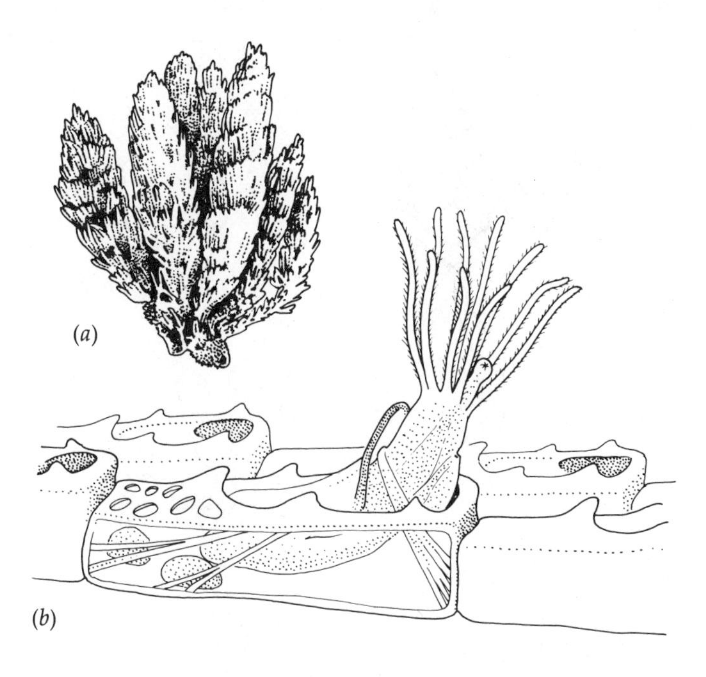

5–26 *(a) A colony of the bryozoan* Bugula, *found in Pacific coast tide pools. Colonies are up to 15 cm high and purple. (b) An extended zooid of the encrusting bryozoan* Electra. *The anus is outside the ring of tentacles—hence the name ectoproct.*

Entoprocta

The entoprocts are sessile, colonial organisms that live primarily in marine waters. They do not have a true coelom. They have a circle of ciliated tentacles for food capturing but this structure is not a lophophore, because it does not have the fold-like morphology and because it surrounds both mouth and anus. Since, in general, they are structurally similar, entoprocts were lumped with ectoprocts in a phylum Bryozoa, but the major differences in morphology have made that arrangement invalid. Entoprocts attach to solid substrate by a long stalk (Fig. 5–28), and rarely secrete a case. New members are added to the colony by sexual reproduction or by budding.

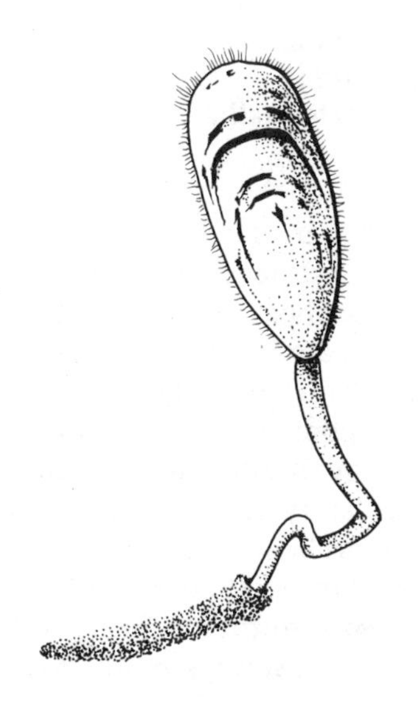

5–27 Glottidia *(a brachiopod) lives in a sand burrow.*

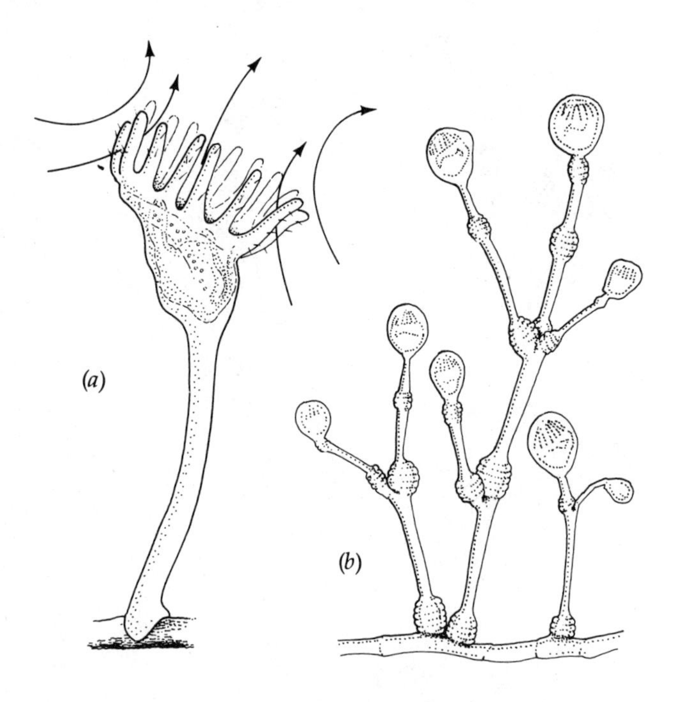

5–28 *(a)* Loxosoma, *a solitary entoproct. Arrows show the direction of water currents passing through the crown of tentacles. (b)* Gonypodaria, *a colonial entoproct.*

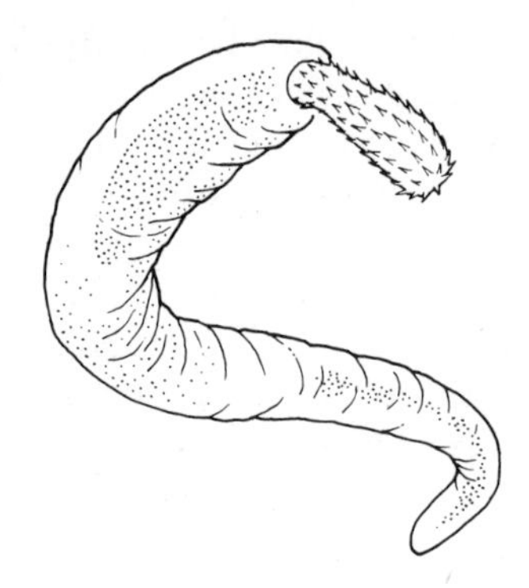

5–29 Leptorhyncoides, *an acanthocephalan parasite in the digestive tract of Atlantic fishes.*

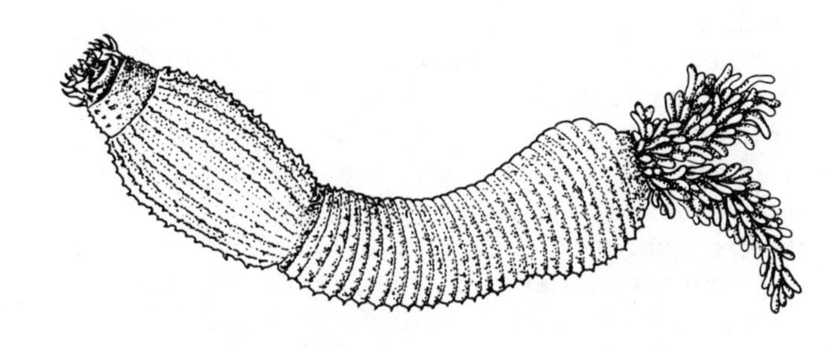

5–30 *A priapulid* (Priapulus).

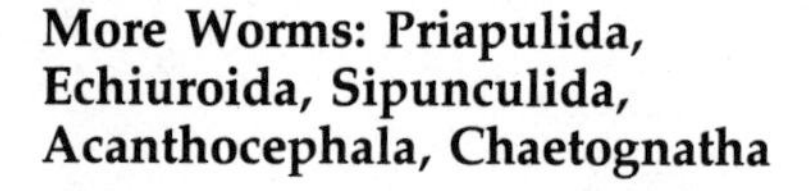

More Worms: Priapulida, Echiuroida, Sipunculida, Acanthocephala, Chaetognatha

There are a number of small phyla of worm-like marine organisms that differ in many respects from the worms already described. The Acanthocephala, or spiny headed worms, are internal parasites—the adults live inside vertebrates (mostly fish) and the juvenile stages inside arthropods. They are found throughout the world. The spiny headed worms do not have a true coelom, a circulatory system, or a respiratory system. They have a proboscis armed with hooks that attaches to the host (Fig. 5–29), but they have no digestive system and absorb nutrients through the skin, as do tapeworms. They do serious injury to their hosts.

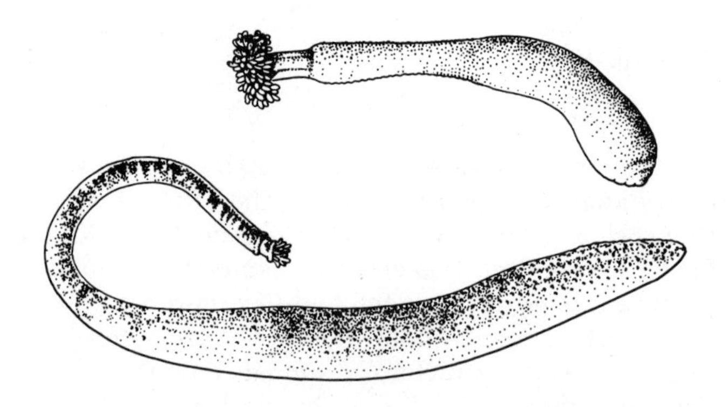

5–31 *Sipunculids.*

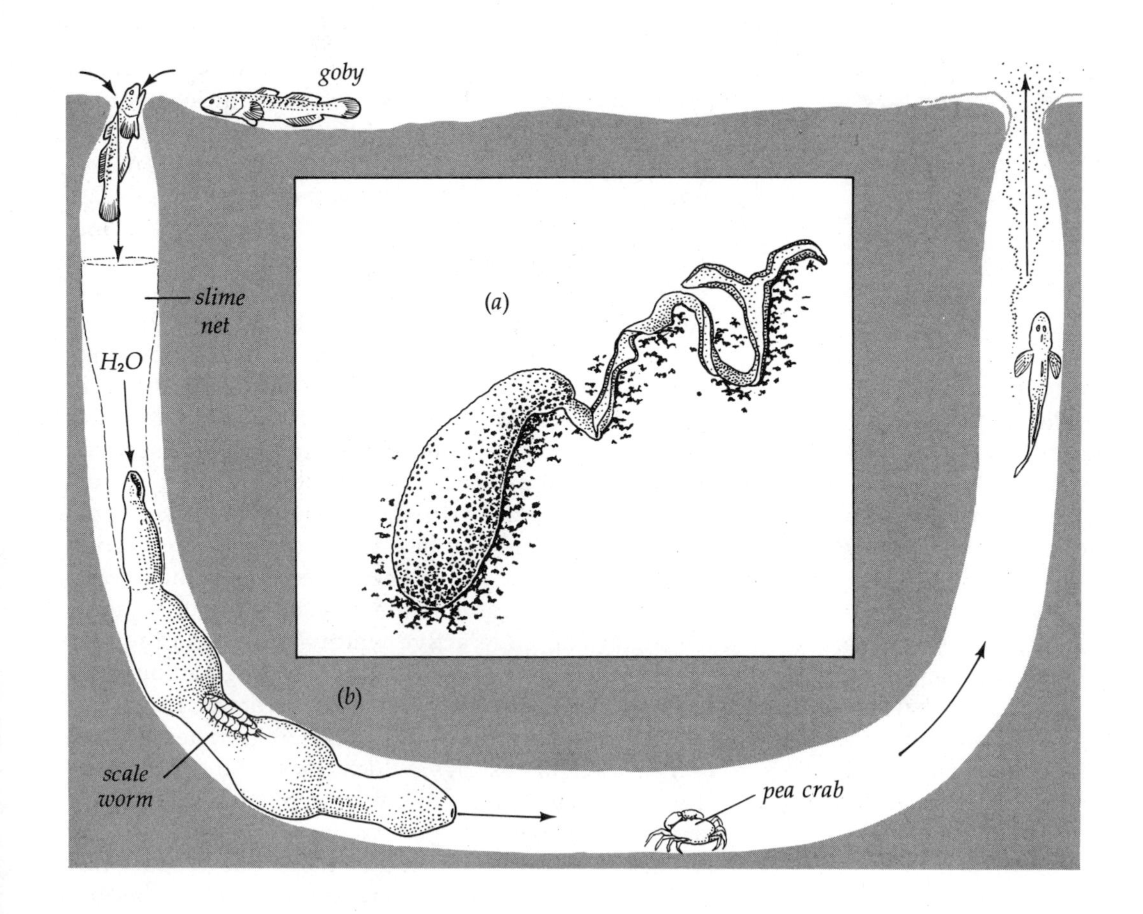

5–32 *Echiuroids. (a)* Bonellia *female; (b)* Urechis, *the fat innkeeper, and the commensals that live in and around the burrow.*

The phyla Priapulida, Echiuroida, and Sipunculida are all "worms" with coeloms that live in burrows in sand and mud. Most live in shallow water, though some occur in very deep water. Priapulids were long thought to be relatives of aschelminths, but the presence of a true coelom is a major distinguishing character. Priapulids are active predators and the proboscis is armed with teeth (Fig. 5–30). Little is known of their embryology, and that knowledge would probably aid in determining the evolutionary placement of the phylum.

Echiuroids and sipunculids are filter or detritus feeders. They ingest particles of food from detritus carried by water currents to the proboscis. Aspects of morphology and embryology relate both phyla to the annelids, but the two phyla are not closely related to each other. Sipunculids are often called "peanut worms" because those that dwell in rock crevices contract to a shape that looks like a peanut (Fig. 5–31). This contraction of the head end is important for defense and during feeding. The echiuroid genus *Bonellia* is of great biological interest. It shows great sexual differences—the females are up to 1 m long; males are about 1 mm long, and are parasites on (or inside) the females. In fact, sex determination is made by that relationship. Larvae that develop independently become females; larvae that attach to the proboscis of a female become males.

The echiuroid *Urechis* (Fig. 5–32) lives in a permanent burrow that has two entrances. It does not leave the burrow unless the burrow is destroyed. *Urechis* traps small food particles in mucus, and larger particles fall out into the burrow. These large food particles are eaten by commensal organisms that also live in *Urechis'* burrow. A polychaete worm, two species of pea crabs,

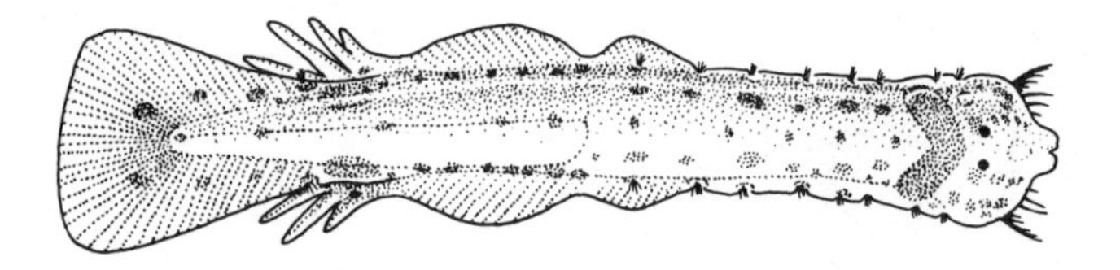

5–33 *A chaetognath* (Spadella).

a small clam, and, at the entrance, a goby fish enjoy *Urechis'* hospitality. *Urechis* provides them with protection from enemies in the burrow, pulls in water currents laden with food and oxygen, and even separates out food particles that are eaten by the commensals. In addition, *Urechis* tolerates species of internal parasitic protozoans.

The arrow-worms (Chaetognatha) are free-swimming, transparent worms that live near the surface of offshore waters. Many are less than an inch long. They have hair-like structures around the mouth that help them grab and seize prey (Figs. 5–33 and 5–5), and they are very active swimmers. They eat small animals in plankton, and in turn, are eaten by coelenterates and other planktonic animals.

Annelida

The phylum Annelida includes animals that have refined the "worm-like" construction. The body plan is based on a series of segments that have much the same arrangement of organs in each unit, a plan called metamerism. One great advantage of metamerism (also found in arthropods and chordates) is the promotion of controlled movement. Muscular movement compresses the fluid in the coelom differentially in each segment, lengthening, contracting, or curving it. There are three classes of annelids. Oligochaeta are primarily terrestrial and fresh-water, and include the "typical" earthworms; Hirudinea, or leeches, are primarily parasites of fresh-water and terrestrial animals, though some species are marine, parasitizing sharks and rays. Polychaeta are mostly marine, so let us consider them in greater detail. They have a distinct head that has eyes and tentacles. Respiratory, excretory, nervous, and circulatory systems are present and well developed. Each segment has a pair of sideways projections called parapodia (Fig. 5–34). These are used in locomotion, reproduction (for gamete discharge), respiration (via gills), and in sensing the environment. There are two main groups, those that move about actively (subclass Errantia), and those that live in tubes and expose only their heads (subclass Sedentaria).

The tubes built by Sedentaria are highly variable (some Errantia build them too, but either leave them for periods of time or carry them about). The tubes range from simple hardened mucus "shells", to those in which mucus secretions cement sand particles together to form the tube, to tubes of calcium carbonate secreted by special glands. The tube dwellers are filter feeders and capture food on their tentacles. The microscopic food particles are trapped in mucus and are carried down ciliated grooves on the tentacles into the mouth. Food then is processed in the long, unsegmented digestive tract. The more active worms can extend the proboscis and suck in food or swallow huge amounts of soil and extract the organic material from it. Those that take in large quantities of indigestible material often excrete piles of castings, which are found on the substrate. Respiration is highly developed, and the gills or other respiratory structures are located on the head or the parapodia. The blood contains one or more respiratory pigments, which are sometimes in cells but usually free. Reproduction is of interest in these worms. Sexes are separate, and when the male and female gametes mature, the worms may gather and swim about, shedding gametes into the water. Fertilization then takes place outside of the worms' bodies.

These worms are incredibly diverse in morphology and life style. They may be minute or very long, nondescript or bright red, green, or orange, simple or complex in structure. Exploration of the sandy bottom can mean the discovery of a tube from which a head with a bright red crown of ornate, branching tentacles is waving in the water current.

Echinodermata

The Phylum Echinodermata includes classes Asteroidea, the starfishes; Ophiuroidea, brittle stars; Echinoidea, sea urchins and sand dollars; Holothu-

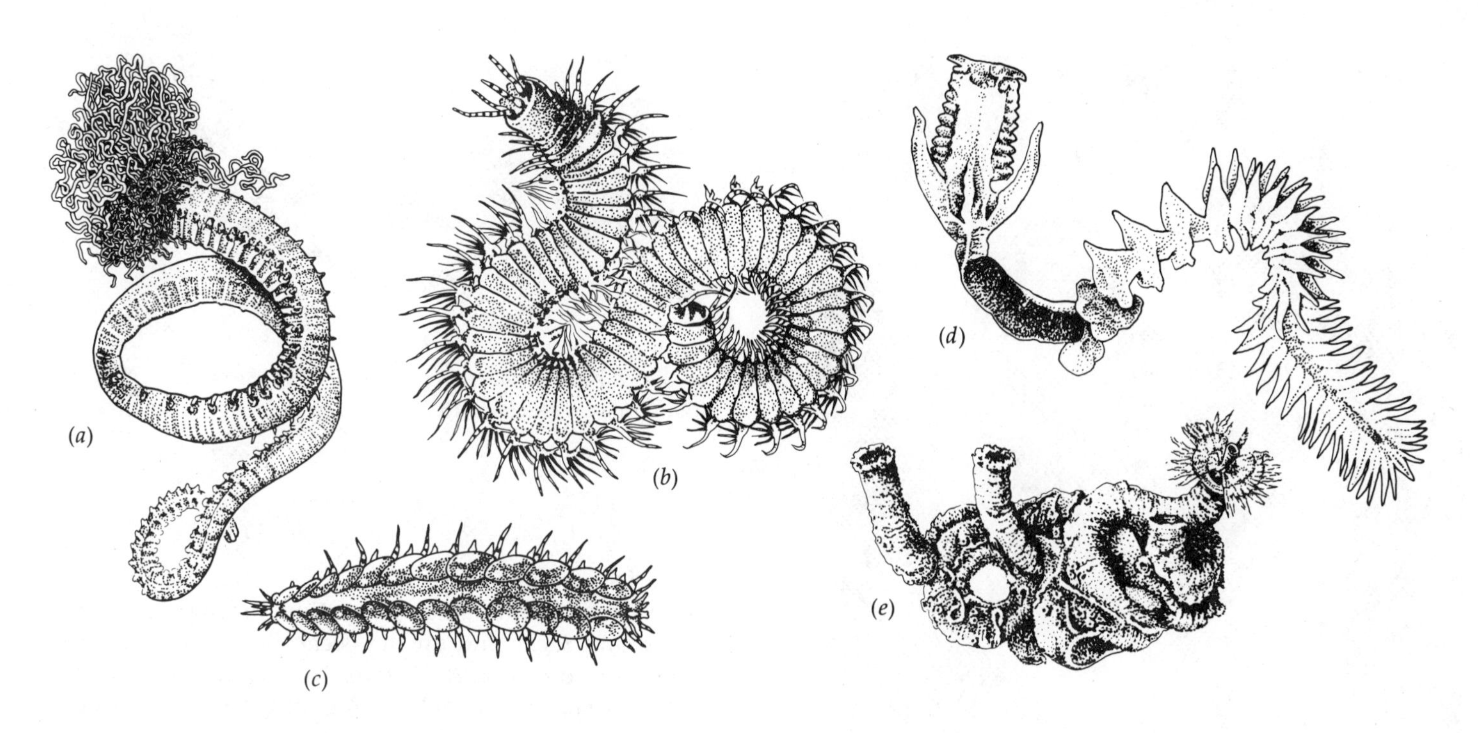

5–34 Diversity among polychaetes. (a) Amphitrite; *(b)* Eunice, *an active crawling and swimming predator; (c)* Lepidonotus, *a scale worm, free-living or commensal; (d)* Chaeopterus, *a tubeworm that produces water currents with its fans through a U shaped burrow; (e)* Serpula, *which lives in limey tubes on the sides of rocks.*

roidea, sea cucumbers; and Crinoidea, sea lilies. All echinoderms are radially symmetrical with a central mouth. They have a calcareous skeleton of plates or spicules. All are marine, and none are parasites. They are bottom-dwellers, from the deeps of the oceans to tidepools. Their mode of development makes them more closely related to chordates than to other invertebrate phyla.

The Asteroidea include the familiar starfish. Most have five arms (Fig. 5–35), though one species has four and one has forty. The mouth is on the under surface, the anus on the upper. A starfish moves around by forcing its tube feet, which end in sucking disks, out of grooves on the arms. These are exserted and retracted by water currents forced through channels in the body —the so-called water vascular system. Tube feet on the ends of the arms are sensory and help to locate food; some tube feet are respiratory. Some starfish are carnivores and feed on clams, mussels, oysters, and other molluscs. The starfish wraps itself around its prey, forces open the shell, and extrudes its stomach around the prey. Digestive juices dissolve the flesh, and the resulting fluid is ingested by the starfish. Sexes are usually separate in starfish, and eggs and sperm are spawned into the water. The larvae are free-living and planktonic for two weeks to two months.

The body pattern of echinoids is similar to that of starfish, but without "arms." Sea urchins have spines that hinge to the skeleton (Fig. 5–36). The spines may be used in locomotion, but tube feet are the primary means of movement. Some species live on rocky shores, others on sandy, muddy, or ocean bottoms. Sand dollars are flattened and some forms live standing on end, turned crosswise to the direction of tidal flow. When the tide goes out, the dollar flops down and covers itself with sand. Echinoids may use their spines to propel small food particles to their mouths.

The ophiuroids (serpent stars or brittle stars) have long arms around a small central disc (see Fig. 5–64). The arms are the primary means of locomotion, and

5–35 A bat star. (Photo by Steve Renick. Courtesy of Steinhart Aquarium, California Academy of Sciences.)

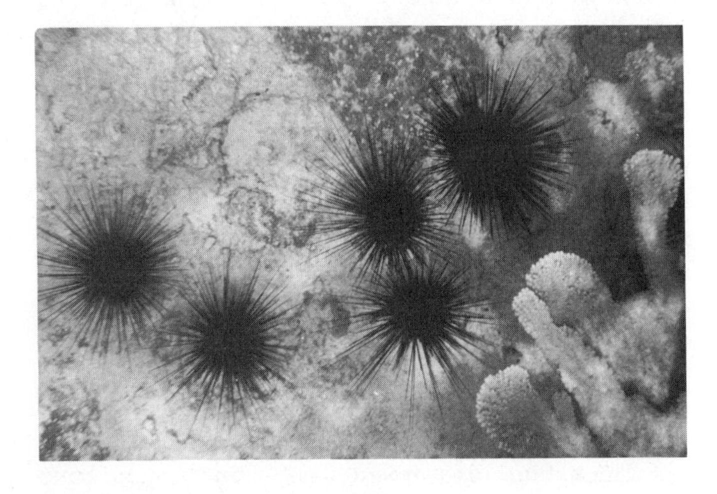

5–36 Sea urchins (Diadema) *from the U.S. Virgin Islands. (Courtesy U.S. Department of the Interior, National Park Service.)*

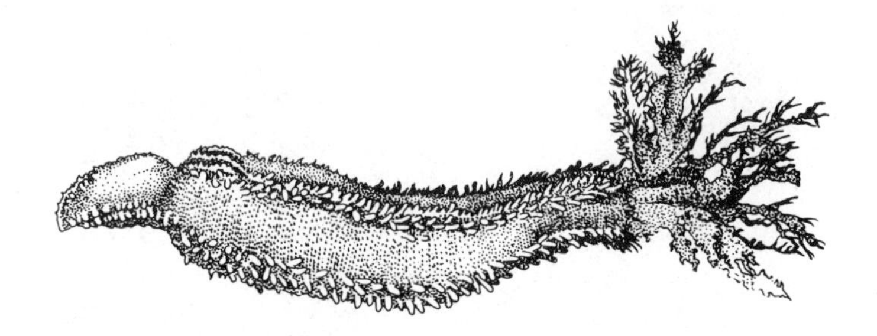

5–37 A sea cucumber (Cucumaria) *from the California coast (about 15 cm long).*

tube feet are used mostly for sensing and respiration. These fragile animals are secretive, hiding under seaweed and rocks and on the ocean floor. They live primarily on detritus. An interesting defense mechanism of brittle stars is that of breaking off an arm if disturbed. They are able to regenerate structures in a very short time, so leaving a predator holding an arm while the rest of the star escapes is quite effective.

The sea cucumbers, or holothurians, are soft bodied, elongate echinoderms (Fig. 5–37). The mouth is at the front end of the body, the anus at the back end. Sea cucumbers have five body regions that are homologous with starfish arms, and they have tube feet. They live on rocks, sand, or mud. They have tentacles that create water currents and trap small food items, such as diatoms, and detritus. Sea cucumbers have a large space, the cloaca, at the end of the digestive tract. Water currents are brought into the cloaca, and oxygen is distributed to the body from there through the so-called "respiratory tree"—an unusual respiratory mechanism. The cloaca also houses commensals, among them crabs and fish. Their reproductive biology seems to be that typical of echinoderms, as described for starfish.

Sea lilies, or crinoids, are abundant deep sea animals. They have the echinoderm rayed body plan, with feathery arms surrounding a mouth on the upper surface. They may have a stalk that anchors them to the bottom (Fig. 5–38), especially in younger animals. They feed on detritus or small organisms. Many fossil species are known, and they were abundant in the sea during Paleozoic times. Fewer species are alive today, and, unfortunately, little is known of their way of life.

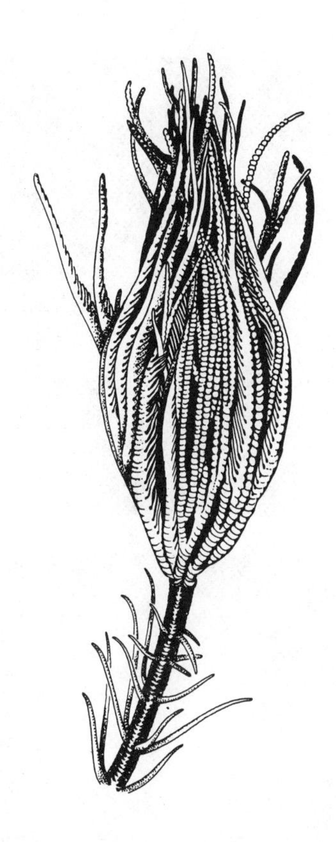

5–38 A deep sea crinoid.

Mollusca

The phylum Mollusca is a large and variable one. Their diversity is well known to shell collectors, for nearly all of the shells that wash ashore are those of molluscs. The animal that lives in the shell is bilaterally symmetrical, usually with a developed head region, has a ventral muscular foot for locomotion, a complete digestive tract with a mouth that has a rasping organ, a circulatory system that includes a heart, a well-developed nervous system, kidneys for excretion, and usually gills for respiration. The body is contained in a fleshy mantle and usually is sheltered in a calcareous shell. Most species of molluscs are marine and live on or under rocks, sand, or mud. They are found in all depths of the ocean. There are five classes of molluscs: Amphineura, the chitons; Pelecypoda, the clams, mussels, oysters, and scallops; Scaphopoda, the tooth shells; Gastropoda, the snails and sea slugs; and Cephalopoda, the squids and octopuses.

All of the Amphineura, or chitons, are marine forms. They have a broad, flat, muscular foot by which they crawl about and attach firmly to solid substrate. Their shell is on the back and usually has eight plates (see Fig. 5–64), though some species have only one or two plates or none at all. Most chitons scrape minute algae and diatoms from rock surfaces to obtain food, though some are detritus feeders. Their shells are often covered with algae, hydroids, or bryozoans.

The pelecypods have a double shell with right and left halves. Each pelecypod species has a different shell shape or color, and in most cases, each of the 11,000 living and 15,000 fossil species can be identified by shell alone (Fig. 5–39). They have a strong muscular foot that is usually used for digging. The foot may also be used to fasten the animal to its substrate, though some pelecypods, such as mussels, secrete "byssal fibers" to anchor themselves in place. Respiration occurs as water is pumped in through one "siphon," over the gills, and out another siphon. Pelecypods use the water current to obtain food, as well—they trap food particles in sheets of mucus over the gills, and then transport the food-laden mucus to the mouth. Pelecypods are very important in the marine food web. They are eaten by many other forms, from walruses to starfish. One pelecypod, the shipworm (which is really a clam), does millions of dollars of damage each year to wooden ships, pilings, and other structures. Why they have gone from efficient filter feeding to digesting wood is an unanswered biological question.

The scaphopods, or tooth shells, live in sand or mud below water level, and occur at all depths. Their shell is a tapered tube that is open at both ends (Fig. 5–40). The foot is extended through the larger opening to dig in the sand. Scaphopods have no heart or gills. They feed on small plants and animals living in the sand.

Gastropods include snails, slugs, abalone, limpets, cowries, nudibranchs, and many other forms. They typically have a distinct head with eyes and tentacles, and have their bodies twisted into a spiral shell. These animals, too, can be distinguished at the species level by shell structure, except for the forms whose shells are very reduced or lost. Molluscs extract calcium from seawater to build their shells. The shell is modified according to the life style of the animal (Fig. 5–41). The snail has a thick, protective shell; the limpet's shell is flattened and fitted to provide less resistance to wave action; the adult abalone has a flattened shape (see Fig.

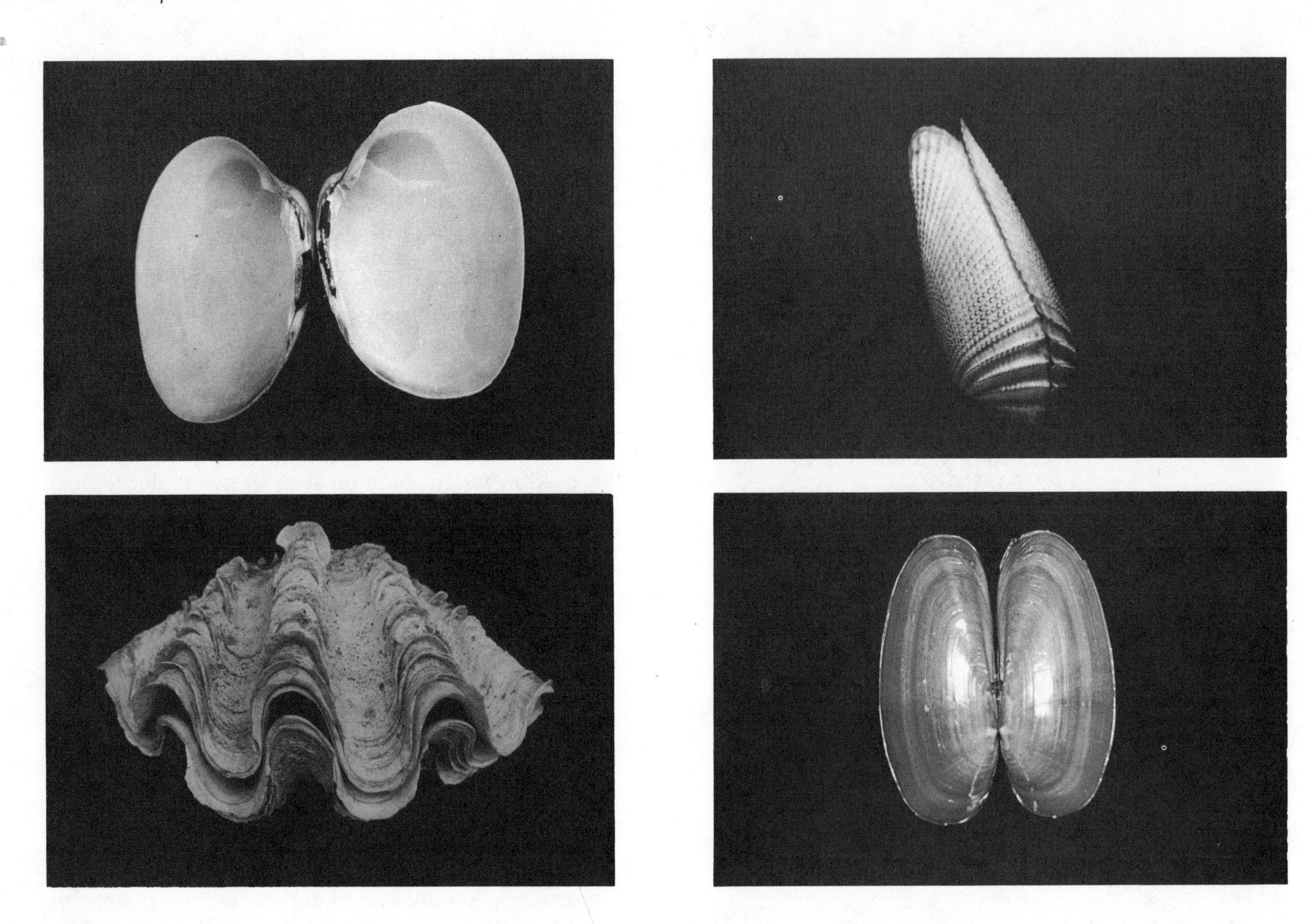

5–39 Various pelecypod shells. (Photos by Steve Renick. Courtesy of Steinhart Aquarium, California Academy of Sciences.)

5–64) that allows its broad foot to produce the great force that holds the abalone onto a rock. The foot is also used to glide along the substrate. Gastropods, too, have a rasping organ that is used to scrape the algae film from rocks in order to feed. Some gastropods, however, are among the few marine animals that feed on the larger algae. The sea hare literally "crops" great masses of algae. Some snails are scavengers, feeding on the dead flesh of many animals (performing a valuable cleaning function). Others prey on clams, barnacles, or other gastropods. Birds, fish, and man also use gastropods as food.

The shell-less gastropods are interesting simply because they have so many beautiful forms. The nudibranchs and tectibranchs (Fig. 5–42) are often brightly colored and patterned; they often have swimming flaps extending from the foot, and have tentacles or fingerlike projections for respiration. These structures all contribute to the unusual appeal of these gliding, brightly colored animals.

The cephalopods include the familiar octopus, squid, and nautilus. Most have a well-developed head, often with eyes, and the foot is divided into eight or ten arms that have suckers for locomotion and prey-

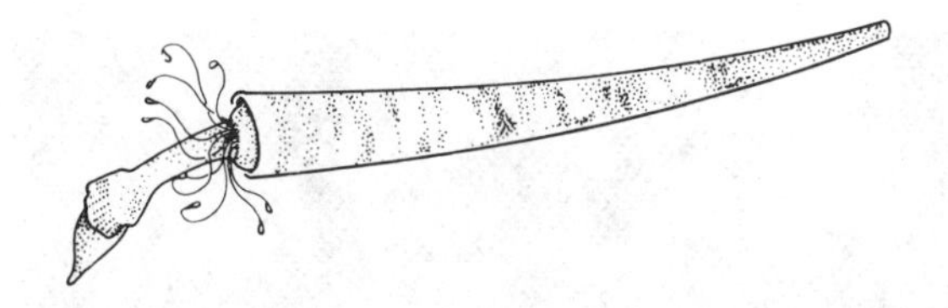

5–40 Dentalium, *a tooth shell.*

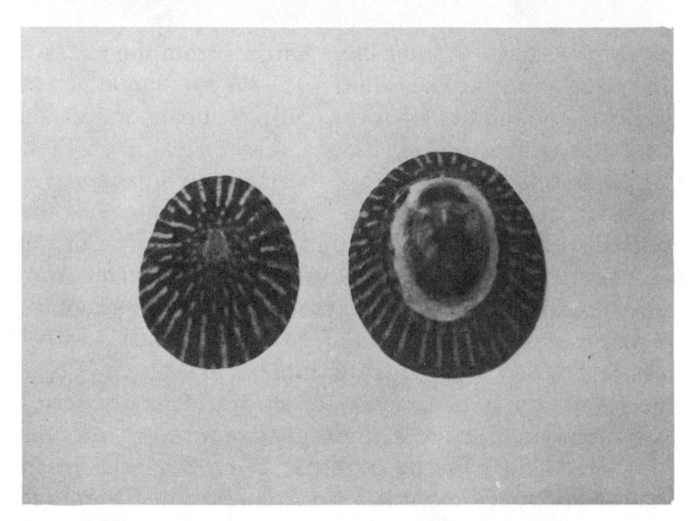

5–41 *Snail and limpet shells. (Photos by Steve Renick. Courtesy of California Academy of Sciences.)*

5–42 *A nudibranch. (Photo by Penny Hermes.)*

grabbing (Fig. 5–43). Only *Nautilus* has a large external shell (Fig. 5–44); in other genera the shell is internal, often greatly reduced or even lost in some forms. The mantle cavity contains the gills, and a mouth with biting mouthparts is below the "head." Locomotion is via movement of the arms, or by taking water into the mantle cavity and forcibly ejecting it through the siphon. This "jets" the animal through the water. All octopuses and squids prey on other animals, including fish, worms, other molluscs, and crustaceans. A number of species have poison "glands" in the mouth that secrete a substance to kill the prey, which is then torn apart and eaten. In turn, cephalopods are preyed on by moray eels and some pelagic fish, and by man. The cephalopods live in crevices in rocks, caves, under rocks, on the ocean bottom in the deeps, and in open water.

Molluscs reproduce by laying eggs, either before or after fertilization; the larvae hatch and swim about freely, often near the water's surface; then the nearly adult form settles to the bottom and completes the change to adulthood. Some octopuses are particularly interesting, for they exhibit maternal behavior. They guard their eggs and brood them, refusing to eat during the brooding period. The mothers clean the eggs to prevent fungal growth and fight off all approaching animals. It is truly a means of assuring survival—the female may lay 45,000 eggs, and with this meticulous care, nearly all will hatch. However, numbers of larvae are eaten by fish shortly after hatching, so relatively few members of each "litter" reach adulthood.

5–43 A 20 kg Pacific octopus. The more than 200 sucking discs are used for locomotion, food capture, and manipulation. Each disc reportedly exerts four ounces of pressure. (Courtesy of Scripps Institution of Oceanography, University of California, San Diego.)

Arthropoda

The amazingly successful arthropods (jointed feet) have a number of marine representatives, most of which are in the class Crustacea. Crustaceans have undergone a great adaptive radiation in the oceans. They all have an external skeleton, jointed legs, a body with several segments, jaws, and antennae. They respire through gills and have an open circulatory system.

The subclass Copepoda (Fig. 5–45) contains many species of minute animals. They are very important in the food web, for they eat diatoms in plankton, and are in turn eaten by larger animals. They are an abundant and rich food source. Copepods live in plankton or in the warm water of tidepools. Many copepods are commensals of molluscs and coelenterates; others are parasites on the skin or gills of fish.

Barnacles are members of the subclass Cirripedia. Though they look little like copepods, their embryology is similar, and a similar larva hatches from the egg. All barnacle adults are sessile (Fig. 5–46) and can attach to almost any substrate—rocks, pilings, boats, the shells of other animals. They have a cement gland on their first antenna that produces an adhesive substance. A barnacle feeds by extending its cirri (homologs of the legs of other crustaceans) through the opening of the divisions of its shell. The cirri sweep the water, trap small particles of food, and stuff them into the mouth. Starfish, worms, and snails feed on barnacles, and occasionally fish and birds do, too. Several kinds of barnacles are parasitic on crabs or shrimp. Parasitic forms are often small, have many organs reduced disproportionately, and have no digestive tract. Barnacles often have very specific habitat preferences when they live freely. They form great clumps, but some species occur in highly prescribed strips according to the depth of the tide.

The subclass Ostracoda includes tiny crustaceans that are enclosed in a bivalved shell (Fig. 5–47). They

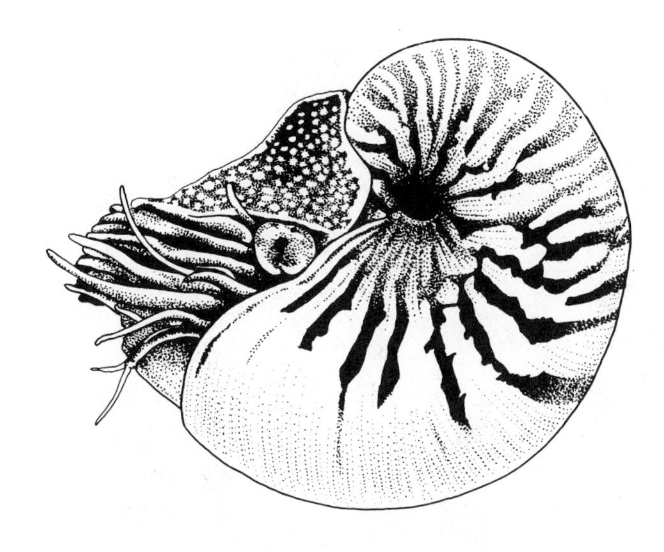

5–44 *A chambered nautilus* (Nautilus macromphalus) *from New Caledonia.*

5–46 *Stalked barnacles, with the cirri exposed. (Photo by Steve Renick. Courtesy of Bodega Marine Laboratory, University of California, Berkeley.)*

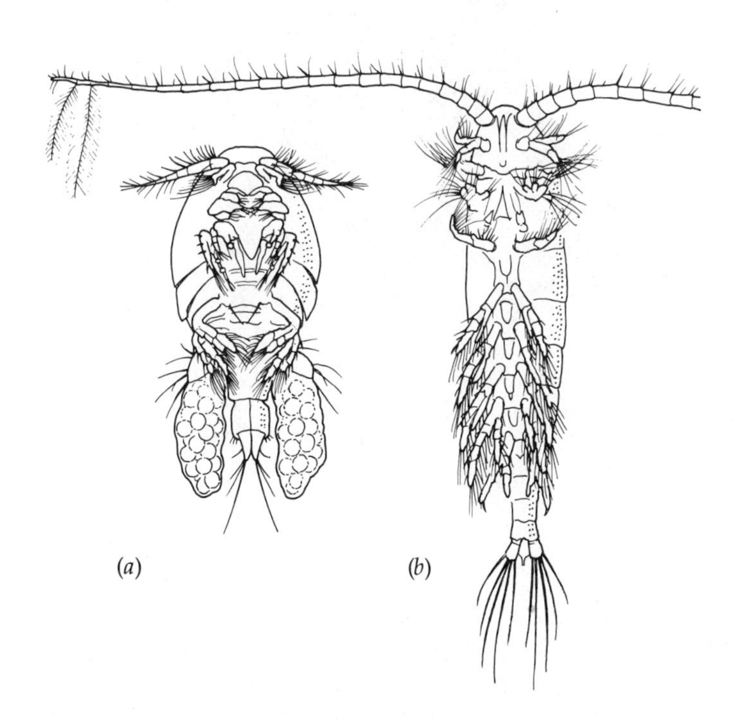

5–45 (*a*) Clausidium, *a copepod commensal on the surface of some kinds of burrowing shrimps.* (*b*) Calanus, *a planktonic copepod.*

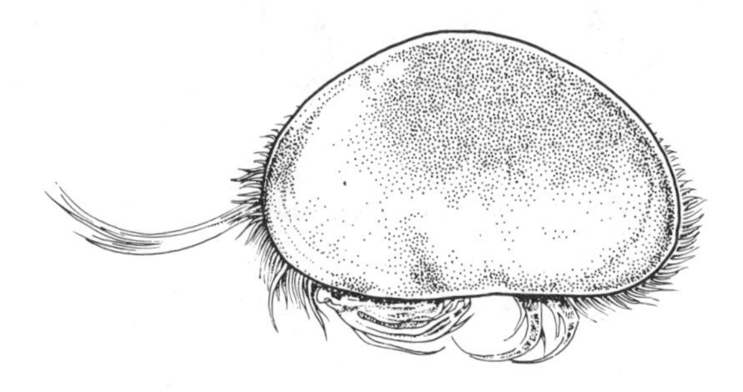

5–47 *An ostracod.*

use their antennae for swimming. Many species are planktonic, and some live on ocean bottoms.

Subclass Malacostraca includes several orders that have marine forms. Isopods (Fig. 5–48) are flattened, and have seven pairs of legs of equal size. They live on seaweeds and on rocky shores. They are scavengers, and you know that the flesh doesn't have to be dead if you've had your toes bitten by some. Some isopods are commensals, others are parasites on the gills and mouths of fishes. Amphipods are flattened from side to side (Fig. 5–49), as opposed to top-to-bottom in isopods. Many amphipods are predators that feed on smaller crustacean larvae. Some amphipods live on algae, in hydroid colonies, in tidepools, and on rocks.

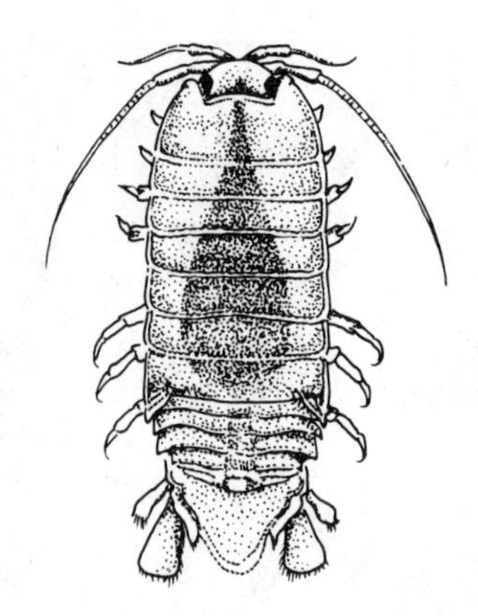

5–48 *An isopod* (Cirolana) *about 20 mm long.*

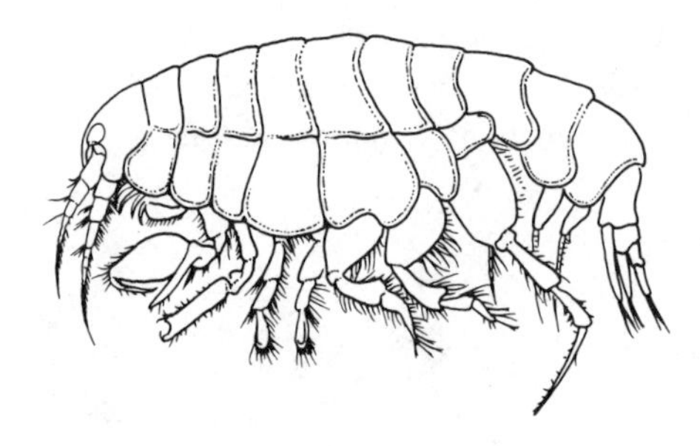

5–49 Pontocrates, *a pelagic amphipod about 16 mm long.*

5–50 *An American lobster* (Homarus), *which ranges along the east coast of North America. (Photo by Steve Renick. Courtesy of Bodega Marine Laboratory, University of California, Berkeley.)*

Others live on the ocean bottom and build themselves a tubular case, which they carry around with them. They cement grains of sand, mud, and debris together to form the case.

The Decapoda include all the crabs, hermit crabs, lobsters, and shrimps. They all have five pairs of legs, and their bodies have two regions—a head-thorax and an abdomen. There are three groups of decapods: the Macrura, or large tailed forms, such as shrimp and lobster (Fig. 5–50); the Anomura (odd shaped tails), such as hermit crabs; and Brachyura, short tailed or true crabs. The Macrura include all sorts of species that man eats. The commercial shrimp and lobster catches are million-dollar enterprises. Some shrimp occur in great numbers in schools, and can be netted easily. There are all sorts of other shrimps, however. Pistol shrimps (Fig. 5–51) live in tidepools, and use their pistol "hand" to stun prey that are swimming by; then they run out and eat the shrimp or fish thus caught. Some shrimp live in burrows, which they keep scrupulously clean.

The Anomura have shorter abdomens; they include burrowing shrimp, hermit crabs, stone crabs, and sand crabs. The burrowing shrimp are constantly maintaining their burrows and shift huge quantities of sand out of the burrow opening. The burrow houses many kinds

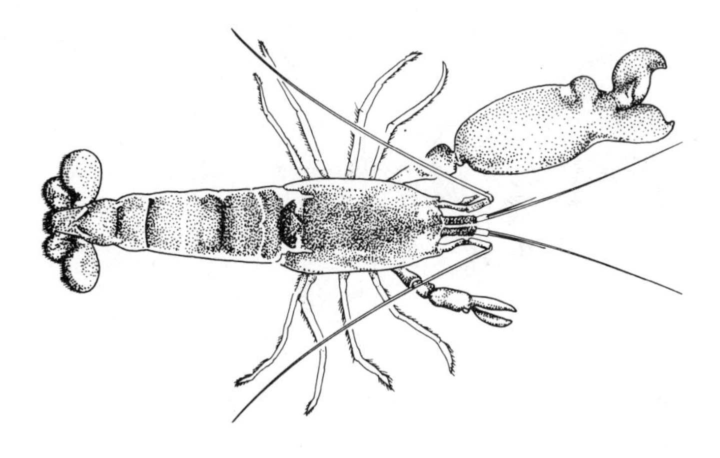

5–51 Cragon, *a pistol shrimp.*

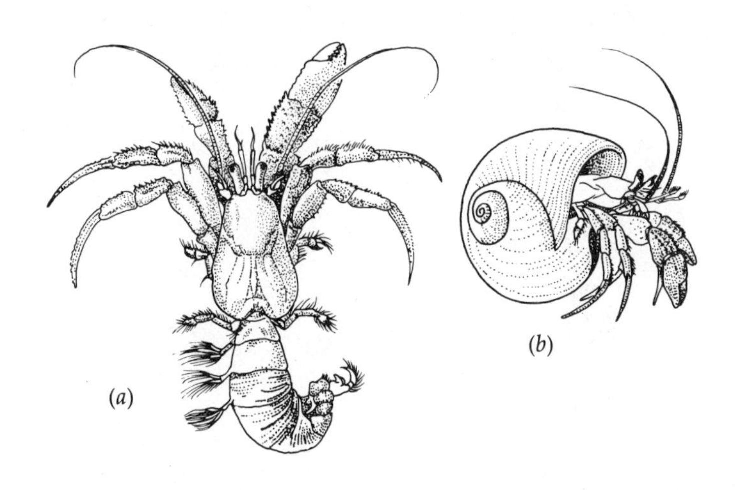

5–52 *A hermit crab* (Pagurus): *(a) removed from snail shell; (b) as it is usually seen, within an empty shell.*

of commensals—goby fish, pea crabs, copepods, a small clam, and scale worms—which all may occupy the same burrow for the protection it affords. Hermit crabs (Fig. 5–52) have soft asymmetrial abdomens, which they insert into the snail shell in which they live. The legs on their right sides are reduced because of the whorl of the snail shell. As a hermit crab grows, it must find larger shells in which to live. It will actively fight other hermit crabs to get a shell of the desired size. Many hermit crabs are scavengers, picking up bits of food as they scurry about, and some are detritus feeders that fan up mud with their claws and strain detritus from it. There are many commensals of hermit crabs—the shell may be covered with barnacles, sponges, or hydroids, and limpets and worms may live inside the shell.

The brachyurans have a large carapace (shell that encloses the head and thorax), and the abdomen is a small flap tucked under it. Their first pair of legs always has pincers. Some crabs eat detritus or plankton, although most are scavengers or predators. Spider crabs, blue crabs, fiddler crabs, and cancer or edible crabs are all brachyurans (Fig. 5–53). Many crabs seem to have a "masking instinct"—they cover their external skeletons with seaweed, hydroids, tunicates, or debris, much of which continues growing and further disguises the animal. The pea crabs (see Fig. 5–32) have been previously mentioned as common commensals. They get their name because they are about the size of a pea and live in all sorts of cavities, usually one crab on each host animal. This poses problems for mating, since females often grow so large that they cannot escape their hosts. Males, however, are usually very tiny, and migrate among hosts, so mating can be effected. Females stay in their refuges and produce many eggs. Shore, or grapsoid, crabs are squarish with eyes at the front corners of the head. They scurry about the rocks, mud, and seaweed. Omnivorous, they feed on dead flesh, small live animals, and the algal film on rocks. Fiddler crabs are also common at the shore. The males have one claw that is greatly enlarged and carried in front of the body. They live in burrows and emerge to get balls of mud, from which they eat the organic content.

The order Euphausiacea is a small one, containing tiny pelagic animals (see Fig. 5–5) whose legs are divided into two parts of equal length. They are the krill eaten in vast quantities by whales.

The order Stomatopoda includes the mantis shrimps, which live on the offshore bottoms in sand or crevices (Fig. 5–54). Mantis shrimps can be quite large, up to a foot long. They have eight pairs of legs modified as mouthparts, hence the name of the order. One leg ends in a large claw that they use to slash prey, such as other shrimp. Some mantis shrimp add to the beauty of the marine world—*Hemisquilla* is banded and spotted in shades of blue-green, orange, and gold.

(*a*)

(*b*)

(*c*)

5–53 Crabs. (a) Cancer. *(Photo by William J. Wallace.) (b) King crab* (Paralithodes) *of the northern Pacific. (Photo by Steve Renick. Courtesy of Steinhart Aquarium, California Academy of Sciences.) (c) A crab that lives among the gulf weed in the Sargasso Sea. (Photo by Penny Hermes.)*

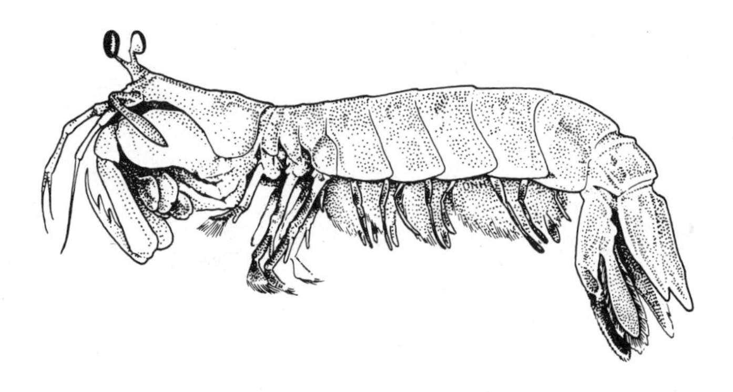

5–54 A California mantis shrimp, Pseudosquilla.

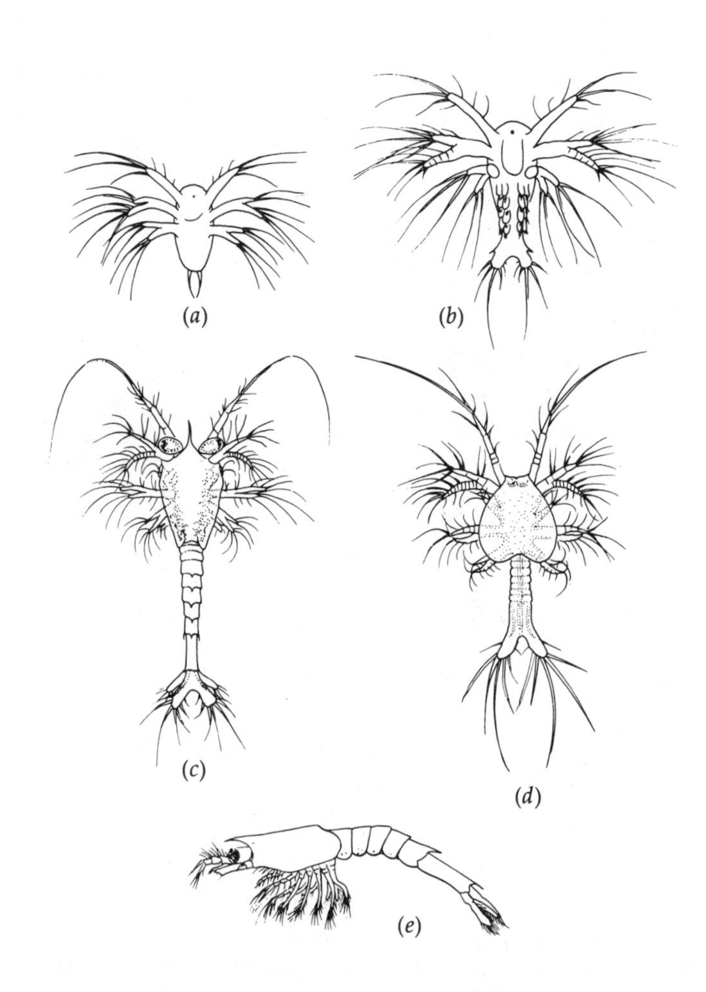

5–55 Larval development in the decapod shrimp Trachypeneus. *Each larval stage has a different name. (a) nauplius (0.26 mm long); (b) metanauplius (0.42 mm); (c) early protozoea (0.7 mm); (d) late protozoea (1.9 mm); (e) mysis (2.8 mm).*

5–56 Horseshoe crab (Limulus). *(Photo by William J. Wallace.)*

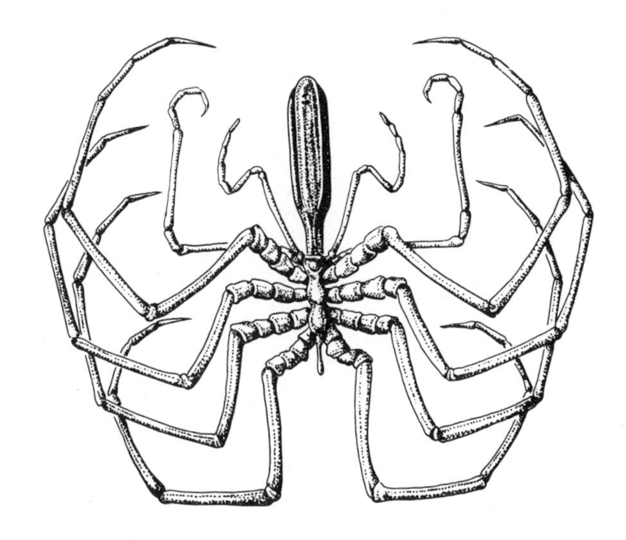

5–57 The pycnogonid Collosendea, *from the North Atlantic.*

Development in crustaceans is significant, for the adult and the larva fill different niches in the marine environment. Many crustaceans brood their eggs until they hatch, and some carry the eggs in sacs on their legs. On hatching, many forms have a lengthy larval period that is free-swimming and often planktonic. The larvae are preyed on by different organisms than the adult, eat different food (usually microorganisms), and are generally subject to very different environmental pressures. As the larvae grow, they must shed their external skeleton and grow a new one. This process, called molting, must take place many times during the growth and development of the organism. The larvae go through many stages before adulthood, each having a specific morphology and its own name (Fig. 5–55). Crustaceans that are more evolutionarily advanced undergo the earlier stages while still in the egg, but most spend some of their lifetime as a free-swimming larva before adopting the adult way of life.

Members of two other major classes of arthropods are marine. The Xiphosura, or horseshoe crabs, are interesting because of their long fossil record. These animals have an arched horseshoe-shaped carapace over the head-thorax region, and a wide unsegmented abdomen (Fig. 5–56). They eat worms and other small animals. They live in shallow waters on the east coasts of North America and Asia. The Pycnogonida, or sea spiders, are usually very tiny animals with small bodies and long legs (Fig. 5–57). They often live on algae and coelenterates, or in the mantle cavity of molluscs. They have sucking mouthparts and eat the tissue of the organisms on which they live.

The Protochordata

The Protochordata are three phyla of animals that have a notochord (a rod-like support structure), gill slits, and a hollow nerve cord along the back. These characters are shared with vertebrates, but protochordates lack a backbone and other truly chordate features. Some authors include a fourth phylum, the Pogonophora —a group of tube worms that has a similar method of coelom formation (Fig. 5–58). The Enteropneusta (Hemichordata), or acorn worms (Fig. 5–59), live in mud flats or on the ocean bottom. They have a long proboscis and ingest mud. They extract organic content from the mud and excrete casts of the nonusable materials.

The Tunicata, sea squirts and salps (Fig. 5–60), typically attach to rocks, floats, boats, and pilings. They have siphons through which they take in and expel water. Tunicates have a large cavity, called the branchial basket, that strains food from the water by trapping it in mucus. The sheets of mucus then are carried to the digestive tract. Tunicates feed on plankton or detritus. Some forms are colonial, and food-bearing water cur-

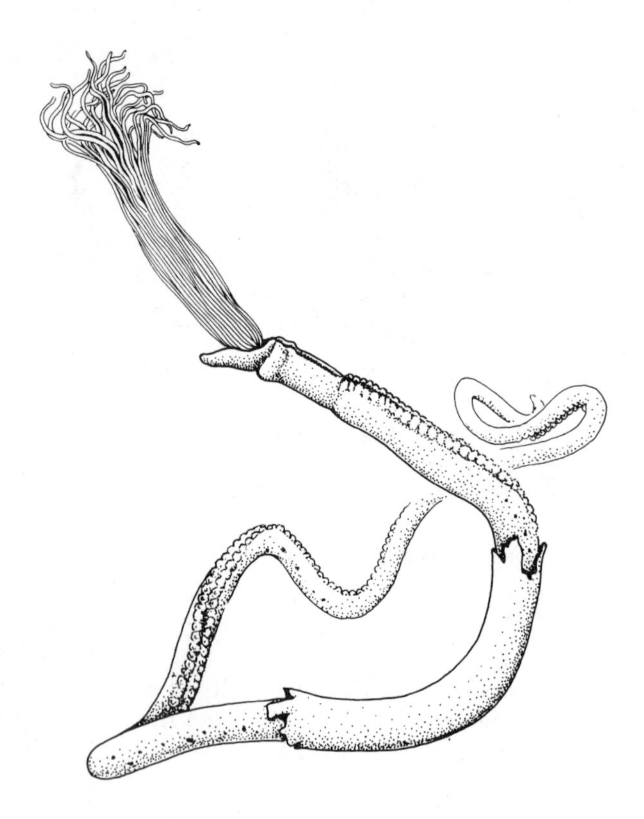

5–58 The poganophoran Spirobrachia, *dredged from the abyssal mud bearing a portion of its coiled cellulose tube.*

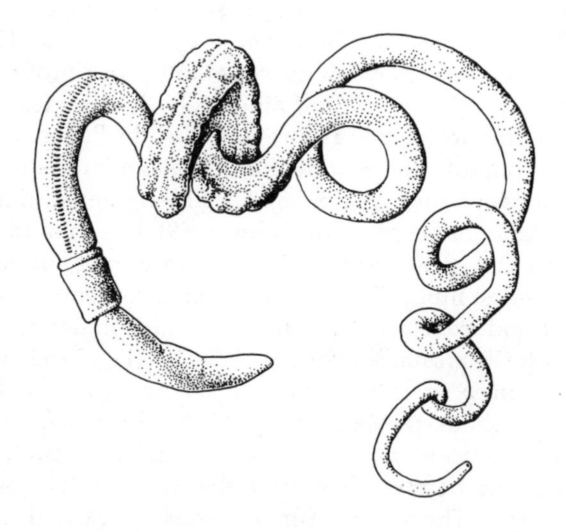

5–59 Saccoglossus, *an acorn worm.*

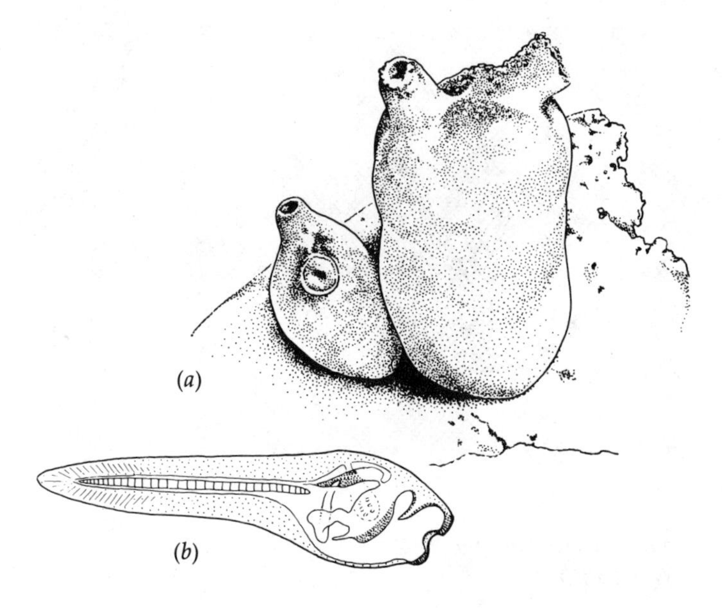

5–60 (a) A tunicate, the North Atlantic sea peach (Tethyum pyriforme); *(b) a tunicate larva* (Clavelina).

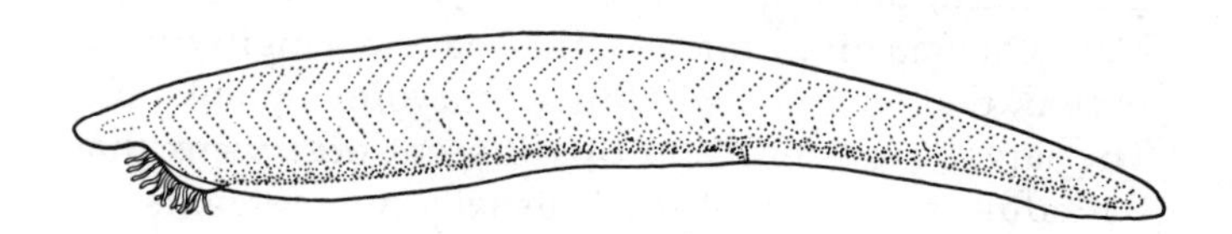

5–61 Amphioxus (Brachiostoma).

rents pass through many members of the colony. They form masses of brown, white, yellow, red, or transparent animals that cover their substrate. Their name is derived from the horny test or tunic that covers them, and the test itself forms a substrate for many animals. Sponges, bryozoans, mussels, tube worms, and other tunicates all fasten to individual animals.

The tunicate larva shows chordate characters. Eggs hatch into "tadpoles" with a long tail, gill slits, a notochord, and a nerve tube (Fig. 5–60b). The tadpoles swim freely for a time, then attach to the substrate and change into the sessile adult form, which shows little resemblance to a vertebrate.

The Cephalochordata, lancelets or amphioxus (Fig. 5–61), bury themselves near the surface on sandy beaches. They strain detritus and plankton from the water in much the same way that tunicates feed. The members of this group were long thought to be a "missing link" between vertebrates and invertebrates because the larvae have the characteristic gill slits, notochord, and tubular nerve cord, and the body form of the adults superficially resembles that of the jawless fishes.

Vertebrates

A great diversity of vertebrates lives on, in, or around the oceans. Various fish exploit nearly every niche possible in the sea; sea snakes and sea turtles, among the Reptilia, spend much of their lives on the ocean's surface, turtles going on land only to lay eggs; many birds live on sandy or rocky shores and fly far out to sea; a number of mammals use marine food sources, and some are adapted for life in a marine environment—sea otters, seals and sea lions, manatees and dugongs, dolphins and whales are all primarily or totally dependent on the oceans for maintenance of their ways of life.

The lampreys and hagfish, class Cyclostomata, are evolutionarily primitive cartilage skeletoned fishes without jaws. Lampreys have a large, circular mouth disc, with which they attach themselves to rocks during their migration from fresh to salt water, and to a host fish or mammal during adulthood. They are very effective parasites, using a rasping organ to cut through the flesh of the host. The migration is a seasonal one, for eggs are laid and fertilized in fresh water. Offspring migrate to the sea, then return to fresh water to reproduce. Hagfish are scavengers that subsist on recently dead fish.

The Chondrichthyes are cartilage skeletoned fish that have jaws, pectoral, dorsal, anal, and tail fins, and open gill slits. Sharks, skates, and rays are included in the class (Fig. 5–62). The skates and rays are flattened in body form, and live on the ocean bottoms, especially along the continental shelf. They primarily eat detritus and small invertebrates, although some eat larger animals that they uncover in the bottom sand. Some rays, however, such as the manta rays, swim near the surface and eat planktonic animals. The main enemies of rays are sharks, many of which are active predators. Sharks

5–62 Ray cruising the bottom at 20 m, near Los Coronados Islands, Mexico. (Photo by Cal Messner.)

(see Fig. 5–5), which have a more fish-like body form, are active swimmers from the moderate deeps to the surface of the ocean. They often feed on schools of fish, and some eat cephalopods and crustaceans, while some large species will attack seals or porpoises. The largest species of sharks, however, are not predators. Basking sharks and whale sharks do not have seizing teeth, but are equipped to strain huge volumes of water from which to obtain plankton. In spite of nets wrecked, fish stolen, and general harassment, man makes considerable use of sharks and rays. They are a food source, their skins are processed for use as sharkskin leather, and their livers are still a source of oil containing vitamin A.

The bony fish, class Osteichthyes, have undergone a fantastic adaptive radiation in the seas (Fig. 5–63). It is impossible to detail the variety of types, but they range from finless eels to blind cave fish to huge tunas to velvet black deep sea forms (see Figs. 5–5 and 5–16), and there are many other bizarre forms. Without discussing the details of morphology that make it possible, let us consider some of the ways of life of bony fish in the ocean. Most marine fish are carnivores who feed on smaller fish, invertebrate larvae, and such forms. However, some are herbivores, eating algae, others are detritus feeders who live on the bottom and pick up decaying matter as food, and some are plankton feeders

5–63 *A school of small tropic fish. (Photo by Penny Hermes.)*

5–64 *A Pacific coast tidepool.*

1 *bushy red algae,* Endocladia
2 *sea lettuce, green algae,* Ulva
3 *rockweed, brown algae,* Fucus
4 *iridescent red algae,* Iridea
5 *encrusting green algae,* Codium
6 *bladder-like red algae,* Halosaccion
7 *kelp, brown algae,* Laminaria
8 *Western gull,* Larus
9 *intrepid marine biologist,* Homo
10 *California mussels,* Mytilus
11 *acorn barnacles,* Balanus
12 *red barnacles,* Tetraclita
13 *goose barnacles,* Mitella
14 *fixed snails,* Aletes
15 *periwinkles,* Littorina
16 *black turban snails,* Tegula
17 *lined chiton,* Tonicella
18 *shield limpets,* Acmaea pelta
19 *ribbed limpet,* Acmaea seabra
20 *volcano shell limpet,* Fissurella
21 *black abalone,* Haliotus
22 *nudibranch,* Diaulula
23 *solitary coral,* Balanophyllia
24 *giant green anemones,* Anthopleura
25 *coralline algae,* Corallina
26 *red encrusting sponges,* Plocamia
27 *brittle star,* Amphiodia
28 *common starfish,* Pisaster
29 *purple sea urchins,* Strongylocentrotus
30 *purple shore crab,* Hemigrapsus
31 *isopod or pill bug,* Ligyda
32 *transparent shrimp,* Spirontocaris
33 *hermit crab,* Pagurus, *in turban snail shell*
34 *tidepool sculpin,* Clinocottus

that strain minute organisms from the water with their gill rakers. The barracuda is a formidable top-carnivore—a few barracuda can decimate a large school of sardines.

Certain fish have modified their ways of life in other regards—schooling behavior, care of young, and defense of territory being only a few. Most fish lay eggs that either stay on the bottom or float on the surface, often as a component of plankton. In some species, a single female can lay several thousand eggs. The larvae also form an important plankton component, and both eggs and larvae are food for other species. Some fish build nests and guard their young, others brood them in various ways (the male seahorse, which has a brood pouch, is a notable example), and others bear live young (the embiotocids, or surf perch, are examples).

Of great interest, but about which little is known, is the large group of deep sea fish. Most of these fish live in the zone of the ocean where light penetrates only slightly or not at all. They may be silvery in color, or, if from very deep, a velvet black, and many have unusual shapes and huge mouths. Some deep sea forms migrate to the surface each night to feed. Several species of deep sea fish (the myctophids, see Fig. 5–5), have bioluminescent organs on their bodies arrayed in a species-specific pattern. These organs are thought to help in species and mate recognition. Others (angler

DARWEN HENNINGS 1972

5–65 *A green turtle* (Chelonia). (*Photo by Steve Renick. Courtesy of Steinhart Aquarium, California Academy of Sciences.*)

fish) have elaborate "lures" that they dangle from their heads in front of their mouths, presumably to entice prey directly there. One species of angler fish has a large female, and a very tiny male. The male is parasitic on the female, and is essentially reduced to a gonad.

Warm, tropical coral reefs provide habitats occupied by a great diversity of fishes (see Fig. 5–16) that have complex niche structures. The coral, with other organisms growing all over it, provides crevices where eels, cleaning shrimp, and schools of small fish may hide. Since coral reefs support so many kinds of living organisms (algae, corals, bryozoans, bacteria, worms, crustaceans, and many other kinds), a great diversity of food is available, and the abundance is divided among many kinds of fish. Schools of large numbers of small, brightly colored fish characterize coral reefs. Their markings are thought to be either species recognition signals or patterns that aid in disguising the fish from larger fish who might prey on them.

Tidepools also may be occupied by several species of fish, each having its own area and using its own food resources. Cottids and rockfish (Fig. 5–64) may swim through the pool, eating smaller animals; blennies and gobies hide in crevices, as do eels; clingfish attach to rocks with their adhesive organs, and yet other fish swim in and out.

There are no amphibians that are truly marine; a few have invaded brackish water. Reptiles, though, have two major marine representatives—sea turtles and sea snakes. The sea turtles (Fig. 5–65) live in warm tropical waters, returning to land only to lay eggs. They travel many miles each year to return to a particular breeding site. They were once in great demand for their hides, but many are now protected. The sea snakes (Fig. 5–66) also live in warm tropical waters, and school in great numbers. These brightly colored yellow and black reptiles are venomous, and prey on small fish.

Many birds use continental shores, islands, and the waters around them, to search for food, to find nesting sites, and to guide migration. Loons, grebes, and sandpipers work along the shore to find small invertebrates to eat; gulls, pelicans (Fig. 5–67), and albatrosses fly

5–66 *Pelamis, the yellow-bellied sea snake, which ranges from the Sea of Cortez to South America. (Courtesy William A. Dunson.)*

5–67 *Pelicans at Pebble Beach. (Photo by Steve Renick.)*

far out over the oceans to pursue their diets of fish. Many of these birds help man by cleaning up the garbage thrown from ships, though their efforts are in vain against ever-increasing quantities of metal and plastic. Perhaps the birds best adapted to marine life are the penguins. Living in the Antarctic, these flightless birds have their forelimbs modified as small paddles for swimming, and they actively seek the fish that they eat. Their general body form and smooth feathers also are adaptations for a swimming life.

A number of mammals are adapted for a marine existence. Several are members of the order Pinnepedia; they include seals, sea lions, and walruses. The seals, sea lions, and walruses (Fig. 5–68), have spindle-shaped bodies, short tails, and limbs formed as paddles. They feed primarily on fish, though walruses also use their tusks to dig molluscs and crustaceans from the bottom. They live and travel in great herds, and in some species one adult male controls a harem of 60 females. Several such "families" gather on rocky islands for the birth and nursing of young. Man still shoots seals and sea lions for their skins, and walruses for their ivory. Sea otters (see Fig. 6–9) belong to the family of weasels, skunks, and mink. They live among the kelp in the surf zone off the western United States. They eat marine invertebrates, especially the purple sea urchin.

Manatees and dugongs (members of the order Sirenia) live in warm waters and feed on plants. These big, bulky animals with fluked tails and paddle-like forelimbs were rarely seen in olden times and were among the "monsters of the deep" reported in many legends.

The cetaceans—whales, dolphins, and porpoises (Fig. 5–69)—have acquired the most fish-like adaptations. They have long bodies, fluked tails, and forelimb paddles. They live exclusively in water, and die if stranded on land. They do, however, mate, nurse their young, and breathe air through lungs as do their more terrestrial counterparts. Dolphins, porpoises, and many toothed whales feed on fish, as does *Physeter*, the sperm whale, who also adds squid to his diet. The killer whale, *Orcinus* (see Fig. 6–12), attacks seals, porpoises, and other whales, as well as large fish. The nontoothed whales have plates of "whalebone" (or baleen) in their mouths and use these to strain small animals, primarily crustaceans, from the water. They process literally tons of water to get their food each day (see Fig. 7–3).

The diversity of life in the sea is great, and the sea has many different kinds of habitats. The incredible variety of living organisms that the sea supports is of great interest, but largely unstudied, in comparison to terrestrial inhabitants. A tremendous number of interesting problems await the attention of marine biologists.

5–68 Elephant seals (Mirounga) *at Guadalupe Island off Baja California. (Courtesy of Scripps Institution of Oceanography, University of California, San Diego.)*

Giants of the Sea—Sharks, Porpoises, and Whales

Among the most exciting animals of the sea are the giants—sharks, porpoises, and whales. Both legends and horror stories have developed about them. Only rather recently has substantial research into their biology been done, for their great size, their wide-ranging movements, and the nature of their habitat have made them difficult subjects to pursue.

The sharks are primitive fishes with skeletons of cartilage. They have five, six, or seven open gill slits on the sides of their heads; this distinguishes them from bony fish, whose gills are covered. Some of the largest sharks are friendly; some of the smaller species are killers. Great white sharks are perhaps most widely known. These large, deep-bodied sharks reach 12 m in length and 3700 kg. They swim constantly and are very fast-moving. They live primarily in the open ocean and apparently attack anything that might be food, even humans. They may take their prey in one gulp; intact 2 m long sharks and 45 kg sea lions have been recovered from white sharks' stomachs. A close relative, the mako,

5–69 A porpoise. (Photo by Steve Renick. Courtesy of Steinhart Aquarium, California Academy of Sciences.)

is streamlined and an even faster swimmer than the white shark. It offers tremendous resistance to being captured and attacks readily.

Thresher sharks have an unusual means of gathering food—their long tails give them such great speed that they "round up" schools of fish. They frighten the fish into tight groups by slapping the water with their tails, and then feed on these schools. These medium-sized sharks reach a length of about 6 m. The strange-looking hammerhead shark is particularly vicious. It attacks readily, and the approach of an animal whose eyes are on pedestals 1 m apart would frighten anyone! It is thought that the "hammer" may have developed as a pair of stabilizers for the shark.

Numerous smaller sharks—leopard sharks, dogfish sharks, saw sharks, and many others—feed on fishes, crustaceans, and small mammals. Some information on the biology of sharks is coming to light—they reproduce by internal fertilization; some then lay eggs in cases, while other forms retain the eggs during development and give birth to living young. The physiology of sharks is interesting—they have solved the problem of osmotic balance in salt water by keeping urea, a nitrogenous waste product usually eliminated by the kidneys, in their bloodstreams. They then have enough "salts" in their bodies to keep them from losing body water to the saltier marine environment.

Curiously, the two largest kinds of sharks are the most docile, and neither eats large fish or other vertebrates. They reach lengths of nearly 17 m. The basking shark is so named for its habit of spending much time just floating at the surface. The basking shark eats plankton, straining huge volumes of water filled with planktonic organisms through long, slender gill rakers on each gill arch. The rakers form a mesh-like trap, while the water flows on out the gill slits. Enough plankton to feed a 4000 kg animal is obtained in this way. The other large species, the whale shark, also strains its food. It feeds on schools of very small fish (such as sardines), squid, and crustaceans. Not only do the two giants prefer slightly different food, but they are geographically separated. The basking shark is usually found in temperate waters and the whale shark in warmer tropical seas.

Sharks cause considerable destruction in commercial fisheries, eating catches and tearing nets. Sharks have been commercially fished to obtain their hides for sharkskin leathers and their livers for oil for tanning and, during World War II, as a source of vitamin A. When shark livers became the primary source of vitamin A, more was learned of their biology than at any time until the most recent ten years.

Let us turn our attention to the mammalian "giants of the sea"—the porpoises, dolphins, and whales. In many ways, their body form is shark-like—large head, pectoral "paddles," flattened tail with two struts. However, they lack scales, gill slits, dorsal and pelvic fins, for they are mammals that have evolved to live in the sea. Their tails are greatly modified hind limbs; they have hair, mammary glands, and other mammalian features. They must breathe air directly, although even this process has been modified tremendously. The whale's nostrils (called blowholes) have migrated from the snout region to the top of its head. The whale closes the blowholes when it submerges and, when it returns to the surface, the animal "blows" warm moisture-laden air, sometimes causing a vapor cloud 5 m high. Early whalers could distinguish different species of whale by the shape and size of the cloud—from whence came the whaling expression "thar she blows." Perhaps the most unusual thing about the whale's respiratory apparatus is the way he "rigs for dive." A whale is able to shut down many bodily functions that are not needed while diving and in so doing conserves oxygen.

Whales are the largest creatures in the sea; in fact, the blue whale is the biggest animal ever to have lived on this planet, reaching lengths of 35 m and weights of 140,000 kg. (This represents approximately the mass of

12 elephants.) Some species of the order, however, are only about 26 kg in weight. One of the anomalies of nature is that the largest whales, the baleen whales, eat some of the smallest animals in the sea—tiny, shrimp-like zooplankton called krill. Whales such as the blue, finback, sei, and California gray all filter krill to live, and each species of whale seems to feed only on organisms of a certain size. The baleen whales have large horny plates on the roof of their mouths. The inner side is frayed and mat-like and hangs vertically. The whale takes a large mouthful of water as he passes through an area with much krill, forces the water up through the "whalebone" (which is composed of chitin rather than true bone) and back out the mouth. The krill then drops as a mass to the whale's tongue and is swallowed. The blue whale takes the largest krill. A blue whale could probably swallow five to ten barrels of krill. The toothed whales, ranging in size from the sperm whale to the common porpoise, eat a variety of things. The sperm whale feeds largely on the giant squid. The killer whale eats virtually anything available, from seals to penguins to dolphins, and may have a dozen or two of these in its stomach at any given time. The dolphin is a fish eater.

Many whales migrate from their feeding grounds to warmer water. Primarily, this migration allows some females to calve and others to copulate. The migration may take six months. The gestation period is between 11 and 16 months. The baby whale is born alive underwater and is immediately pushed to the surface for air by its mother. A baby blue weighs about 7,000 kg. Its mother must provide it with 50 gallons of the richest milk in the natural world each day. The baby whale feeds in the same way as almost all mammals, but its lunch counter is underwater. To speed up the process the mother whale has muscles in her breasts that allow her to force-feed the baby. She simply pumps him full of milk in only a few seconds. With this rich diet, the baby blue grows at the rate of 5 kg per hour or 1,000 kg each nine days.

Unlike a fish, whose body temperature is that of the water it is in, marine mammals maintain a much higher body temperature than their surroundings. To conserve their body heat they must have insulation. This is accomplished by blubber, a thick fatty layer just below the skin surface. Blubber is not a solid fat but an oil that is confined in a tissue network, giving it a solid appearance. In addition, whales have a countercurrent blood vessel system in the tail, which conserves heat at the body temperature of the whale.

Hydrodynamically, whales are quite adept in their water environment. From snout to tail or fluke, they possess smooth, streamlined bodies. The muscle power that drives the fluke is tremendous and capable of propelling a 100,000 kg animal at speeds approaching 20 knots.

Little is known about the senses of whales. Sight, especially underwater, appears to be quite good and important to the animals. Their hearing is superb. Not only do whales, dolphins, and porpoises possess a great ability to pick up errant sound, but some may emit sounds much like clicking. When returned after bouncing off an object, the sound is, in effect, sonar.

Whales emit other sounds. The squealing sounds that porpoises use to communicate with one another are being investigated in great detail. In fact, Dr. John Lilly, perhaps the foremost researcher on dolphin and porpoise communication, is structuring his experiments toward the exciting possibility of man's being able to communicate with these other species. Not only is he evaluating communication patterns among dolphins, but he is analyzing the evolution of communication. He suggests that the only reasons communication between man and dolphin might *not* be established are (1) the dolphin's brain may be so different that we could not understand its processes (but Lilly's evidence suggests similarities, not differences); (2) the dolphin brain may have no speech centers; (3) they may be "stupid," using the large brain only for complex motor control; (4) differences in vocal apparatus may prevent either man or dolphin from communicating directly. Lilly thinks communication will be established. Whales are also known to "sing." Some of these songs reach a musicality and plaintiveness that is both amazing and sad, amazing because these creatures seem so advanced, and sad because man may soon have killed all the great whales. Like the dolphins, many of the whales engage in play or frolic having no discernible purpose but joy. Apparently, only advanced animals express what seems to be emotion.

The porpoises and dolphins, too, vary appreciably in size in their many species. A few achieve 10 m in length, but most species are 2 to 3 m long. Although the vast majority of species are oceanic, four species of dolphin have been found to live in the Amazon, Ganges, Yangtze, and La Plata rivers. The distinctions between dolphins and porpoises have caused considerable discussion. Some have said that a dolphin's snout protrudes in front of his head, whereas the porpoise has a rounded head, but taxonomists do not all agree on this distinction.

Unlike the more solitary whale, the porpoise seems attracted to man. Legends have grown up concerning these remarkable marine relatives of man. Much remains to be learned about these animals, but several things are quite certain. They possess extreme intelligence—perhaps second only to man. Anyone who has seen their acts at marine centers or on television can attest to their trainability. And there has never been a recorded instance of a porpoise trying to injure a man. Indeed, there are many stories in which men have been pushed and kept at the surface by these mammals. Because porpoises are air breathers, they push their young to the surface and have been known to hold other weak or injured members of their species at the surface. Perhaps they have taken pity on less well-adapted man and have assisted his feeble attempts to move about in the ocean.

Further Reading

Abbott, R. T., *American Seashells.* New York: Van Nostrand-Reinhold, 1954.

Barnes, H., *Oceanography and Marine Biology: A Book of Techniques.* London: Allen and Unwin, 1959.

Barnes, R. D., *Invertebrate Zoology,* 2nd ed. Philadelphia: W. B. Saunders, 1968.

Bayer, F. M., and H. B. Owre, *The Free-Living Lower Invertebrates.* New York: Macmillan, 1968.

Buchsbaum, R., *Animals without Backbones.* Chicago: University of Chicago Press, 1948.

Darwin, C. R., *On the Origin of Species by Means of Natural Selection or the Preservation of Favoured Races in the Struggle for Life.* New York: New American Library, 1958.

Fraser, James, *Nature Adrift.* London: G. T. Foulis and Co., 1962.

Hardy, A. C., *The Open Sea: Fish and Fisheries.* New York: Houghton-Mifflin, 1959.

Hardy, A. C., *The Open Sea: The World of Plankton.* New York: Houghton-Mifflin, 1957.

Hedgpeth, Joel, and Sam Hinton, *Common Seashore Life of Southern California.* Healdsburg, Calif.: Naturegraph, 1961.

Herald, E. S., *Living Fishes of the World.* Garden City, N.Y.: Doubleday, 1961.

Hickman, C. P., *Biology of the Invertebrates.* St. Louis: C. V. Mosby Co., 1967.

Isaacs, J. D., "The Nature of Oceanic Life." *Scientific American,* September 1969, pp. 65–79.

Jessop, N. M., *Biosphere: A Study of Life.* Englewood Cliffs, N.J.: Prentice-Hall, 1970.

Jorgensen, C. B., *Biology of Suspension Feeding.* New York: Pergamon Press, 1966.

MacGinitie, G. E., and Nettie MacGinitie, *Natural History of Marine Mammals,* 2nd ed. New York: McGraw-Hill, 1968.

Marshall, N. B., *The Life of Fishes.* London: Wm. Clowes & Sons, Ltd., 1965.

Morton, J. E., *Molluscs: An Introduction to Their Form and Functions.* New York: Harper Torchbooks, 1960.

Murphy, R. C., "The Oceanic Life of the Antarctic," *Scientific American.* September 1962, pp. 186–210.

Nichols, David, *Echinoderms.* London: Hutchinson University Library, 1962.

Ricketts, E. F., J. Calvin, and J. Hedgpeth, *Between Pacific Tides.* Stanford, Calif.: Stanford University Press, 1968.

Scammon, C. M., *The Marine Mammals of the Northwest Coast of North America.* New York: Dover Publications, 1968 (original publication date 1874).

Schmitt, W. L., *Crustaceans.* Ann Arbor: University of Michigan Press, 1965.

Southward, A. J., *Life on the Seashore.* Cambridge, Mass.: Harvard University Press, 1965.

Sverdrup, H. U., M. W. Johnson, and R. H. Fleming, *The Oceans: Their Physics, Chemistry, and General Biology.* Englewood Cliffs, N.J.: Prentice-Hall, 1942.

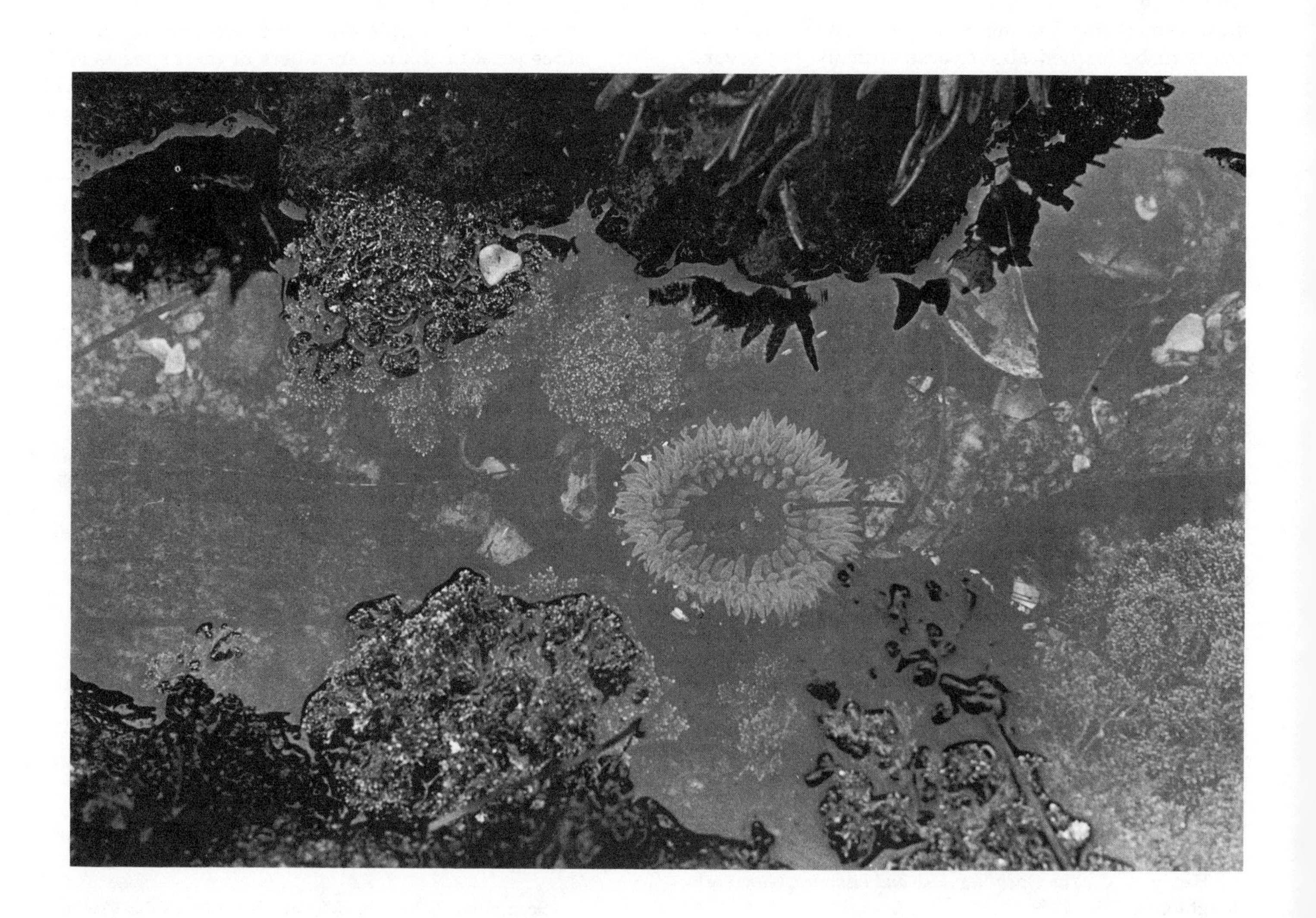

6 The Marine Environment

The world below the brine,
Forests at the bottom of the sea, the branches and leaves,
Sea-lettuce, vast lichens, strange flowers and seeds, the thick tangle, openings, and pink turf,
Different colors, pale gray and green, purple, white, and gold, the play of light through the water,
Dumb swimmers there among the rocks, coral, gluten, grass, rushes, and the aliment of the swimmers,
Sluggish existences grazing there suspended, or slowly crawling close to the bottom,
The sperm-whale at the surface blowing air and spray, or disporting with his flukes,
The leaden-eyed shark, the walrus, the turtle, the hairy sea-leopard, and the sting-ray,
Passions there, wars, pursuits, tribes, sight in those ocean-depths, breathing that thick-breathing air, as so many do,
The change thence to the sight here, and to the subtle air breathed by beings like us who walk this sphere,
The change onward from ours to that of beings who walk other spheres.

Walt Whitman

I. Classification of Marine Environments

II. Littoral Zone

- *A. Definition*
- *B. Conditions*
- *C. Zonation*
- *D. Littoral environments and organisms*

III. Estuaries

- *A. Definition*
- *B. Processes and description*
- *C. Key organisms*

IV. Sublittoral Zone

- *A. Definition*
- *B. Conditions*
- *C. Zonation*
- *D. Great Barrier Reef*

V. Euphotic Zone

- *A. Definition*
- *B. Photosynthesis*
- *C. Nutrients*
- *D. Life*
- *E. Water mass types*
- *F. Fisheries*

VI. Aphotic Zone

- *A. Definition*
- *B. Euphotic/aphotic transition*
- *C. Conditions*
- *D. Life*

VII. Integrating Ideas

- *A. Food webs*
  1. *definition*
  2. *sample webs*
- *B. Nutrient cycle*

Classification of Marine Environments

Environments have been classified by whatever variable is perceived to be significant by individual scientists. These have included different chemical, physical, geological, and biological variables. For instance, some scientists have found a system based on salinity to be useful (Table 6–1). The most common criterion for the classification of marine environments, however, is depth. In addition, most scientists superimpose other relevant variables on the basic system. The most commonly superimposed parameter for the ocean is light penetration, which adds two zones—the euphotic and aphotic—to the original system.

Probably the most widely accepted classification of marine environments was proposed by the Committee on Marine Ecology and Paleoecology of the Geological Society of America (see Hedgpeth, 1957, p. 18). This zonation has been widely adopted although it is not universally accepted (Fig. 6–1). The question marks indicate that the limits or boundaries are either variable, as in the case of euphotic and aphotic, or uncertain, as in the case of bathypelagic and abyssopelagic. Zones that will be discussed in detail in this chapter are: littoral, the intertidal region between high and low water; sublittoral, the bottom between low water and 200 m—that is, approximately from low water to the edge of the continental shelf; aphotic, below the maximum depth of light penetration; euphotic, area of sunlight penetration.

Marine ecologists often use a biological system for classifying localized marine environments, especially for nearshore benthic areas. This system is based on the recognition of a dense population of a particular organism. C. G. J. Petersen designed sea bottom samplers for acquiring quantitative data on the number and distribution of bottom-living organisms. After analyzing thousands of samples, Petersen discovered remarkably uniform combinations of macrofauna species. A relatively few species were found to be extremely common. This uniformity and dominance by a few species led to the concept of animal communities. Since the concept of communities was introduced as a descriptive–statistical unit, geological, physical, and chemical properties of the communities have been measured. The uniformity extends to these nonbiological properties also. The concept of community can now be considered a fundamental ecological unit. Communities are usually classified by a common genus, such as *Tellina, Macoma, Venus,* or *Modiola.*

Tellina communities are typically found from the intertidal zone to a depth of about 10 m in areas where the bottom is fine, closely packed sand. The most common genera include the pelecypods *Tellina, Donax* (coquina shell or little bean clam), and *Dosinia;* the echinoderm *Astropecten* (sand dollar); and the crustacean *Emerita* (sand crab).

Littoral Zone

As previously defined, the littoral zone is the area along the shoreline between high and low tide. This is the zone that is probably the most familiar to many people. The physical and chemical conditions in this zone are without a doubt the most variable and rigorous found in the marine environment. First, tidal fluctuation alternately produces submarine and subaerial conditions. Second, wave action and longshore currents cause constant movement and the transfer of energy. Third, even geologic characteristics are subject to rapid alteration, mainly as a result of wave action and currents. Fourth, this area is subjected to much pressure from the behavior of humans, especially from dredging, filling, and waste disposal. Fifth, the combination of these factors also causes major fluctuations in the chemical conditions.

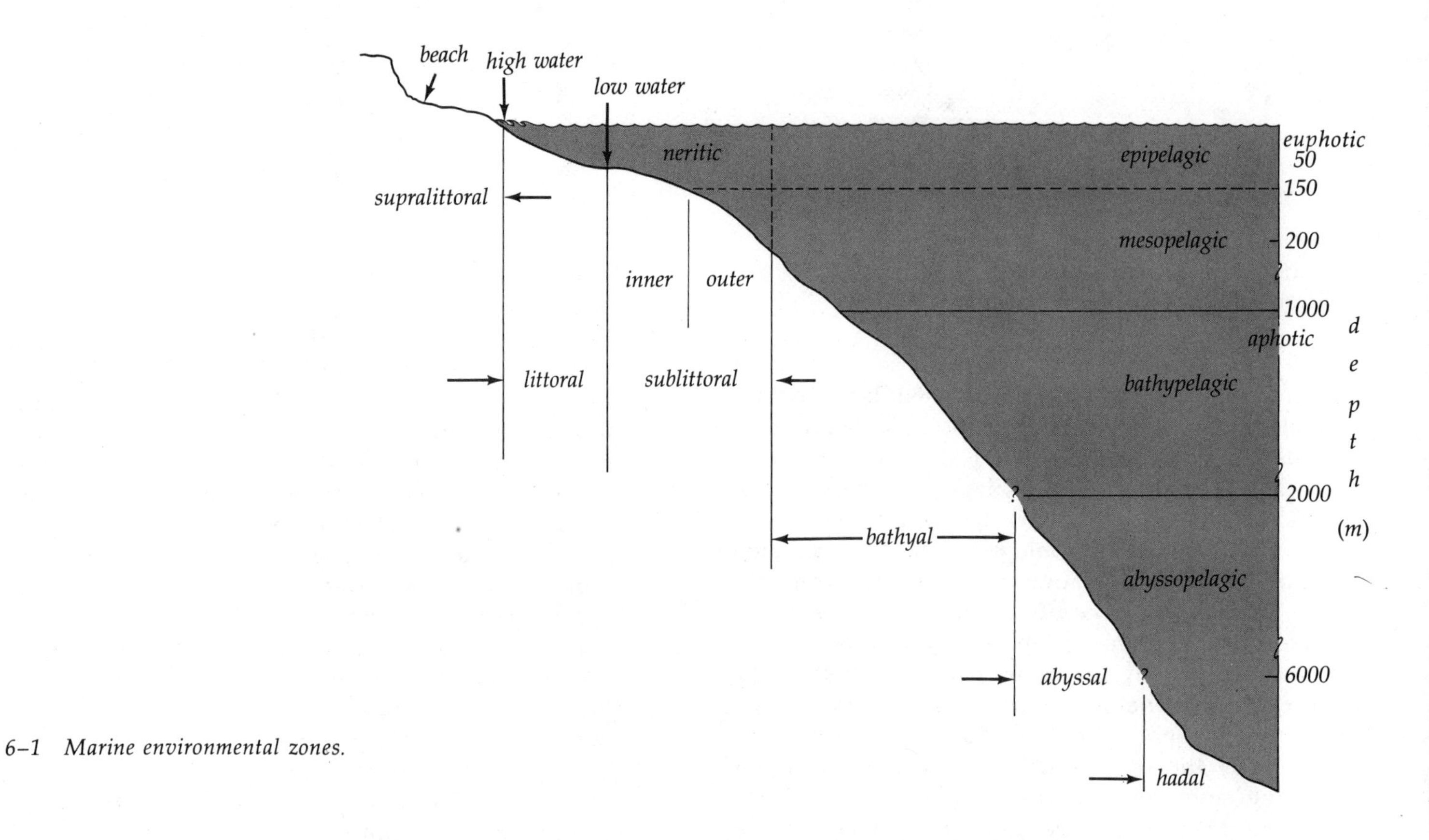

6–1 *Marine environmental zones.*

In a simple division, consider the following four zones within the littoral rocky coast.

Zone 1: wave splash zone, rare submersion during storms
Zone 2: submerged regularly during the highest tide of a normal daily cycle
Zone 3: exposed only during the lowest tide of a normal daily cycle
Zone 4: exposed only during especially low tides as occur at full or crescent moon phase

Each zone is progressively more submerged from zone 1 to zone 4. There is some overlap of organisms across zones; however, the most common organisms visibly differ in each zone. This differentiation is most obvious on pilings. Along rocky coasts, the zonation is less obvious because of the irregularity of the rocks and the presence of isolated tidepools, which allow perpetual submergence above zone 4. It may be interesting to think about the differences in conditions in these zones and to see what effect these differences have on the types of organisms found in the zones in southern California (Figs. 6–2, 6–3, 6–4, and 6–5).

Zone 1 receives relatively little ocean water and is subjected to relatively long atmospheric exposure. As a result, the organisms living there must be capable

Table 6–1 Marine Environments Classified by Salinity Level

Name	Salinity (‰)
Infrahaline	0.5
Oligohaline	0.5–3.0
Mesohaline	
Meiomesohaline	3.0–8.0
Pleiomesohaline	8.0–16.5
Polyhaline	16.5–30.0
Ultrahaline	30.0 up

Information from J. Hedgpeth (ed.), *Treatise on Marine Ecology and Paleocology,* vol. 1. Washington, D.C.: GSA Memoir 67, 1957.

6–2 Zone 1 at La Jolla Shores, California. (*Photo by John P. Baker.*)

6–3 Zone 2 at La Jolla Shores, California. (*Photo by William J. Wallace.*)

of enduring such conditions. The most common organisms found in this zone on rocky shores in southern California are snails, barnacles, limpets, and green algae.

Zone 2 is submerged for longer periods than zone 1 and is subjected to more wave action. In this zone are found different species of snails, limpets, a number of crabs, some clams, chitons, and algae. Of these only the snails and crabs are not attached to the rocks. They are both able to cling tightly or hide when wave action is the greatest. For these organisms, curious humans "collecting" are the most serious predators.

Zone 3 contains a wider variety of organisms than zones 1 and 2 and some relatively large individuals. Most conspicuous in this zone are the mussels, goose barnacles, and starfish. The large clumps of mussels are found toward the top of the zone, with several species of barnacles, algae, crabs, sea anemones, and starfish just below. In some remote areas where tidepools are scattered among the rocks, an occasional abalone or octopus may be seen. Along most of the southern California coast today, few abalones are found in the littoral zone. Humans—and not sea otters, as abalone hunters have charged—have reduced the abalone population tremendously, since they are delicious food. Divers, however, may find them at depths of from 3 to 8 m.

Zone 4 exhibits a still greater variety of marine organisms. Some of the most conspicuous are green surf grass, snails, and larger limpets. Also, occasionally, the California spiny lobster is found here. Like the abalones, the lobster population has dwindled tremendously in the past ten years. They are still fairly common offshore but are hunted heavily by scuba divers.

Littoral Environments. The major intertidal environments are rocky shores, sandy beaches, and muddy bays. Any of these may experience high or low energy conditions depending on local wave action and currents. Bays, however, rarely experience high energy conditions. Sandy beaches and muddy bays will usually be inhabited by organisms that burrow, or walk, swim, or crawl along the bottom. Some attached organisms are also found, but they do not dominate as they do on rocky shorelines such as southern California's.

One of the most unusual inhabitants of a rocky littoral environment is the unique marine iguana of the Galapagos Islands. Although these lizards spend much of their lives on land, they go into the water to graze on algae, often at a depth of 10 m or more, and are capable swimmers. Charles Darwin described these animals in his book *The Voyage of the Beagle.* Part of his description follows:

The nature of this lizard's food, as well as the structure of its tail and feet, and the fact of its having been seen volun-

6–4 Zone 3 at La Jolla Shores, California. (Photo by William J. Wallace.)

6–5 California mussels clump on the pilings in zone 3 at the pier at Scripps Institution of Oceanography. (Photo by William J. Wallace.)

tarily swimming out at sea, absolutely prove its aquatic habits; yet there is in this respect one strange anomaly, namely, that frightened it will not enter the water. Hence, it is easy to drive these lizards down to any little point overhanging the sea, where they will sooner allow a person to catch hold of their tails than jump into the water. They do not seem to have any notion of biting, but when much frightened they squirt a drop of fluid from each nostril. I threw one several times as far as I could, into a deep pool left by the retiring tide; but it invariably returned in a direct line to the spot where I stood. It swam near the bottom, with a very graceful and rapid movement, and occasionally aided itself over the uneven ground with its feet. As soon as it arrived near the edge, but still being under water, it tried to conceal itself in the tufts of seaweed, or it entered some crevice. As soon as it thought the danger was past, it crawled out on the dry rocks, and shuffled away as quickly as it could. I several times caught this same lizard, by driving it down to a point, and though possessed of such perfect powers of diving and swimming, nothing would induce it to enter the water; and as often as I threw it in, it returned in the manner above described. Perhaps this singular piece of apparent stupidity may be accounted for by the circumstance, that this reptile has no enemy whatever on shore, whereas at sea it must often fall a prey to the numerous sharks. Hence, probably, urged by a fixed and hereditary instinct that the shore is its place of safety, whatever the emergency may be, it there takes refuge.

Along sandy beaches, the littoral zone will include shore birds, crabs, and even some terrestrial visitors such as foxes and raccoons. The burrowing organisms that are most common are usually found in zones 2, 3, and 4. They include the coquina clam, the Pismo clam, the "steamer" clam, and many other edible clams. In addition, burrowing worms like the beach bloodworm (an annelid), sand dollars, and sand crabs are extremely common. Less often seen, but important inhabitants include microscopic plants and animals. Unicellular algae and foraminifera are basic sources of food for larger organisms.

In protected bays with muddy bottoms, a surprising variety of life exists. Often low circulation and polluting drainage from adjacent land areas apply undue pressure on these localities. Among the common organisms here are marsh grass, fiddler crabs, snails, and a host of different birds. Here again the presence of unicellular algae and foraminifera is essential to support the other organisms.

The ecological relationships between the organisms, the substrate, and the physical and chemical conditions in the littoral zone are exceedingly complex. It has been difficult for marine ecologists to identify the relationships and even more difficult to measure them after identification. The role of bacteria in the littoral zone, for example, is only at an early stage of investigation. Small changes such as the addition of DDT in a concentration of less than 2 parts per billion in the water in some areas have even led to the near extermination of some organisms. Understanding the ecological relationships in the littoral zone is a major problem that needs much more research.

Estuaries

Not always considered part of the marine environment, estuaries, or river mouths, are particularly important because they are under such heavy use by man. The most polluted parts of the marine environment, they are the primary channels through which pollutants flow into the ocean (Fig. 6–6). In addition, many organisms migrate to or through estuaries at some time in their life cycles.

Many estuaries in the United States are familiar in name; for example, San Francisco Bay and Chesapeake Bay are estuaries. An estuary extends from a region that is invariably marine to a region that is invariably nonmarine. In *Estuaries,* D. W. Pritchard proposed this definition: "An estuary is a semi-enclosed coastal body of water which has a free connection with the open sea and within which sea water is measurably diluted with fresh water derived from land drainage" (Lauff, 1967). If estuaries are analyzed using the main marine water parameters—depth, salinity, temperature, and light penetration—the distinction from normal ocean water is obvious.

The salinity in an estuary such as Chesapeake Bay or San Francisco Bay is highly variable. Near the open ocean, the salinity is often the same as that of ocean water, 35‰. The salinity along one shoreline may decrease rapidly to less than 5‰ in only a few kilometers while remaining over 20‰ on the other side of the bay. This and other variations are caused by complex current patterns resulting from the interaction of such physical factors as tides, river drainage, ground water seepage, and the Coriolis effect. In parts of an estuary, particularly when river drainage is low, evaporation causes dissolved salts to precipitate from the water, and reduced circulation causes stagnation and the accumulation of pollutants.

The temperature of estuaries also fluctuates greatly. In tropical climates, the water may exceed 30°C for long periods of time and in mud flats may exceed 40°C. In colder climates, many estuaries freeze over completely. Localized springs or power plant effluents may also have a tremendous effect on the water temperature. In the Crystal River estuary on the west coast of Florida, springs emerging from the Florida aquifer produce a relatively constant temperature of about 20°C. In the southern part of San Diego Bay in southern California, power plant effluent produces temperatures exceeding 35°C in appreciable parts of the bay.

Estuaries also vary in depth depending on their geological history. Some may be extremely shallow, except where dredged. The St. John's River estuary near Jacksonville, Florida, is shallow because the low gradient of the river from its source to mouth precludes much erosion. The Amazon River estuary, however, has channels more than 100 m deep.

Because estuaries have a free entrance to the ocean and at the same time are protected from heavy wave action by adjacent land, humans have used them as shipping centers. Thus, cities have developed around estuaries. Before the explosion of population and industry, pollutants dumped into the water were sufficiently diluted so that organisms were not adversely affected. In some cases, organisms benefited from the increased nutrients. However, pollution pressure by sheer volume in the case of sewage, by toxicity in the case of lead, and by concentration in the food web in the case of DDT and other pesticides has begun to endanger natural estuarine organisms. Countless examples can be cited. In San Francisco Bay, market crabs are endangered and, where found, contain pollutants that render them inedible for humans. Oysters in many parts of Chesapeake Bay are likewise endangered. People have been slow to realize their effect on the environment as a whole but have been especially negligent with estuaries, where economic interests have been dominant in preventing the adoption of conservation principles.

6–6 Estuarine tidal flats with mudflat grasses at the mouth of the Housatonic River near Bridgeport, Connecticut. (Photo by Dale E. Ingmanson.)

In spite of the pollution pressure on some estuaries, many are still relatively unaffected (Fig. 6–7). A myriad of organisms, usually uniquely characteristic of estuaries but occasionally a mixture of marine and nonmarine, are found. Reasonably common to estuaries are mud flat grasses, oysters, barnacles, mussels, snails, crabs, mullet, and sea trout. Many other organisms, varying in size and complexity from bacteria to killer whales, can also be found in estuaries.

A few of the permanent residents are members of the order Sirenia, which includes the manatees and dugongs. These large, slow-moving herbivores graze on algae and other plants in coastal tropical and subtropical estuaries, lagoons, and rivers near the sea. Manatees can still be found along the coastal lagoons and in the canals of south Florida. They even live in canals within the city limits of Miami. The name Sirenia would appear to be a misnomer as these mammals do not match the normal person's image of a siren. They do, however, float vertically in the water and suckle their young at their breasts in such a position. These bizarre mammals are thought to be the source of mermaid legends translated from ancient Greek literature. Perhaps early Greek sailors on board ship for two or three months looked longingly at some female manatees.

Among the animals that migrate through estuaries are salmon, which as adults leave the ocean and head upstream to their spawning grounds. Eels swim through estuaries to live their adult lives in rivers after having spawned far out to sea. Estuaries also temporarily contain many other transients.

The zonation patterns of the organisms in an estuary are often striking and are interesting to trace. Each species has its own characteristic limits. The barnacle has different limits than the mussel. In addition, species within a genus will have discernible patterns of location. Usually these patterns are determined by salinity variations and the tolerance of species and genera for these variations.

Spring-fed estuaries offer an unusual mixture of marine and nonmarine organisms. In the Crystal River estuary, marine fish such as sheepshead and snook mingle with fresh water fish such as bass and bream.

6–7 Marine marsh and estuary, looking toward the ocean. Drake's Estero, Point Reyes National Seashore, California. (Photo by Steve Renick.)

For some reason, the marine organisms are able to adapt to fresh water conditions. The most widely accepted theory explaining this is that many fish can alter the efficiency of their kidneys to maintain a suitable osmotic balance for sustained life. In other words, they can increase or decrease the rate at which water can be pumped out of their bodies and so adjust to varying salinity.

Another unusual estuary is the Everglades region of south Florida. The drainage is spread thinly over a broad area instead of in a deep channel. Its ecologic characteristics are essential to a wide variety of organisms (Fig. 6–8).

Sublittoral Zone

This zone consists of the bottom extending from the low tide area at the shoreline to the edge of the continental shelf or to a depth of about 200 m if the shelf edge is not a clear separation zone. The physical conditions are determined mainly by the geologic substrate and by the extent of wave action and currents in particular regions. Off southern California and the east coast of Florida currents on occasion move with speeds up to 4 knots. Most of the Gulf of Mexico shelf, however, is usually unaffected by currents. Wave action affects the shallower parts of the sublittoral zone, especially where the continent faces long open ocean stretches. Subsurface currents associated with surges in submarine canyons and internal waves moving on the thermocline or other density layer interfaces also affect the shelf. The temperature at a particular area of the sublittoral zone usually remains relatively constant except in regions of upwelling, where it can vary as much as 5°C in a few days. Temperatures in different geographic regions in the sublittoral zone vary from −2°C in the Arctic to more than 30°C in the Red Sea and Persian Gulf.

Chemical conditions in the sublittoral zone are, for the most part, relatively stable. Like the physical conditions, the chemical characteristics are largely

6–8 White Shrimp (Penaeus) *spawn in estuaries of the Everglades mangrove swamps. (Courtesy of the United States Department of the Interior, Fish and Wildlife Service.)*

affected by the open ocean. Locally, some important variations should be mentioned. Sewage effluent pipes are now being extended out several kilometers into the sublittoral zone as an alternative to polluting the littoral zone and estuaries. This continuous flush of nutrients has produced abnormal populations of algae, foraminifera, and even sea urchins, all of whom are apparently capable of absorbing the nutrients directly from seawater. Extensive studies are presently in progress near Los Angeles. What long-term effect this pollution will have is presently unknown.

A few anoxic basins are located within the sublittoral zone. They are created where natural geological dams block circulation of the water, allowing stable stratification and deoxygenation by bacteria to occur. The sublittoral zones in fjords are classic examples. Two other anoxic basins within this zone are found off Puerto Vallarta, Mexico, and off the north coast of Venezuela.

On the continental shelf off Florida, fresh water springs alter the chemical conditions in the sublittoral zone. These areas of low salinity are very localized, however. Where they occur, abundant algae grows attracting green turtles and loggerhead turtles. Because the flowing spring water causes turbulence, visibility is reduced and scuba divers have been known literally to bump into 160 kg turtles!

The geology of the sublittoral zone is as varied as the geology of the continents. The zone, by definition, is part of the continent and has been exposed above sea level at least during continental glaciation, so all the erosive processes observed on rocks above sea level can also be observed on the rocks of the sublittoral zone. In large regions of this zone, however, extensive sedimentation is actively in progress, especially in deltas. In addition, beautiful limestone banks are developing in tropical sublittoral zones off Florida, Yucatan, Australia, Kenya, and many other locations. Submarine canyons are a dominant feature of sublittoral zones especially off southern California and the Congo.

Some attempts have been made to establish benthic community designations that correspond to those of the littoral zone and estuaries. The main division point is the maximum depth of light penetration. Some indiscriminate scavengers such as lobsters cruise the whole sublittoral zone. Most other types living in this zone, however, are either light dependent or not. One type of benthic organism that apparently does lend itself to zonation from the shoreline out across the sublittoral zone is the foraminiferan. This zonation is especially useful in helping paleontologists identify the water depth at which rocks containing fossil foraminifera were formed. To date, foraminiferal zonation has not helped marine ecologists understand the relationships on the shelf as a whole. Perhaps with more access via scuba, submersible vehicles, photography, television, and better sampling techniques, the use of foraminiferal zonation will prove invaluable.

Numerous organisms are found in the sublittoral zone. Of the algae, one of the most spectacular is the kelp. Because of their forest-like beauty and the abundance of other life associated with them the kelp beds are a scuba diver's paradise. A few divers, however, have drowned because of entanglement in the kelp. Some of the marine life found around kelp beds is also potentially dangerous, especially barracuda and occasional sharks and killer whales.

Kelp is important commercially as a source of iodine, fertilizer, agar, and ingredients in bread, candy, ice cream, jelly, and many other items. Large commercial "mower" ships cruise daily over the kelp beds off southern California cropping the tops of the plants.

The giant kelp, *Macrocystis*, is common in the sublittoral of southern California. The plant usually called giant kelp is the multicelled stage of a complex life cycle. This large multicelled stage produces microscopic spores that grow into a microscopic, planktonic stage. This tiny plant produces sperms and eggs, which after random fertilization, settle to the bottom and begin to grow into new giant kelp.

The success of the growth and development of kelp beds is dependent on the extent of predation and chemophysical environmental variables, which may differ at different parts of the life cycle. Sea urchins are known to eat the giant kelp plants. Predators on the microscopic stage are not well known but probably include copepods and foraminifera. Storms that occur when free sperms and eggs are in the water may prevent fertilization. Recent studies have indicated that although kelp is not directly affected by sewage from outfalls, sea urchins take up nutrients directly from the sewage thus multiplying the sea urchin population and resulting in increased kelp predation.

Since kelp is adapted to cool water, the beds in the vicinity of thermal outfalls from power plants are endangered along the California coast. Cycles of kelp die-off, and proliferation has been observed in some areas. No satisfactory explanation has been given for these cycles although sewage may play a significant role in causing die-off.

The giant kelp beds are referred to as the kelp habitat. The kelp provides an environment for a complex set of marine organisms including foraminifera, copepods, polychaetes, shrimp, crabs, amphipods, serpulid worms, many fishes, and several marine mammals. Although many organisms are found in the kelp habitat, the presence of the kelp is essential to only a few. But the kelp is visually dominant, so its name is associated with the habitat.

One of the more conspicuous inhabitants of the kelp beds is that remarkable creature the sea otter, *Enhydra lutris* (Fig. 6–9). The sea otter could live on land but almost never comes in from the sea. Originally, the natural range of the sea otter extended from the Bering Straits to Baja California. By the early 1900s, fur hunters brought this furry creature to the brink of extinction. Strict protection by international agreement has allowed its slow return, and today there is a herd of about 1,000 animals off California and a more sizable one in Alaskan waters. The sea otter is approximately 1.3 m in length; the males weigh between 22 and 36 kg and the females about 20 kg. They are completely covered with the thickest, richest, most durable fur known.

6–9 An Alaskan sea otter. (Photo by Calvin B. Harris)

An adult sea otter uses its nimble-fingered forepaws to tie a leg to the kelp or wrap a frond about its body to keep it in the seaweed's shelter before going to sleep. It is not uncommon for a female to tie her pup's rear leg to the kelp while she goes off to hunt for food, letting the kelp babysit.

Apparently, the bulk (probably over 50 percent) of the sea otter's diet is the purple sea urchin. Very little else in the ocean ever attempts to eat the urchin. Blooming urchin populations are taking such a heavy toll that kelp beds and sea otters should be welcomed if they can keep the urchin population in check. Recently, abalone gatherers have been complaining that the sea otter decimates the abalones. The abalone gatherers are unwilling to admit that their own lack of conservation is probably the primary cause of the extreme decrease in abalones in California waters. In reality, no one knows how many abalones sea otters eat.

Also among the numerous organisms found in the sublittoral are the microorganisms including bacteria,

6–10 Many kinds of corals require relatively warm water (above 26°C), well-circulated water, and low amounts of suspended sediment in the water. They have many forms, including finger coral, brain coral, staghorn coral, and palmate coral—all of which look somewhat like their namesakes. (a) Black coral at 4 m depth at Panapompom, New Guinea. (Official United States Navy Photograph.) (b) Acropora in an interisland channel at 2 m depth. (c) Table coral at a depth of 20 m. (d) Three types of coral: Acropora *(left),* Dendrogyra *(center), and* Monastrea *(right) at Lameshur Bay. (b, c, d: courtesy of Tom Dana.)*

foraminifera, and diatoms. Some of the more well-known larger invertebrates are lobsters, corals, sea urchins, scallops, and sea anemones. Cruising close to the sublittoral zone and eating organisms on or in the bottom are many vertebrates including sharks, rays, catfish, and sea otters.

Great Barrier Reef. One of the most extensive coral reef areas is found in the sublittoral zone on the continental shelf off Queensland, Australia, roughly between 10°S and 26°S. This reef system differs appreciably from the Indo-Pacific island reefs (see p. 52). Many of the organisms found on the reefs are the same

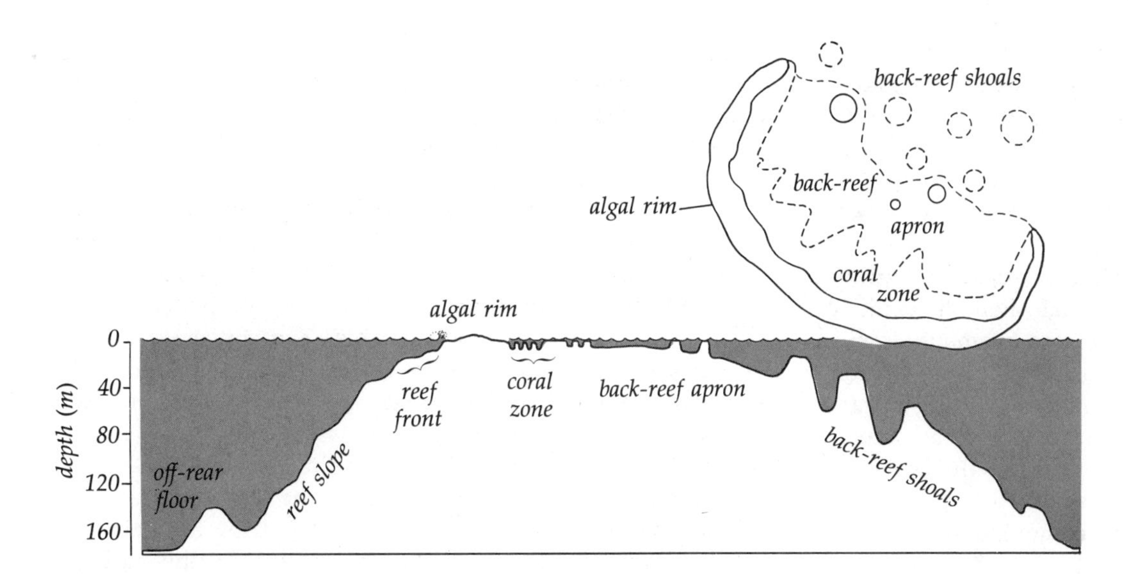

6–11 Open ring and composite apron reef.

species but the general shapes and geologic evolution are very different because the reef is on the continental shelf and not in mid-ocean. Even the name, Great Barrier Reef, is a misnomer because it is not a barrier reef in the sense of the term applied to oceanic island reefs for which it was coined.

Corals are one of the predominant life forms found on tropical reefs, the Great Barrier Reef, as well as other reefs (Fig. 6–10). Everyone has heard of these animals, but few know much about them. Most species are colonial and are connected by an external calcium carbonate casing. A single polyp lives in each small chamber. This polyp is the living coral animal—it secretes the casing. The coral polyp is also interesting for a number of less well-known reasons. It feeds mainly at night on planktonic organisms captured by its small tentacles. In addition, the cells of the coral polyp contain many unicellular algae. These algae, zooxanthellae, are actually dinoflagellates that carry on photosynthesis in the daytime while providing the coral with oxygen and organic nutrients as a diet supplement. The coral in return prevents the algae from being eaten by herbivores and supplies carbon dioxide.

A number of reef shapes have been described by scientists. The open ring, composite, apron type is a common example (Fig. 6–11). Reef shapes are determined by the direction of current flow and the configuration of underlying bedrock. Because the bedrock on the continental shelf anywhere is less apt to have a circular pattern than a volcanic oceanic island, the reefs are less likely to be circular. The geologic evolution of reefs on continental shelves will be determined by what is happening to that part of the continent which, in many cases, will differ from what is happening to a volcanic island in the oceanic province.

The reef slope is comprised of a very few reef building corals. Other organisms are common, including unattached mushroom corals. Between depths of 1 to 10 m is the reef front, where spectacular reef building corals are found. Although many types of corals grow in the reef front, species belonging to the genus *Acropora* dominate, especially the broad palmate species. These species require extensive water circulation, aeration, and relatively uniform temperature and salinity. The algal rim is located between the surface and 1 m deep and is so called because calcareous encrusting algae grow on the remnant coral skeletons. This part of the reef absorbs most of the wave energy that is dissipated on the reef. In the lee of the algal rim is the reef flat. Many subzones are discernible in this region. Because foraminiferal species are especially responsive to subtle environmental changes, the microzonation is based on their population patterns. Corals found on the reef flat include the brain corals, platy corals with intricately foliated skeletons, and several encrusting species. Currents and wave action are not strong on the reef flat, so many delicate or unattached organisms are found, including various mollusks, echinoderms, crustaceans, sea squirts, polychaete worms, and algae such as *Halimeda*. Further to the lee of the algal rim is an area of patchy shoals composed mainly of large isolated algal buttresses, brain corals, and broad palmate corals.

On the Great Barrier Reef, coral does not grow actively over the entire continental shelf. Instead, it forms patch reefs, which are separated by channels and broad limestone banks. The banks may have some scattered corals, but other organisms including calcareous algae such as *Penicillus* and *Halimeda* are much more abundant. The sediment found on the banks is principally composed, therefore, of the calcareous shells and debris of the algae and common animals such as the mollusks.

One of the few reef front coral predators is the crown of thorns starfish *Acanthaster planci*. For some reason currently being investigated, this starfish population seems to have increased since 1964, appreciably diminishing living corals. This was recently noted with alarm on the Great Barrier Reef and has now been observed on several reefs bordering Pacific islands. Guam, for instance, has lost about 30 percent of its living coral on the reef front.

Euphotic Zone

The euphotic zone is the part of ocean water and nearshore bottom exposed to solar radiation, especially visible light. Sunlight may penetrate ocean water to a depth of more than 200 m. The depth of penetration depends on at least four major factors: (1) cloud cover, (2) the angle of inclination of the sun's rays to the surface of the ocean, (3) the amount of suspended inorganic material, and (4) the population density of planktonic organisms. Light should penetrate deepest on a cloudless day near the equator during an equinox with no waves in an area far from land, so that a minimum of sediment and organisms would be present. At a given geographic location, the maximum depth of light penetration can be highly variable even during a short timespan.

All life in the euphotic zone depends on sunlight as an energy source. Photosynthetic plants, primarily, convert this energy into other forms usable to animals. Unicellular algae are both the most common organisms and the most essential for the survival of the other organisms. As do all photosynthetic plants, they contain chlorophyll pigments, which absorb solar energy. This energy is used by the plant cell in photosynthesis to produce carbohydrates, proteins, and lipids (fats) that can be metabolized by other organisms. In the euphotic zone, sunlight, carbon dioxide, and water are readily available. The concentration of dissolved carbon dioxide varies. Its principal sources in the euphotic zone are respiring animals and migration from the atmosphere across the air–ocean interface. The exchange of carbon dioxide across the air–ocean interface is determined by the temperatures and pressures of the air and water. Respiring animals exhale carbon dioxide. The variability of the carbon dioxide concentration depends on photosynthesis and the concentration of bicarbonate and carbonate ions. The latter enter seawater mainly in two processes. In one, water on land dissolves limestone rocks.

$$\underbrace{2H^+ + H_2O}_{\text{acidic ground water}} + \underbrace{CaCO_3}_{\text{calcite}} \leftrightharpoons H_2O + Ca^{+2} + \underbrace{HCO_3^-}_{\text{bicarbonate ion}}$$

$$+ H^+ \leftrightharpoons H_2O + Ca^{+2} + 2H^+ + \underbrace{CO_3^{-2}}_{\text{carbonate ion}}$$

Bicarbonate, carbonate, and calcium ions are then carried in solution to the ocean. In the other process, shells of marine organisms dissolve, freeing carbonate ions.

$$\underbrace{CaCO_3}_{\text{shell}} \rightarrow Ca^{+2} + \underbrace{CO_3^{-2}}_{\text{carbonate ion}}$$

The bicarbonate and carbonate ions and carbon dioxide then become available to enter one of the important buffer reactions that help maintain the acidity (*p*H) of ocean water. Carbon dioxide dissolved in seawater will react to produce bicarbonate and carbonate ions if there is an excess of carbon dioxide and vice versa.

$$CO_3^{-2} + 2H^+ \rightleftarrows HCO_3^- + H^+$$

$$HCO_3^- + H^+ \leftrightharpoons H_2O + CO_2$$

Oxygen, one of the products of photosynthesis, is, of course, essential for animal respiration. Estimates of the net amount of oxygen produced by marine plants range from 40 to 90 percent of the total oxygen produced by plants on the earth each year.

Photosynthetic plants produce organic compounds, attracting herbivores and omnivores to the euphotic zone to eat the plants. And as a result of the presence of herbivores and other animals, carnivores are also attracted there to prey on them.

The materials essential for life in the euphotic zone can be summarized as water, carbon dioxide, sunlight, and inorganic nutrients, mainly phosphate and nitrate. The first three are essential for photosynthesis; the last are essential nutrients. Other materials, such as iron and silicates are also important. These essential materials are not uniformly distributed through the euphotic zone. Consequently, populations of marine life are not uniformly abundant through the zone.

Conditions favoring abundant plankton growth are found mainly along coastlines where nutrients and carbon dioxide are readily available due to mixing, upwelling, and river runoff. Even though light penetration in these areas may be less than 100 m, high productivity is possible in this narrow surface water zone because of the presence of the other essential materials. Two organisms that undergo high population expansions because of favorable conditions are the diatoms and the dinoflagellates (several species of which are the cause of red tide).

In the northwest Gulf of Mexico, the dinoflagellate *Gymnodinium* reproduces so rapidly that population densities of up to 100 million individuals per cubic meter are not uncommon. The organisms produce a toxin that is lethal to man when ingested and lethal to fish, which receive the toxin through their gills. Thus, fish kills are grim evidences of dinoflagellate populations that have run wild. Not all red tides are deadly to marine life. In southern California waters, *Prorocentrum* (another dinoflagellate) blooms occur in summer months. A mild toxin is produced which does not kill marine life. Because clams and mussels eat dinoflagellates by the thousands, the toxins are concentrated in their bodies making them potentially deadly to humans between May and October. The factor that triggers red tides is not fully understood. In fact, different factors may be involved in different areas. Two possibilities are increased nutrients provided by river water and upwelling along coasts.

Diatoms can be found in fresh water as well as the ocean. In the ocean, they are most common near shore and in the open ocean in latitudes above 40°N and 40°S, although they are ubiquitous geographically. Because diatoms are photosynthetic, they need sunlight and therefore live only in the upper 100 m or so of the sea. The reproduction of diatoms is rapid. They divide into two new cells in 6 hours to 10 days depending on the temperature, the availability of nutrients like phosphate, nitrate, and silica, and the amount of sunlight. This rapid rate of reproduction enables tremendous blooms to occur where physical and chemical conditions are suitable, thus insuring survival of some individuals by statistical chance even when conditions are unsuitable. Many diatoms are planktonic. They are only slightly more dense than ocean water, tend to sink, and have evolved a number of adaptations to retard this tendency. The major adaptations are low surface-to-volume ratios via elongate shapes and storage of oil within the frustule. The significance of diatoms in marine geology was discussed in Chapter 2. Their significance in marine biology is that they are basic organisms in the food web. One of the organisms most responsible for converting inorganic nutrients to organic nutrients, they provide the energy supply of most herbivores and carnivores in the ocean.

Especially abundant euphotic zone life can presently be found bordering Antarctica, off Peru and Ecuador, extending westward along the equator for 2,000 km, off the west coast of North America, and in a broad band extending across the North Atlantic from Nova Scotia to Iceland and on to Norway. Except for the equatorial current system and Antarctic waters, the major parts of the oceans are not conducive to plant growth and are, therefore, relatively barren of life. One exception is the Sargasso Sea, southeast of Bermuda in the Atlantic Ocean. *Sargassum* is a pelagic brown alga found in the open ocean unattached to the bottom. Beneath, between, or on the stipes and floats live numerous animals, including fish, hydroids, pycnogonids, amphipods, and many others.

The most common organisms in the euphotic zone are microplankton including unicellular algae, flagellates, foraminifera, and radiolaria. The most common multicellular invertebrates include pteropods and copepods, and the most common vertebrates are probably anchovies, herring, and sardines.

Based on comparative communities of marine life and the properties of the respective water masses, at

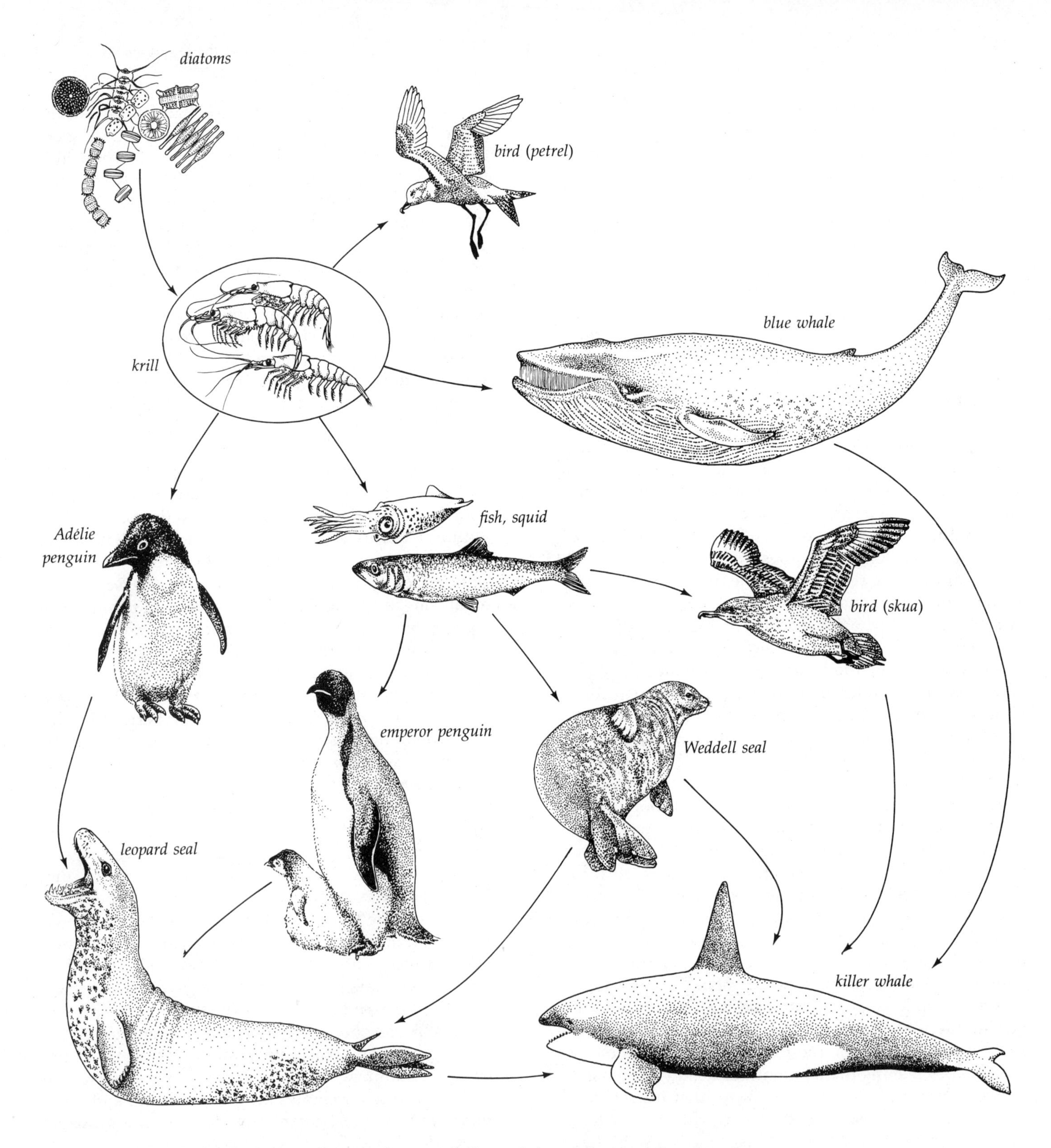

6–12 The shortest food web in the ocean is the diatom → krill → whale web in Antarctic waters. In addition, the Antarctic holds the most productive waters of the world. Although whales have been hunted in the Antarctic for at least a century, fish and krill are just recently being exploited.

least six general categories of euphotic zone types can be identified: (1) mid-oceanic, (2) tropical reef, (3) Antarctic, (4) temperate oceanic, (5) continental border, and (6) Arctic. The mid-oceanic type is characterized by relatively warm, saline water with low biologic productivity. The tropical reef is warm and saline but more productive than the mid-oceanic because of its proximity to land. Because of extensive mixing, the Antarctic has abundant marine life. The water mass is much colder and less saline than the mid-oceanic type. The temperate oceanic type is somewhat colder and less saline than the tropical types but has few organisms. Because of the abundance of nutrients and upwelling, continental border areas vary in temperature and salinity. An abundance of life is also present where overfishing or pollution has not depleted the normal productivity. The Arctic type is categorized separately because little is known of its marine life. The water is cold, has low salinity on the surface, and is well stratified. As sea ice covers most of the Arctic Ocean during the whole year, presumably the euphotic zone is greatly reduced, reducing the abundance of marine organisms. This guess, however, is not based on much evidence.

Special mention should be made here of the incredibly high productivity of the waters surrounding Antarctica. Most people associate warm, clear water with abundant life. Observations of the distribution and abundance of life in the ocean indicate that cold, nutrient-rich water contains the most life in terms of the number of living individuals present. Antarctic waters contain the most individuals of any part of the ocean.

Diatoms are particularly common in the Antarctic. They are food for the krill, red, shrimp-like crustaceans that are also very common. Krill in turn are fed on by a host of predators, including blue whales, squids, penguins, and some types of seals. The squids, penguins, and seals are fed on by leopard seals and killer whales (Fig. 6–12).

The Antarctic waters are so productive because cooled sinking water is constantly being replaced by warmer water from the Pacific, Atlantic, and Indian oceans (see pp. 160–161). This circulation of waters brings abundant nutrients into the region. In addition, the low temperatures allow up to 95 percent saturation of carbon dioxide and oxygen. Although penguin eggs possibly have been reduced in thickness due to DDT, pollution of Antarctic waters is less than in other oceans.

6–13 Fishing with a stern trawler net in the North Sea. (Courtesy of Dr. F. Krügler, Hamburg.)

The variety of phytoplankton is limited mainly by the lack of carbonate and bicarbonate ions which, with calcium, form the tests of the common dinoflagellates of tropical and temperate seas. The lack of carbonate is due to the absence of river runoff, the major source of carbonate in ocean water.

Because the sinking of cooled water occurs mainly near the Antarctic continent and the coastal sea ice reduces light penetration, diatom populations and the populations of other directly or indirectly dependent organisms is richest between the Antarctic convergence and the ice barrier. The penguins and seals therefore prefer the offshore waters and go onto the ice or shore only for temporary escape from predators or for reproduction.

Fisheries. Most of the fisheries industry is located in the euphotic zone along continental margins (Fig. 6–13). Tuna, however, may be found in shallow or deep water near or far from coasts (Fig. 6–14). The five major fishing grounds (shown in Table 6–2) harvested approximately 77 percent of the total ocean catch of 55 million metric tons during 1967. In descending order of the metric tons of fish returned to port, Peru, Japan, China (mainland), the USSR, Norway, and the United States are the countries with the largest fisheries.

6–14 Big-eyed tuna from the Pacific Ocean. (Courtesy of the United States Department of the Interior, Fish and Wildlife Service.)

6–15 Pacific sardines. (Courtesy of the United States Department of the Interior, Fish and Wildlife Service.)

Peru with about 10 million metric tons per year, mostly of Peruvian anchovy, has the world's largest national fishery. The catch by U.S. fishermen has been steady at 2 million metric tons per year for a number of years.

Fish occur in many shapes, sizes, and colors. They are found from rarely submerged tide pools to the bottom of the Challenger Deep. Some are bright orange (garibaldi) and a number of deep sea fish are bioluminescent. A few common schooling fish are significant commercially. One of these, the common sardine, *Sardinops caerulea* (Fig. 6–15), is well known for its schooling behavior and for its appeal to human appetites. Schooling provides at least two advantages for marine life. First, members of the sexes are in constant close proximity. The problem of a male individual finding a female is potentially a serious one because of the great size of the ocean. Schooling eliminates this problem. Second, schooling provides protection against natural predators. When a predator attacks, the sardines move rapidly almost in unison, changing direction often. This behavior often confuses predators who snap their mouths at random because they are unable to identify and chase an individual fish in the school. One predator, man, finds schooling an advantage because one large net can engulf a whole school of sardines. In fact, sardines and anchovies (another schooling fish) provide the largest catch by weight and numbers by commercial fishermen. A third possible advantage of schooling to sardines is resistance of the group to trace toxins in the water. Research on this possibility is presently needed. The mucus of large numbers of goldfish, a fresh water carp, has been described as causing colloidal silver, a toxin in the water, to precipitate harmlessly out of solution. This phenomenon has not been described for marine fish that school.

Table 6–2 The Major Fishing Grounds and Their Catches in 1967

Fishing ground	Catch (in millions of metric tons)
Southeast Pacific: around Peru and Ecuador	11.2
West central Pacific: between the Philippines and New Guinea	10.5
Northeast Atlantic: north of the British Isles	10.2
North Pacific: between Japan and Alaska	6.4
Northwest Atlantic: the Grand Banks	4.0

Statistics from the United Nations Food and Agricultural Organization.

The question has often been asked, "How much food can be taken from the sea without depleting its population?" This question is currently especially pertinent because of the human population explosion and the reduction of tillable soil on the land. It is a most difficult question to answer, however, because marine populations of swimming organisms cannot be easily assessed. A scientist cannot, for instance, count the number of anchovies in a school. Computer analysis based on careful sampling has improved assessment

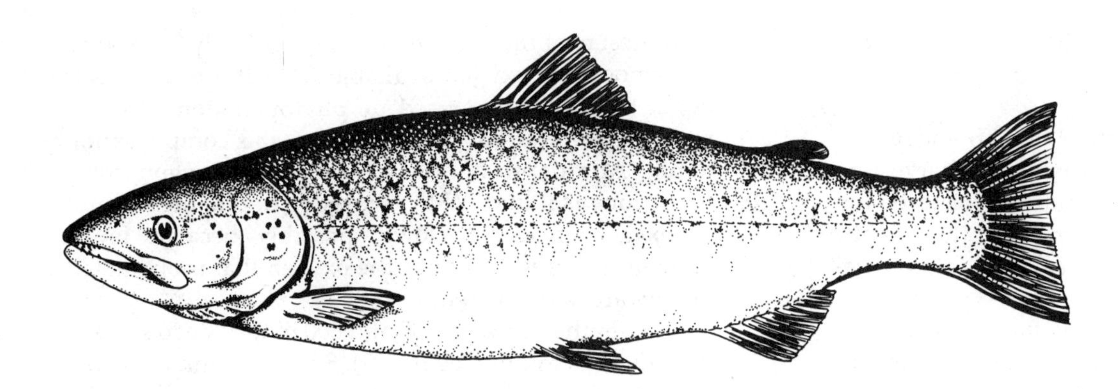

6–16 *Atlantic salmon.*

but schooling and other behavior patterns inhibit careful sampling. Most marine biologists specializing in fisheries research feel that they can assess with 15 percent accuracy at best. Given this limitation, the maximum sustainable catch per year—that is, the maximum amount of fish caught without lowering the following year's fish population—is estimated to be somewhere between 100 and 200 million metric tons. At present rates of fishery growth, especially in underdeveloped countries, this amount could be reached in less than twenty years. The reduction in the Peruvian fishery in 1969 and 1970 indicates that the maximum sustainable catch for anchovies has already been surpassed in that area of the Pacific. Is it any wonder that Peru and Ecuador are sensitive about other countries fishing the southeast Pacific? On the brighter side, oceanic regions not yet tapped appreciably are the Antarctic, the continental border of the Indian Ocean, and the region below the euphotic zone. The Antarctic has, of course, long ago passed the maximum sustainable catch of most whale species, including the largest animal to have ever lived on the earth, the blue whale.

The problem of overfishing has reached many organisms. In addition to the anchovy and blue whale, presently endangered marine life forms are the Pacific sardine, the Atlantic salmon, the Atlantic sturgeon, at least four other species of whales, all of the marine turtles, oysters, and sea otters. The case of the Atlantic salmon is an interesting example of how this can happen.

The Atlantic salmon, *Salmo salar* (Fig. 6–16), spends most of its adult life at sea but returns to fresh water streams to spawn. Spawning occurs in the fall and winter of each year. The eggs are laid by the female in gravel beds of lakes or streams far from ocean water. The eggs hatch in approximately 60 days. When the salmon emerge, they have an odd little yolk sack and are called alevins; when they grow to a length of a few millimeters, they are called parr. The early growth period, which is in fresh water, usually lasts two or three years. During the spring and summer of each year, the two- to three-year-old smolts migrate to the ocean. Once in the ocean, the Atlantic salmon eats voraciously and grows rapidly. It then returns to the stream in which it was born, the females to lay and the males to fertilize the eggs. Although the Pacific salmon dies after its first spawning trip, the Atlantic salmon may spawn a second and possibly a third time.

Surprisingly, the salmon invariably return to the rivers and streams where they hatched, probably identifying them by the memory of some subtle biochemical characteristics. Because fishermen simply stretched nets across the rivers during spawning runs upriver, they tremendously depleted the population of adults capable of reproduction. In addition, chemical pollution of streams by sewage, lumber wastes, and other industrial wastes also affected the fish. International regulations were set up limiting the number of fish caught and preserving the natural character of some rivers and streams. The salmon fishery at this time was limited to harvesting from the rivers because no one knew where the Atlantic salmon went between his stream exodus and return. However, within the past five years, the oceanic salmon grounds have been located in the Labrador Sea. Since this discovery, oceanic fishermen, especially from Denmark, have used special small mesh nets, called purse-seines, to fish the salmon grounds. Again the salmon, after staging a comeback, is below sustainable population. International negotiations are once again being held to save the Atlantic salmon.

In fisheries as well as in other human endeavors, exploitation must be replaced by reasoned harvesting if the natural resources of the world are not to be depleted beyond the point of no return. Must the Atlantic salmon become extinct? Will the Atlantic salmon become extinct?

The role of aquaculture may be important in preserving marine species and in providing food for humans. With some marine fish, being corralled or caged has not been successful. No one has successfully raised tuna. Salmon hatcheries, however, are common with salmon growing up to more than 2 m in length and 32 kg in weight in fenced estuaries. The Japanese have been successful in culturing oysters, seaweed, and shrimp. In many parts of the United States, research scientists at the U.S. Bureau of Commercial Fisheries and other institutions are working on the controlled growth and reproduction of many fish, lobsters, and oysters.

Aphotic Zone

There is no sharp, consistent border between the euphotic and aphotic zones. This transition area may be found at depths anywhere from 100 to 400 m depending on biological, chemical, geological, and physical factors.

In 1843 Edward Forbes, an English marine biologist, suggested the theory that the number of marine organisms decreased as the depth increased. He further suggested the existence of a deep azoic zone without any life. Even before he published this theory, considerable evidence opposing the azoic zone concept was known based on dredging at depths close to 2,000 m. The first part of Forbes' theory has been substantiated, however. Published in the *Challenger* reports are descriptions of 1500 species of animals collected during that historic voyage from depths greater than 2,000 m. In fact, 20 specimens representing ten species were collected from below 6,000 meters.

Because oxygen gas becomes dissolved in ocean water mainly by migration across the air–ocean interface and by photosynthesis, the concentration of oxygen is highest near the surface. At a depth of only 25 to 50 m, the amount of oxygen available from those sources is equal to the amount used by phytoplankton and animals in respiration. This is called the compensation depth. At a depth of about 1,000 m, the oxygen minimum zone is found. With increasing depth, the amount of oxygen increases slowly possibly because fewer animals are present to use the oxygen that is circulated downward with currents.

At depths between 200 and 800 m the "deep scattering layer" is found (see p. 102). Sound within a narrow wavelength band is scattered at this layer. The DSL migrates closer to the surface at night and back downward in the morning. A slight yo-yo movement can be detected as clouds pass overhead. This strange behavior has led scientists to believe that the DSL is caused by plankton and their predators that migrate vertically each day as a feeding and protective behavioral adaptation. The sound wavelength that is absorbed indicates that the organisms most responsible are probably small fish, squids, or shrimp.

Physical, chemical, and geological conditions in the aphotic zone are believed to vary less widely than conditions on the continental shelf. The density of water masses increases with depth and the ocean waters are stratified although moderate, usually slow, mixing occurs. As a rule, the temperature decreases with depth, reaching a typical bottom temperature of about 1.6°C. Drastic exceptions occur in some locations such as the Red Sea. Chemically, little is known about deep water or *in situ* reactions on the ocean bottom. It is believed that complex silicate equilibrium reactions are major buffering agents maintaining the *p*H of deep water. The deep zone is known to have lower concentrations of oxygen and carbonate ions. The subject of deep water chemical oceanography is a major unexplored field at this time. Several geological factors are important on the deep ocean bottom. Submarine volcanoes must have some effect, although just what is unknown. Better known are the effects of submerged ridges and proximity to rivers and canyons. Submerged ridges provide barriers to the distribution of abyssal swimming and benthic organisms. Rivers and canyons usually funnel nutrients, organic materials, and oxidized minerals to adjacent deep water zones. This tends to contribute to increased aphotic zone life.

In the aphotic zone at least three modes of living exist. Best known are the benthic organisms that live on or in the bottom feeding mainly on detritus, which falls from the water above. These organisms include sea spiders, various types of worms, brachiopods, and crinoids. Brittle stars are predators of the attached forms. A second mode is that of wide-ranging swimmers, which eat larger fish or other animals that have fallen from above. Grenadier fish and sable fish seem able to locate fallen carcasses rapidly, as was discovered by continuous photographic monitoring of lighted bait placed at depths exceeding 2,000 m. The third mode consists of fish predators. Because most of the known deep sea organisms are less than 2 m long, many of these predators could only be considered mini-monsters from human perspective. Like the deep sea angler fish the predators are often fierce looking in spite of their small size. A few large predators such as the giant squid and Arctic shark venture into the mid-water region but are not known to inhabit the area below 4,000 m.

Many large marine animals are known to swim in both the euphotic and aphotic zones. Whales dive into the aphotic zone to feed and to escape enemies. Other marine mammals do likewise. Marlin, swordfish, sailfish, and tuna are known to venture below 500 m. A number of species of sharks also cruise parts of the aphotic zone.

The giant squid lives mainly in deep and intermediate water of the open ocean and is known to reach a length of more than 15 m. This creature is thought to be the cause of a number of tales of sea monsters told by fishermen. It is also one of the principal foods of the sperm whale. Apparently, judging from tentacle scars on harpooned whales, the giant squid puts up quite a battle, but few, if any, men have observed such a battle. The octopus, a shallow water cephalopod related to the squid, also has a grisly reputation. They are not aggressive toward humans, however. When attacked by a human diver, the octopus will try to get away. If he cannot, he is capable of putting up a formidable wrestling match, which may be accompanied by severe and sometimes poisonous bites. The octopus is also recognized as one of the most intelligent invertebrates. Many learning experiments have been performed on these animals, and they have displayed amazing ability.

Because the marine life in the aphotic zone is so dependent on the life in the euphotic zone, its population density is closely related to surface productivity. This operating hypothesis still needs more supporting evidence. The technological difficulties and the expense associated with exploring the marine life of deep ocean waters has been prohibitive until the past five years. Certainly each succeeding year will bring new, interesting, and occasionally startling information to us.

Integrating Ideas

A few broad ideas have been proposed to understand the ecology of marine life. These include biogeochemical cycles such as the sulfur cycle, the phosphate cycle, the nitrogen cycle, and the carbon cycle. A more generalized cycle is the nutrient cycle. Also, intricate food webs tracing organic nutrients from plants to predators are being investigated. To illustrate these major theories, it may be helpful to study generalized models of the nutrient cycle and a food web.

Food Webs. References have been made elsewhere in this text to food webs and the place of specific organisms within some web. A food web is a system for tracing energy from its storage in organisms through complex patterns of distribution of that energy to maintain and perpetuate marine life (Fig. 6–17).

The first level in any food web is converting energy into a form that can be stored and used by organisms. The major way this is accomplished is by the process of photosynthesis. In the euphotic zone, many photosynthetic plants are known. In the mid-oceanic province, the phytoplankton are the primary converters. The most important phytoplankton, according to population magnitude, are diatoms, unicellular algae, dinoflagellates, and coccolithophores. Near the shorelines, a wider variety of plants become important converters. For instance, in the littoral and sublittoral zones numerous groups of plants exist, any of which can provide the first step. Partly because of the variety of sources, nearshore food webs tend to be far more complex.

The second level in food webs are herbivores. The web may end at this point, as when a plant dies. The plant remains are then consumed by bacteria, the major final-step organisms in most webs. Another short but broad web would be:

variety of plants → manatee

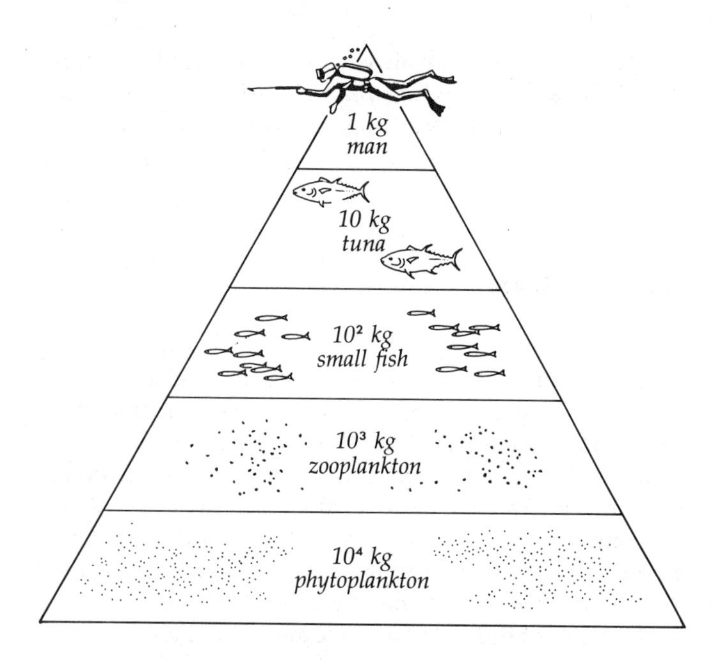

6–17 *Model of mass transfer in the food web.*

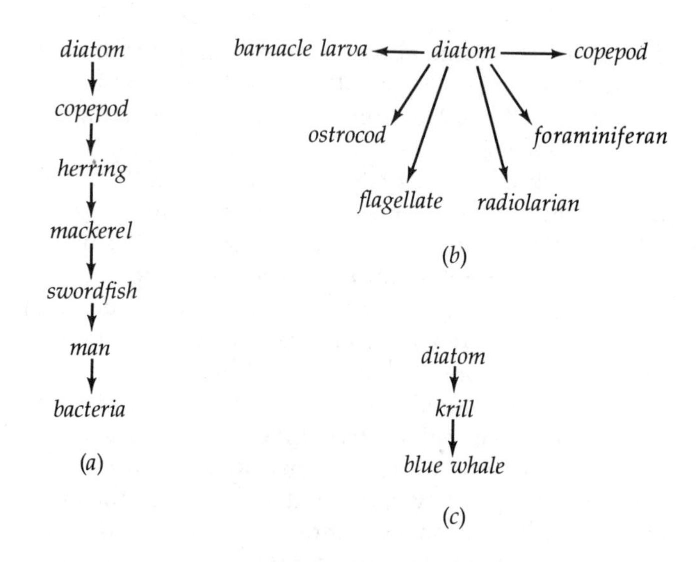

6–18 *Simplified food web models. (a) A seven-level model; (b) a two-level model; (c) an Antarctic food web.*

An adult manatee essentially has no enemies other than man today. Possibly in the past sharks or crocodiles may have preyed on young or injured individuals. The manatee feeds on a host of different types of plants living in estuaries and lagoons. In the oceanic province, the most likely second step would be to some species of zooplankton. The most common zooplankton, on a world-wide basis, are flagellates, foraminifera, krill, copepods, pteropods, and thousands of different types of larvae.

For the third level in food webs, the predator–prey relationships are important. The majority of third-level organisms are carnivores that tend to capture their prey somehow. These are generally organisms that feed on zooplankton or other herbivores. In some instances, large zooplankton may feed on smaller species. Copepods may eat foraminifera. In fact, flagellates may eat other flagellates. Of the larger organisms at this level, the herring, anchovy, sardine, squid, baleen whale, and whale shark are most familiar.

At the fourth level of the food web are the fish eaters. Several species of marine life eat herring, among them sharks, squids, mackerels, and sea birds. In turn, the mackerels may be eaten by bonitos, which are then eaten by seals.

The fifth level is comprised of marine life that essentially have no enemies, at least as adults. These include the great white shark, killer whale, sperm whale, tuna, swordfish, and marlin.

If the food web is defined as the distribution of energy through marine life to a "climax" organism—that is, one that has no predators—the fifth level ends the food web. There are, however, other levels that are part of a more complete trace of energy through marine life as a whole. The sixth level consists of organisms that feed on detritus or dead animals that may be falling through the water column, floating in it, or resting on the bottom. In the sublittoral and littoral zones, such organisms are lobsters, crabs, gulls, snails, and worms. In the oceanic zone, many species of sharks and deep sea fish such as grenadiers are examples. The final level of organisms in any food web contains the decomposers. Principally, bacteria are at this level. Dead organisms from any of the previous levels may directly supply nourishment to bacteria.

Examination of simplified food web models (Fig. 6–18) shows that complete and detailed tracing of energy through marine life is extremely complex. The complexity becomes even more apparent with the realization that many species are species-selective in their choice of food. Barnacle larvae of a particular species, for instance, will digest only certain individual species of other organisms. When the hundreds of thousands of species of marine organisms are taken into account, a

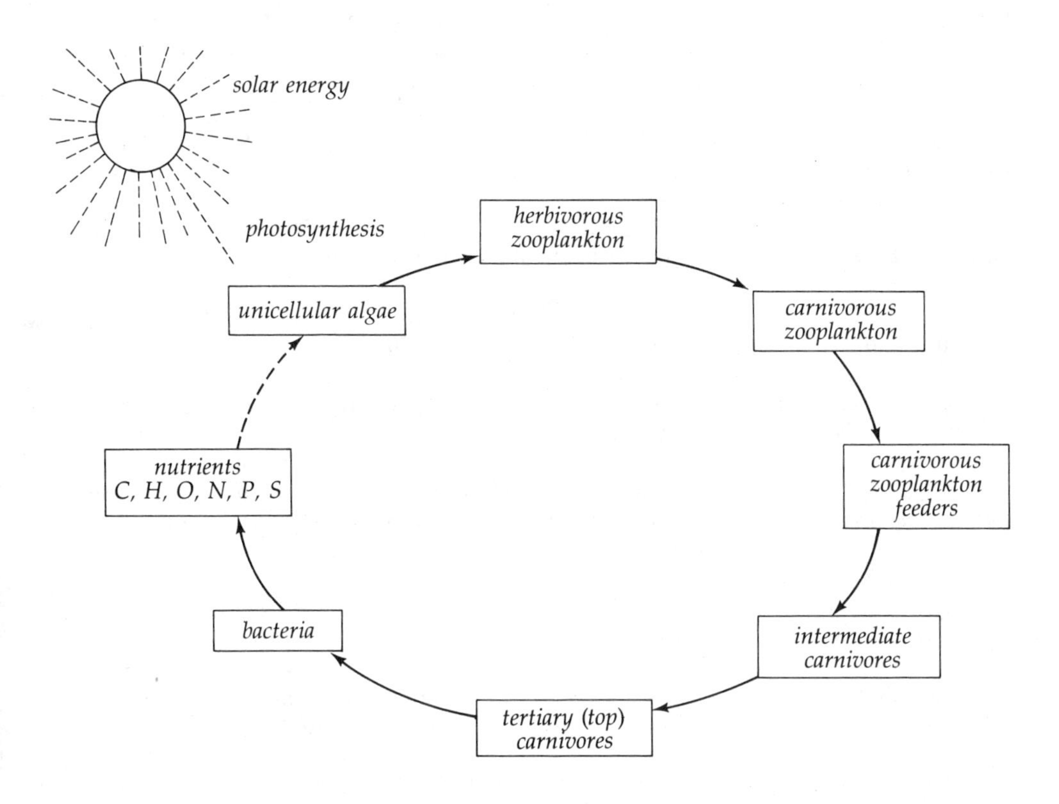

6–19 Nutrient cycle model.

complete food web description is impossible. With further research, particularly using systems analysis techniques and computer programming, our levels of understanding can be greatly increased.

Nutrient Cycle. The nutrient cycle traces, generally, the route of oxygen, phosphates, nitrates, and carbon from plants through the food web to bacteria and back to plants. The major parts of the nutrient cycle consist of the process of photosynthesis of the food web, the conversion of organic substances to nutrients by bacteria, and the route of nutrients from bacteria back to plants (Fig. 6–19).

Some of the most important inorganic nutrients are carbon, oxygen, nitrogen, hydrogen, sulfur, and phosphorus. These may be in single element molecular form or combined with one another. The significance of these nutrients for life in the ocean can be conveyed when their roles are simplified into a few statements. Carbon dioxide, water, and solar energy are used by plants to produce sugar and oxygen. Phosphate forms the high energy bonds in adenosine triphosphate (ATP), the substance which enables plants to synthesize protein molecules. Ammonia is essential in the synthesis of proteins. Sulfur links complex chains of carbon, nitrogen, oxygen, and hydrogen in proteins.

In the oceanic province, the gradual overturn of water slows the return of nutrients to the euphotic zone. In the littoral, the sublittoral, and estuaries, mixing and overturning of water by convection, upwelling, shoreline processes, and estuarine processes greatly increase the availability of nutrients in the euphotic zone. For this reason, abundant phytoplankton blooms are most likely to occur in those areas.

To understand how these nutrients are cycled in the ocean requires an understanding of the food web, the chemical processes in organisms and in ocean water, and the physical movements of the currents that distribute the nutrients. It is not difficult, however, to appreciate that marine biology involves interacting, complex processes between the chemical, physical, biological, and geological domains of oceanology and thereby to gain some insight into the interdisciplinary nature of oceanology.

Further Reading

Allee, W. C., et al., *Principles of Animal Ecology*. Philadelphia: W. B. Saunders, 1949.

Barnes, H., *Oceanography and Marine Biology: A Book of Techniques*. London: Allen and Unwin, 1959.

Barnes, R. D., *Invertebrate Zoology*, 2nd ed. Philadelphia: W. B. Saunders, 1968.

Bayer, F. M., and H. B. Owre, *The Free-Living Lower Invertebrates*. New York: Macmillan, 1968.

Bolin, Bert, "The Carbon Cycle." *Scientific American*, September 1970, pp. 124–132.

Christy, F. T., Jr., and Anthony Scott, *The Common Wealth in Ocean Fisheries*. Baltimore, Md.: Johns Hopkins Press, 1965.

Cloud, Preston, and Aharon Gibor, "The Oxygen Cycle." *Scientific American*, September 1970, pp. 110–123.

Cushing, D. H., *Fisheries Biology: A Study in Population Dynamics*. Madison: University of Wisconsin Press, 1968.

Darwin, C. R., *On the Origin of Species by Means of Natural Selection or the Preservation of Favoured Races in the Struggle for Life*. New York: New American Library, 1958.

Darwin, C. R., *The Voyage of the Beagle*. New York: Bantam Books, 1958.

Deevey, E. S., Jr., "Mineral Cycles." *Scientific American*, September 1970, pp. 148–158.

Delwiche, C. C., "The Nitrogen Cycle." *Scientific American*, September 1970, pp. 136–146.

Ekman, Sven, *Zoogeography of the Sea*. London: Sidgwick and Jackson, 1953.

Hardy, A. C., *The Open Sea: Fish and Fisheries*. New York: Houghton-Mifflin, 1959.

Hedgpeth, Joel, and Sam Hinton, *Common Seashore Life of Southern California*. Healdsburg, Calif.: Naturegraph, 1961.

Hedgpeth, Joel, ed., *Treatise on Marine Ecology and Paleoecology*, vol. 1. Washington, D.C.: GSA Memoir 67, 1957.

Holt, S. J., "The Food Resources of the Ocean." *Scientific American*, September 1969, pp. 93–106.

Hutchinson, G. E., "The Biosphere." *Scientific American*, September 1970, pp. 44–53.

Isaacs, J. D., "The Nature of Oceanic Life," *Scientific American*. September 1969, pp. 65–79.

Jessop, N. M., *Biosphere: A Study of Life*. Englewood Cliffs, N.J.: Prentice-Hall, 1970.

Jorgensen, C. B., *Biology of Suspension Feeding*. New York: Pergamon Press, 1966.

Lauff, G. H., ed., *Estuaries*. Washington, D.C.: American Association for the Advancement of Science, Publication 83, 1967.

Moore, H. B., *Marine Ecology*. New York: John Wiley & Sons, 1958.

Murphy, R. C., "The Oceanic Life of the Antarctic." *Scientific American*, September 1962, pp. 186–210.

North, W. J., and C. L. Hubbs, *Utilization of Kelp-Bed Resources in Southern California*. Sacramento: California Department of Fish and Game, Fish Bulletin No. 139, 1968.

Scammon, C. M., *The Marine Mammals of the Northwest Coast of North America*. New York: Dover Publications, 1968 (original publication date, 1874).

Southward, A. J., *Life on the Seashore*. Cambridge, Mass.: Harvard University Press, 1965.

The State of World Fisheries. New York: UN Food and Agricultural Organization, World Food Problems, No. 7, 1968.

Sverdrup, H. U., M. W. Johnson, and R. H. Fleming, *The Oceans: Their Physics, Chemistry, and General Biology*. Englewood Cliffs, N.J.: Prentice-Hall, 1942.

Trager, William, *Symbiosis*. New York: Van Nostrand-Reinhold, 1970.

Wiens, H. J., *Atoll Environment and Ecology*. New Haven, Conn.: Yale University Press, 1962.

Wood, E. J. F., *Marine Microbial Ecology*. New York: Van Nostrand-Reinhold, 1965.

The Contaminated Ocean

7

The sea was wet as wet could be,
The sands were dry as dry.
You could not see a cloud, because
No cloud was in the sky:
No birds were flying overhead—
There were no birds to fly.

Lewis Carroll

The otherwise smooth water surface
was ruffled over a broad area
by myriads of small breaking bubbles
of gas accompanied by rising
globules of light oil that quickly
spread as they surfaced to form
iridescent films streaked with brown.
On the sea floor the oil issued
as streamers from pinpoints
and larger openings in siltstone outcrops
and soft sediment alike.
Pits and elongate craters on the
sea floor ranged from a few
feet to many feet in length
and up to several feet in depth.
In this vicinity the sea floor
was littered with angular rock debris, . . .
presumably ejected forcefully from
the sea floor depressions
at the time of the blowout eruptions.

Thane McCulloh

I. Natural Changes in the Sea

A. Physical

B. Biological

II. Unnatural Changes in the Sea

A. Fish

B. Mammals

1. *Atlantic walruses*
2. *dugongs*
3. *seals*
4. *sea otters*
5. *whales*

III. The Sea as a Dump

A. Sewage

B. Chemicals

1. *mercury*
2. *red tides*
3. *nerve gas*
4. *lead*
5. *DDT*

C. Oil

D. Solid refuse

1. *trash*
2. *land fill*

E. Radioactivity

F. Heat

IV. The Future

Man has never been particularly neat and tidy in his treatment of the environment. In the earlier stages of his evolutionary development, when man's ancestors were tree dwellers, he simply dropped his wastes out of the trees to the ground beneath. When he moved down to the ground, there was not much change in his habits. With the coming of agriculture, man no longer needed to lead a nomadic hunter's existence to survive. Agriculture allowed him to settle in one locale and, as agriculture progressed, large numbers of people could be supported on the land in one region. Man's accumulated wastes became harder to dispose of but they were natural, and sooner or later—perhaps after a plague—nature could accommodate them. By the Middle Ages, however, garbage in city street gutters was evidence of man's lack of concern for the removal of his wastes.

Streams have always been convenient places near which to live; they are a readily available source of water and in many parts of the world the easiest paths for travel. It was fairly simple for man to throw most of his garbage and wastes into the river. By 1750 parts of the Thames River in England, which had been a dumping ground for cities such as London for many years, were foul with pollutants.

After the invention of pipe, man was able not only to pump water to other areas for his own needs, but also to share his wastes with his fellow man on a larger scale. But nature has always been a large help. Through the course of time natural processes have absorbed most of man's wastes. Occasionally, a river became algae-choked because too many nutrients were present from waste deposited in the river. In some areas, especially where the water flow was not too rapid, this could have been a source of disease. When man became aware of the relationship between bacteria and disease, he had strong impetus to clean up his environment, but in many areas of the world not sufficiently. Even now, large cities such as San Diego, San Francisco, and Los Angeles each pump more than 100 million gallons of only primary treated sewage into the ocean daily.

Pollution, then, is not a particularly new problem but a rather old one. Man has always contaminated his environment, but until fairly recently the pressures he has exerted on the environment have not been great enough so that nature could not recover. The pressure on the environment—not only on the air and land but also on the ocean—really began when man's population grew so large and his amount of waste so great that nature was no longer able to absorb all the slop (Fig. 7–1).

By the early 1800s a tremendous network of canals had been developed in the central and northeastern United States. One can still see fragments of this canal network in northern New York and Pennsylvania. Many of the canals were artificial, but most of the network used existing rivers and lakes. Lake Erie was a vital link in that canal traffic. Most people of those days probably would have thought it impossible for Lake Erie to become so squalid that fish would vanish and most of the life in the lake would die off. A stream perhaps might become polluted, but not something as big as Lake Erie; and until heavy industry began in the United States this pretty lake was too big to pollute. On the assumption that the waste would dissipate and that the capacity of a lake like Erie to absorb wastes was infinite, large amounts of industrial sewage were dumped into rivers and natural bodies of water. As population increased and with it the demand for material goods, the amount of waste pumped into the rivers continued to grow. Most of these industrial by-products do not occur naturally in the physical world and, depending on their composition, could take an extremely long time to be absorbed. As the amount of poisons increased, many eastern rivers and lakes became polluted. To keep man supplied with creature comforts, industry and industrial practices had wiped out the purity of most eastern rivers by the late 1800s. By the early 1900s few rivers in the eastern United States near a population center were unpolluted. With a few exceptions, this process has been continuing constantly into the present. Only recently have attempts been

7–1 A common scene in most of the world's harbors.

made to rectify or curtail the tremendous dumping that the massive industrial complex has been unleashing on the environment. The pollution problem has, of course, proceeded well beyond the original eastern states. In some regions it is almost unimaginable. In the summer of 1970 the Illinois Department of Public Health declared that not a single river or stream in the entire state was safe to swim in, so polluted were the waters. And the problem still continues almost unchecked. The environment is literally being poisoned and cluttered by substances it cannot naturally absorb, from DDT to the lowly though more costly pull-top aluminum can.

Man has always liked to think that whatever he turns his hand or his bulldozers to has a touch of the divine. He has constructed many things and even changed landforms in an attempt to control aspects of his environment, such as water flow, water runoff, and so on. Some have even been beneficial. But man has never been overly concerned with his mistakes, perhaps because he is incapable of admitting that they might be mistakes, or perhaps because he believes that nature can always rectify them. Man has always thought that nature was capable of supplying all his needs and of absorbing most of the stress he has placed on it. In the early days, lakes were always considered inexhaustible. But even a lake has a natural life span. It comes into existence at some time and will eventually dry up. The Great Lakes were created during the Wisconsian glaciation of approximately 50,000 years ago, by melt water runoff and glacial depression. They would naturally fill in during the final stage of eutrophication, about 50,000 years from now. At man's present rate of changing the environment, eutrophication is proceeding at least 20 times faster than normal. Unlike the Great Lakes, the sea is in no danger of drying up—only of dying.

Natural Changes in the Sea

The oceans also change naturally. Unlike lakes, the oceans, of course, do not vanish. But the position of the ocean basins can change, as the global plate tectonics theory indicates. Water may also be added to or subtracted from the ocean to change sea level. In the geological history of the ocean, sea level has changed a number of times, and there is every reason to believe that it constantly changes, though more so during some times than others. Other aspects of the ocean also change naturally. Salinity may be very slowly increasing. Other parameters, such as the temperature of the ocean and the amount of ice cover at the poles, are bound to undergo change. Local regions are changed by forces such as wind erosion and wave activity. Even the color of the seawater in a region may undergo natural change. Natural changes in the physical parameters of the ocean may or may not bring about changes in sea life. A change in upwelling such as

El Niño (see p. 153), for example, causes an accompanying change in sea life. Fossil evidence shows that there are many victims in the struggle for survival in the sea. Numerous species, creatures both scaled and shelled, have ceased to exist in the battle of the fittest. Physical parameters may change so that a species favored at one time is no longer able to live. In the study of marine species, there have been a number of surprises. One fish, the coelacanth, which was thought by scientists to have been extinct for over 70 million years was found to be alive when a Portuguese fisherman netted one off northeast Africa in 1938. Some changes in nature are dramatic. The seasonal changes in plankton and plankton blooms are an example of large-scale rise and fall in biological life. Perhaps the most predictable facet of nature to man is its changeability. As man learns more about the complexities of the ocean—its systems as well as its creatures—he may be able to understand and even predict its changes. Unfortunately, it is becoming all too evident now that a number of changes in the sea are not natural.

Unnatural Changes in the Sea

Although man has dabbled in the sea, charted it, sailed on it, fought on it, he really knows only little about its interactions. This is aptly indicated by the current discussion concerning *Acanthaster*, the crown of thorns starfish. These creatures move across the face of reefs and atolls, devouring living coral polyps. This is now happening along the world's largest reef system, the Great Barrier Reef of Australia, as well as on some western Pacific islands, including Guam. Scientists don't really know whether this is an unnatural, man-caused catastrophe that has to be rectified or a normal cycle of nature. Apparently the only natural predator of the crown of thorns starfish in its environment is the marine gastropod, the great triton. The large, spiral triton shells are attractive, and they have been taken in large numbers by fanciers. Scientists surmise this may be the cause of the crown of thorns starfish infestation. It was recently discovered that the painted shrimp devours the crown of thorns starfish. If the painted shrimp were introduced as a regulatory agent and could decimate the crown of thorns starfish, would the ecology of the reef system change, or would a balance be set by nature? Again, in dealing with the starfish problem man is unsure whether this is a potentially disasterous invasion of the reef system or just a natural cycle that includes a temporary anomaly or imbalance.

Fish. To see some of the changes in the ocean that certainly are not natural, one need only look at fish catches. Man seems always to have thought that the sea can produce an inexhaustible amount of food. He need only go out and get it. Many see it as the solution to the world's food problem—a massive oceanic reserve that could be tapped whenever needed. However, when one measures the fish takes in some regions, another story emerges. Off Los Angeles and San Diego, overfishing for many years and perhaps the proximity of one of the world's largest DDT manufacturing plants have virtually wiped out the sardine and anchovy fisheries that once prospered there. For a good picture of man's voracious assault on these fish, reference to John Steinbeck's *Cannery Row* and *Sweet Thursday* is strongly recommended. Although there is some evidence that the sardine and anchovy population was in a period of natural decrease, this decrease was so great in the area that the fishery based on these two fish simply died out. In the last ten years, according to the National Marine Fisheries Service, the numbers of these fish have increased markedly in the area, presumably because after the pressure on the species by the fishing industry was removed nature could restock the region. The tremendous catches of immature herring off the east coast of Britain in 1968 and 1969 appear to have wiped out the fishing industry in that region.

The North Atlantic salmon (see Fig. 6–16) is in acute danger of extinction. A number of countries surrounding the North Atlantic fish this salmon, but by international agreement nets are not used. Generally, sportsmen fish on shore with poles, especially along the rivers as the salmon come back to spawn. Atlantic salmon spawning runs occur in Maine, Newfoundland, Quebec, Scotland, and a number of other countries in Europe. Recently, the number of salmon coming back to spawn has decreased. The average size of the adult salmon is less, and many of the salmon caught by fishermen show definite signs of net damage. By 1964 and 1965, the number of salmon being sold in Denmark indicated that the Danes had found the area in which the Atlantic salmon spends most of its adult life—the Davis Strait, between Greenland and Canada. Danish and Eskimo fishermen (from Greenland), not adhering to the international fishing agreement on the salmon, are quickly but surely decimating the species. There is no indication that Denmark feels any desire to go to the conference table to discuss the possible extinction of the Atlantic salmon. According to current predictions, at the rate with which Danish and Eskimo fishermen are taking these salmon, the species will be all but extinct in the next five years. Denmark is not a heavy-industry country. Because foodstuffs such as hams and cheeses are her primary export, she would be very susceptible to the economic pressure of individuals' boycotting these products, a step that may be necessary to make her ease off on the salmon.

The North Atlantic haddock has also been overfished for years. In an attempt to allow the haddock population off the Grand Banks of the North Atlantic to increase, areas have been set aside within which no fishing is allowed. This partial moratorium of certain areas rather than numbers taken appears to be working for the haddock, and although the physical limits of the no-fish regions should be increased, the success of the idea so far is promising.

There are a number of other species of fish that are in definite jeopardy. The Northwest Pacific salmon and several bottom fish in that area show signs of depletion. In the North Atlantic there are also decreases in the amount of hake and cod, to mention only a few. But the most hotly contested region in which large-scale fisheries exist is that of western South America, off the shores of Chile, Ecuador, and Peru. These waters have a tremendous plankton population, owing to the large-scale upwelling in the whole region. Thriving on these patches of plankton are numerous anchovies, sardines, and other fish. Stalking these smaller fish are mackerel, shark, squid, and tuna. The tuna is of special interest. Because of the tremendous riches of the water, massive schools of yellow fin tuna, bonito, and skipjack exist.

Before World War II, and for a few years afterward, South American countries had made little attempt to fish these waters. Then, in 1950, Peru began to use anchovies as a source of high protein fish meal for poultry feed. At that time Peru did not contribute significantly to the total tonnage of fish taken in the world. Shortly thereafter, Chile, Ecuador, and Peru extended their international boundaries to 200 miles from the coast, so that they alone would have the "legal" right to fish those waters. (The international boundaries of most other countries extend only 3 miles out to sea.) By then ships from a number of nations, especially the United States, were actively engaged in tuna fishing in these waters. In 1954 Peruvian naval destroyers began seizing American tuna boats off the coast. By 1966 Ecuador had seized 43 American ships and Peru had seized 25. Although these seizures are, of course, piracy, the politics of the situation off Peru is not particularly important here, but only the environmental effects. By 1961 it was becoming apparent that the inexhaustible amount of yellow fin tuna in the eastern tropical Pacific waters was somewhat less than that. Like the fish in so many other regions, the supply of yellow fin began to disappear.

Yellow fin and other tuna had been taken in these waters before World War II. During the four years of the war, virtually no fishing was done in these waters. When fishing resumed in 1946, the amount of yellow fin available was greater than it had been in 1940. It should have been evident to the world's fisheries, or at least to the fisheries in this region, that because fishing pressure had been relaxed for several years, the schools were capable of regrowth. But the total fishing pressure on the yellow fin by 1940 had not been extremely severe. Unfortunately, it is now.

The nations of the world, and primarily Peru, Japan, and Russia, are now engaged in a competition to loot the ocean of its fish. As in the even more serious case of the whales, these countries do not seem to care that they are decimating the stocks of fish. Each only wants to get his share while the fish still remain. With the tremendous increase in fishing by Peru and Russia, the tonnage of fish taken per year between 1953 and 1967 has increased more than ten times. In the past few years American fishermen have often complained about the large-scale fishing activity by Russian fleets and factory vessels off both coasts of the United States. Unlike the countries, the fish themselves are international. A particular skipjack may at one time be found off Mexico and later off Hawaii.

7–2 Virtually all of the shoreline in this especially sea life–rich area of Pacific Grove, California, is a marine sanctuary. This is a much-needed glimmer of sanity when one remembers that annually tons of creatures, shelled and unshelled alike, are pirated from the beaches of Southern California alone, primarily by children. Almost all of these pirated "specimens" are thrown into a bowl of water (often not even seawater) and are then thrown out after a day or two, when they die and begin to stink. (Photo by William J. Wallace.)

The sea supplies only a small percentage of the total calories consumed by the world's population each year. Seafood, however, is rich in protein and currently supplies the world with about one-fifth of its total protein per year. It is obvious to people who can look at the problem rationally that fish catches over the world now are exceeding the ocean's ability to restock. For some years now the catches of yellow fin and other tuna in Peruvian waters have decreased. Other examples have already been cited. Much of the problem is that scientists do not know the maximum sustainable yield that could be taken from the oceans. The figure could undoubtedly be increased from its present value if wanton, indiscriminate fishing procedures by the world's fishing fleets are abandoned. In 1967 about 60 million metric tons of food were taken from the sea, but this probably represents an overtake, largely because of indiscriminate and greedy methods. With care and intelligent conservation, the ocean could be capable of yielding 700 million metric tons of seafood per year. This would involve, of course, not only the aquaculture of seaweed and oysters but fish culture as well. Apparently, the knowledge now exists to predict fishing success from year to year. But many large corporations are stripping the sea of fish and, when these have run out, they will simply switch to other enterprises. They have no economic incentive toward conservation. Therefore, some form of international agreement is needed to regulate the use and resources of the sea and its potential food.

Mammals. Of all the creatures that live in the sea perhaps those affected most by man are the mammals. Although a normal occurrence in nature, the unnatural extinction of a species is a sad thing. Man has completely wiped out numerous species—some perhaps for a reason, most generally not. When colonists first came to this country, clouds of millions of passenger pigeons passing overhead would almost darken the sky. When the last passenger pigeon, Martha, died in the Cincinnati Zoo in 1914, several men who visited the cage shortly thereafter cried. The story of man's relationship with his fellow mammals in the sea is heartbreaking. Steller's sea cow, for example, which weighed approximately 4500 kg (10,000 lb), the largest of the order Sirenia, is now extinct. This peaceful, serene, vegetable-eating animal, much in temperament like the deer, had virtually no commercial value—no horns, no ivory, no fur coat, no leathery hide that was useful; it was simply slaughtered like the buffalo, only for meat. Creatures facing extinction are found in virtually every order of animal classification; besides the mammals there are many, many birds, some reptiles, amphibians, fishes, and even plants (Fig. 7–2).

A number of marine mammals are on the very brink of extinction. Although its Pacific cousin appears to be somewhat safer, the Atlantic walrus, a large animal 3 to 4 m long and weighing up to 1400 kg, is seriously threatened. Originally its range extended from the coast of Massachusetts north. Now only small numbers exist in northwestern Greenland and southern Hudson Bay. It is estimated that a minimum population of 23,000 would be required to sustain the herd at its present level considering the rate of hunting, especially by the Eskimos. In 1956 about 25,000 Atlantic walrus existed but, at the present rate of killing, the population may already have been reduced to the point of extinction. Small herds of Atlantic walrus do exist elsewhere; the species is protected by Dutch and Norwegian law, and in Russia the killing of walrus has been prohibited since 1956. Although there has been some minor legislation, the Canadian government has made little effort to preserve the Atlantic walrus.

In the order Sirenia, the genus *Dugong* has a single species. These inhabit coastal marine waters of tropical regions along the east coast of Africa to the Red Sea, eastward along India and Ceylon to Formosa, Malaya, the Philippines, New Guinea, and the Solomons. They are vegetarians, rather slow-moving, and not very bright. Dugong meat is reported to be tasty (somewhat between pork and beef) and it has been hunted relentlessly primarily for this reason, although its hide makes a good quality leather. Almost no one has attempted to protect the dugong. Most of the hunting is done by primitive peoples, especially in Indonesia, where dugongs are now slaughtered with modern weapons.

Several species of seal are in danger. The ribbon seal, ranging from northern Japan to the Bering and Chukchi seas, has a total population of between 5,000 and 20,000 individuals. The ribbon seal has never been very abundant, and its skin has a low market value. They are hunted mostly by Japanese sealers, primarily for their oil and meat. There has been no attempt to protect these creatures in North America; in fact, Alaska still offers a bounty of $3 for the ring or harbor seal, and ribbon seals are taken along with these. A number of other seals are also in danger, including the Ross seal (the smallest Antarctic seal), the Mediterranean monk seal, the Caribbean monk seal, and the Hawaiian monk seal. Attempts to keep these seals in captivity have not proven successful. None have survived longer than three years.

In 1867 the estimated seal population of the largest fur seal rookery in the North Pacific, Alaska's Pribilof Islands, was 4.7 million. By the early 1900s, only 125,000 were counted. The drastic rate at which the fur seal population decreased led the United States, Japan, Russia, Britain, and Canada into a joint agreement to protect the Pacific fur seals. Before the 1911 fur seal convention, there had been hot disagreement between these countries as to the disposition of the seals. At one point the U.S. House of Representatives attempted to pass a bill that ordered the mass slaughter of all seals and therefore would end any future dispute. Only the failure of the Senate to pass this bill stopped the wholesale extermination of the seals. Since the 1911 agreement, management has proceeded remarkably well. Today the seal herds are in tremendously better shape than they had been in 1910, yet because of proper management, more seals have been taken than ever could have been had the situation not been approached intelligently. This has been man's most successful attempt to manage and preserve marine mammals, or any marine species.

Another marine mammal, the sea otter, was also hunted almost to extinction. The sea otter pelt, which is probably the best and most valuable of all fur, brought as much as $1,000 per pelt in the late nineteenth century, when a prime beaver pelt was bringing no more than $15. By 1900 most sea otters had been slaughtered. In 1910 the United States passed a law prohibiting the capture or killing of sea otters in American waters. For the next 25 years most people thought the sea otter was extinct. Just before World War II, two small colonies were found: one off Monterey, California, and another in Alaskan waters. The sea otter is the most highly specialized of all otters.

A sea otter spends a fair amount of time grooming his coat. Unlike other marine mammals, the sea otter does not rely on a thick coat of fat beneath the skin to insulate him from the cold. He has only the trapped air in his fur to insulate him from the cold water. This makes the sea otter especially vulnerable to oil spills. Recall that the sea otter seldom leaves the ocean. While he never goes far out to sea, he seldom ventures more than a few yards from the sea's edge if he ever comes onto land. The increasing amount of crude oil in the ocean may well become the sea otter's greatest enemy. Crude oil coats the animal's fur causing it to mat and lose its insulating property; the animal becomes chilled by the cold water and will probably die shortly thereafter.

Of all the marine animals, none has been hunted more relentlessly or is in greater danger of extinction than the whale (Fig. 7–3). It is now certain that the world's largest mammal, probably the largest animal ever to have existed on this planet, faces extinction unless the senseless slaughter by a few vested interests is stopped at least for a time. The rate at which the great whales are now being killed is beyond the rate at which they can reproduce.

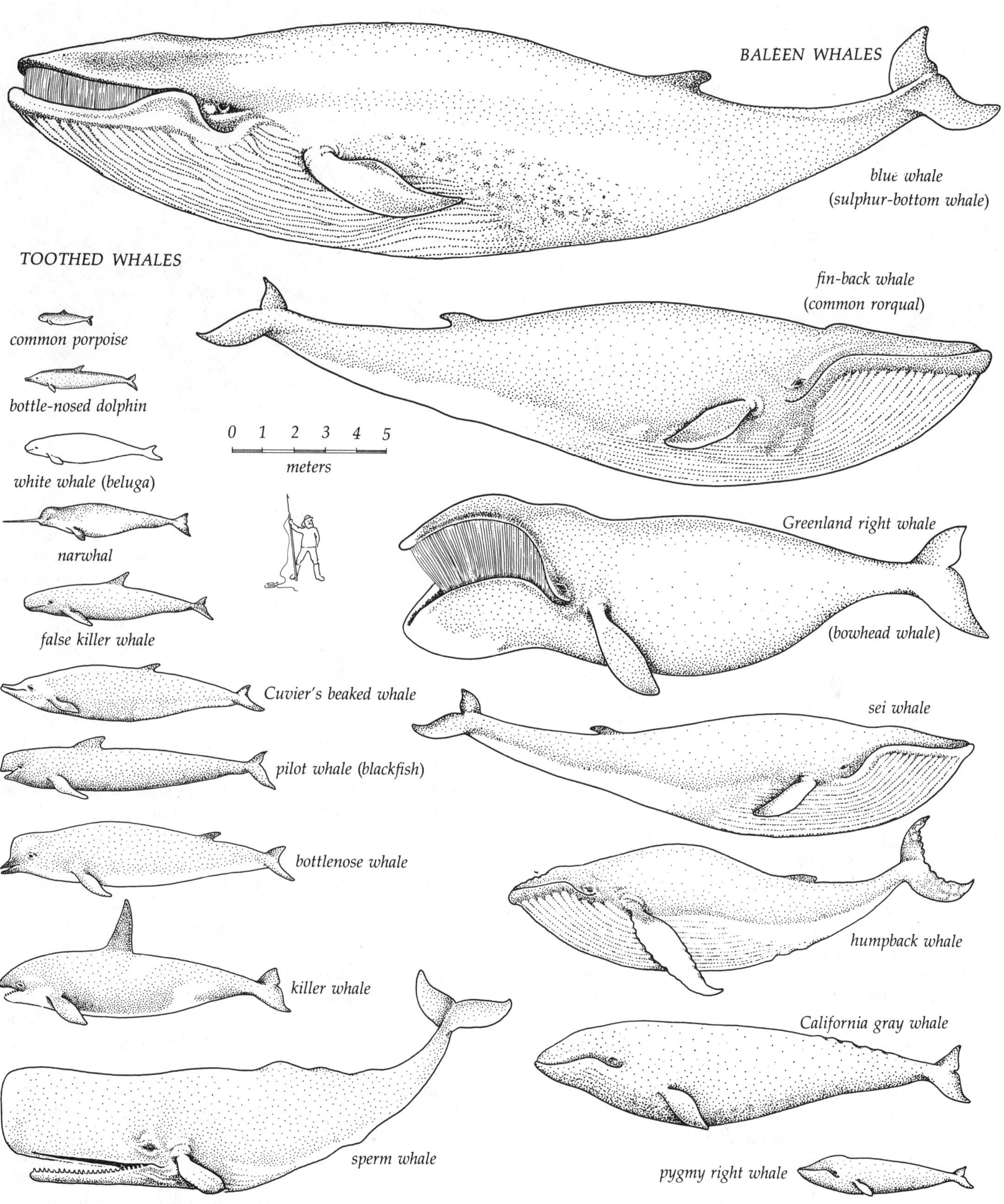

7–3 *Relative sizes of whales.*

As far back as 1851, Herman Melville, in his great novel *Moby Dick,* questioned whether whaling practices of his day might not wipe out the whale, but he decided that these animals would probably survive. In the days of Melville and sail-driven ships, not all large whales were hunted. Some whales were not hunted simply because they sank immediately after being killed. But with steam power, faster ships, explosive harpoon guns, and compressed air to keep the dead animals afloat, by the early 1900s virtually all species of large whales began to be taken. The blue and the finback, which could exceed 20 knots in speed, were now at the mercy of the whalers.

Before World War II, there were virtually no attempts to limit the catches of whales. As early as 1900, when U.S. whalers were withdrawing from the industry because it was becoming unprofitable, the blue whale population was definitely decreasing in the North Atlantic. Deep sea or pelagic whaling then moved to the Antarctic and just before World War II large numbers of whales were taken. In 1931, for example, almost 30,000 blue whales were killed. In 1934 a Japanese whaling fleet joined British, Dutch, and Scandinavian vessels in the Antarctic to pursue the whales. Shortly after the war, eighteen nations formed the International Whaling Commission to conserve whale resources and to develop methods for determining the best use of whale populations. Although the idea that the commission be formed was a good one, as might be expected, the IWC has virtually no powers. It cannot seize or inspect whaling vessels nor can it enforce any of its mandates. At best, it can only recommend the number of each species of whale that might be taken in a year. Unfortunately, the IWC has not set individual quotas for each whale species but set up a unit, the blue whale unit (BWU), in which one blue whale is equated to other species. One blue whale unit, for example, equals two finback whales or six sei whales. Because the blue whale is the largest, it made sense for whalers to seek out and take these largest whales and, of course, the numbers of blue whales decreased rapidly. As one might expect, there are a tremendous number of violations even of these not-too-stringent regulations; and apparently no one has ever served a jail sentence for violating whaling regulations.

By the early 1960s damage to the blue whale population was becoming all too apparent. With the surge in whaling that occurred after World War II, approximately 7,000 blue whales were caught each year until 1950 (Fig. 7–4). In 1952 the catch had dropped to 5,000, in 1956 less than 2,000 were taken, and in 1962 the catch was just slightly over 1,000. Yet the radical decrease in the numbers of blue whales did not stop the whalers and only when the 1963–64 whaling season yielded approximately 110 blue whales did the International Commission begin to take more notice. The hunting of blue whales was officially terminated in 1966, when only 20 blue whales were caught.

The countries engaged in whaling were Britain, Japan, the Netherlands, Norway, and Russia. By 1963, Britain had abandoned its whaling industry, and

7–4 The large-scale factory-like slaughter of the great whales. Included in the group of dead whales to be rendered by the factory ship are a number of blue whales. (Photos by Dr. Ted Walker.)

Holland, in 1964 deciding that it was no longer economical to whale hunt, sold its fleet to Japan. In the last couple of years Scandinavian interests have decided also that whaling is no longer profitable. This has left the bulk of the whaling to Japan and Russia, although Chile and Peru also do some whaling, primarily from shore stations. With the exception of some small advances made by conservationists in these two countries, neither Chile nor Peru have shown any indication that they will stop taking blue whales. The best estimate now is that there are less than 1,000 blue whales left. If this does not appear to be a drastic or a severe reduction, remember how large the ocean is. Experts agree that if the numbers of the blue whale fell below 1,000 the statistical chance of a male finding a female in the vast expanses of the Antarctic Ocean would be smaller than the species' normal mortality rate. So there is a very good chance that the blue whale is already functionally extinct. Probably, the greatest of all the animals ever to have existed on the planet has already passed below the survival level.

Because the numbers of blue whales have decreased so drastically, whalers have begun hunting other species of whale more intensively, and the pressure on these species has increased tremendously. At the current rate of hunting, the finback, humpback, sei, and sperm whales will shortly reach extinction. The number of kills of all the great whales has been decreasing steadily since the early 1960s (Fig. 7–5).

The whaling fleets, now almost entirely Japanese and Soviet, are exterminating all the great whales. In 1967 Japan took 39 percent of all the whales caught, and the Soviet Union took 41 percent. It should also be pointed out that this is not a tremendous heavy industry. In 1967 the entire whaling fleet was less than 300 ships—about 250 catchers and 16 floating processing factories. Compared to the gross national products of Japan and Russia, the whaling industry is a small business. But unlike most of Japan's business, for example, it is moving toward extinction. The Japanese seem to know this, for already they are making plans to convert their whaling vessels to other uses, based on the supposition that in four or five years no more whales will exist and it would be uneconomical not to use an already existing ship for something. The Japanese use whales not only for the oil, which is made

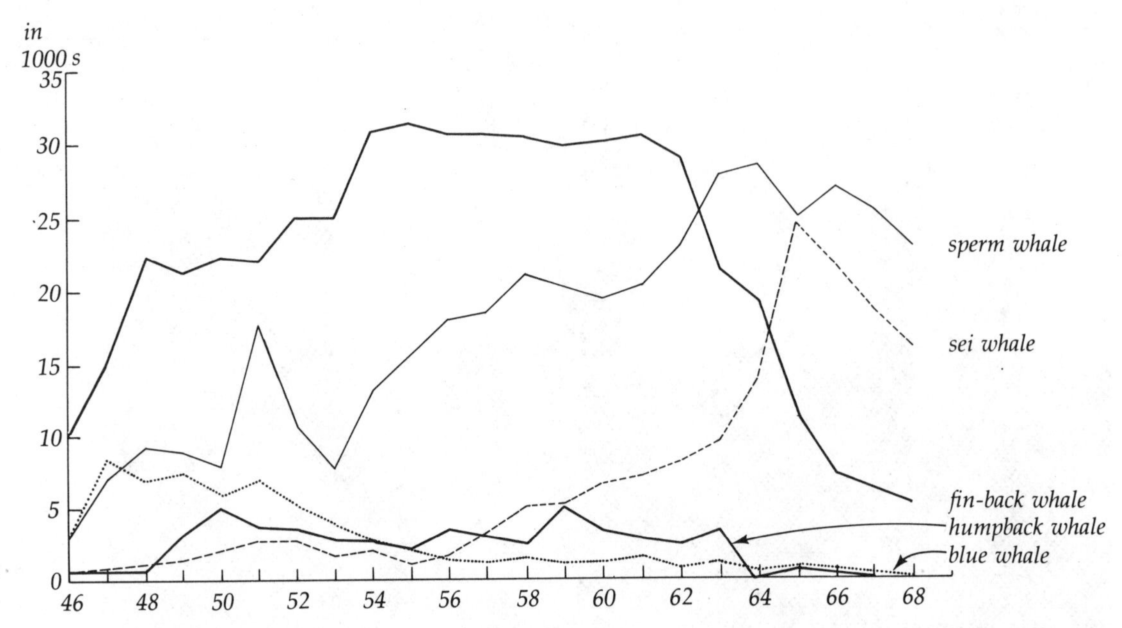

7–5 *Whale kills from 1946 to 1968.*

into soap and margarine, but also for the meat, which is used as animal food and fertilizer. In 1967 the Japanese ate over 176,000 tons of whale meat.

The most important question is a personal one. Does it make any difference to you personally that somewhere in the world's oceans whales exist, whether or not you ever see one? If it makes no difference, they will simply cease to exist. If it does make a difference, then each of us has to work at it. What can one do if these countries do not now begin to solve the problem themselves? Probably the logical place to start is with Japan. Because whaling is such a small part of Japanese industry, there is no reason why it can't be curtailed. All whale hunting would not have to cease. The actual take in whales could increase if the entire whaling industry were managed as man manages a number of the other animals on the planet. Conservation is not only preservation of a species; it can also involve the intelligent use of those species. Intelligent handling could not only preserve the whales but could make the whaling industry profitable for a longer period of time rather than ending in three to five years when all the whales are gone.

The Japanese economic system is capitalistic, and much of it depends on trade with the United States. On a personal basis, then, what can one do? One can write letters, especially to companies like Honda, Datsun, Toyota, and Sony—the heavy industry companies of Japan. Quite probably, letters to these companies explaining that a number of Americans hadn't yet purchased a Japanese car or electronic equipment only because they were concerned about what the Japanese are doing to the whales would be effective. Any form of pressure by the Japanese automotive industry on the Japanese whaling industry would probably result in action by the whalers. What can one do about the Russian destruction of the whales? Little directly, but if Japan were to intelligently attack the problem, the Soviet Union—which is extremely conscious of its world image—might defer to public opinion.

Whale Catch Limit Set

(Associated Press—Tokyo) *Japan, the Soviet Union, and Norway signed an agreement fixing the limit on whale catch in the Antarctic Ocean during the next whaling season, the Foreign Ministry said.*

They said under the agreement Japan is permitted to catch 1,493 whales, Norway 231, and the Soviet Union 976. (Christian Science Monitor, *28 July 1970*)

Very recently a glimmer of sanity *may* have entered the whaling scene. Though no mention of species is given in the article, it is hoped that these countries have agreed to so limit their whale kills. They may herald a true beginning of whale conservation. Time will tell.

As people are becoming aware that damage to the environment cannot be taken lightly, one need only think of Lake Erie, or the oil in the Santa Barbara Channel, or the air over Los Angeles, or the destruction of fertile valleys and forests by bulldozers, or the chemical defoliation and decimation of tremendous forests in Southeast Asia, or the huge losses in the bird population (including eagles and pelicans) through pesticides to begin fearing for the ultimate ruin of the planet. Personally, one can take steps not only to protect his environment but also to help the whales. Many worthy organizations can be supported both financially and with your time.

A disturbing thought has come to the authors more than once concerning not only the whales but all the many endangered and extinct species. One wonders if the end of all these species is not an omen or a warning of the possible end of the human species through too much cleverness, too much ignorance, and too little caring.

7–6 A sewer outfall at Point Piños, Pacific Grove, California. Here sewage, only primarily treated, is dumped directly into the surf zone. The dark mass on the rocks in the background is the normal seaweed or algae cover in this area. The prevailing current is toward the viewer. Note that the sewage effluent, containing too much chlorine, has bleached the rocks in the foreground free of algae. (Photo by William J. Wallace.)

The Sea as a Dump

Using the sea as a dump is potentially more dangerous than any other way man changes the sea. The question of human survival may be involved here.

Sewage. If a river's water is clear and clean and if there is little or no accumulation of green algae on rocks, the river is free of an inordinate amount of nutrients. The rivers close to man in North America and in most countries of the world are contaminated. Japan, Russia, Western Germany, and Britain all have polluted rivers. The commodity most deposited by man throughout history in his waters is sewage. This is hardly new, but the disposal now is much more copious and efficient. The once beautiful Hudson, now beautiful only from a distance—if one covers his nose—is for all practical purposes an open sewer, so bad that a truck could almost be driven on it. The effects of this tremendous stream of pollution can be noted well over 100 miles at sea. Some careful watching at a city's waterfront will reveal barges loaded with garbage and debris heading out to sea and always returning empty. Scientists do not know the effect of sewage on the open ocean, but no serious accumulative effect has yet been discovered. However, in the nearshore waters—the bays and the estuaries—sewage pollution has already closed down thousands of acres of oyster and clam beds in this country alone (Fig. 7–6). The water is so foul in most of this nation's estuaries that people cannot go swimming for fear of disease, and there is always the possibility of epidemic. Twenty years ago, children in New York City used to swim in the Hudson River; now it is often said that if one tried to fish there, the worm would get typhoid. The situation is not unlike that of cities of the Middle Ages, which had raw sewage running right down the gutters of the streets. Coastal towns in Italy still do.

Chemicals. Associated all too often with human sewage are chemical wastes. As man throws sewage in his vital waters, polluting not only streams, lakes, rivers, bays, and estuaries, but also ground water, he generally for reasons of expediency throws in all sorts of chemical wastes. Without question, the most potent environmental danger to man is the chemicals he puts in the water and air.

Much has been said lately about DDT and lead in the environment. These two do present real dangers, and more about them will be said presently. But a tremendous volume of lesser known substances are also released, ranging from other pesticides and herbicides to detergents, industrial solvents such as phenols and benzenes, and strong acids like hydrochloric and sulfuric. Until recently, a stream or river polluted with detergents could be spotted by the foaming. Now modern science and the oil industries have made compounds that take away the foaming. The detergents are still there, but now you can't see them or their action.

The unleashing of chemical poisons is insidious. Numerous deaths about the world have been reported. A classic example of water-borne, industrially caused death is the "Minamata disease." Minamata Bay, on Kyushu Island, Japan, was the scene of over 100 deaths by this strange affliction between 1953 and 1960. The symptoms of the disease varied, but general deterioration of mental and nervous faculties culminated either in death or complete insanity. Investigations showed that people and pets who ate seafood seemed to be the only ones afflicted by this disease. Shellfish taken from the bay and fed to laboratory animals caused similar symptoms. The cause of the malady turned out to be a large poly-vinyl chloride factory located on the headlands of the bay. Waste water from the flushing of catalytic columns that contained mercuric chloride was being pumped into the bay. The mercury was concentrated by the shellfish, and people who had eaten the shellfish were suffering from highly toxic mercury poisoning. All heavy metals, of which mercury and lead are only two, are poisonous to the human system. Mercury disrupts and destroys segments of the central nervous system. Although the plant has stopped dumping its waste waters into the bay, it will be years before this bay can be used by fishermen and clammers because the sediment is still heavily contaminated with mercury. Similar instances of mercury poisoning on a smaller scale have occurred throughout the world.

No one knows why the occurrence of the phenomenon known as red tides has become more common. With the exception of two small outbreaks, before World War II this phenomenon seldom occurred either in the Gulf of Mexico or off California. Now it has become all too common. Massive red tides occurred off Florida in 1946–47 with subsequent outbreaks in 1952–54, 1957–64, and 1966. The red tide is a plankton bloom. The plankton responsible for red tides exists normally in numbers approximating 1,000 per liter of seawater. For unknown reasons, over very short periods of time this number increases to approximately 50 to 60 million organisms per liter. Toxins secreted by the organisms (*Gymnodinium breve* in the Gulf of Mexico and *Gonyaulax polyhedra* in southern California waters) are capable of killing fish. Through the lack of oxygen in the water caused by fish decomposition more fish die, and the decomposition generates more nutrients to further extend the plankton bloom. The red tides of 1946 killed over 50 million fish, which tides and waves piled up onto the beaches of Florida, taking over two months to clean up. Aside from the potential danger of disease and epidemic from the large number of dead fish piled on the shore, red tides cause other problems. Peru is worried about red tides off its coast because the anchovy, the most important fish taken by that country's fishing industry, is very susceptible to red tides. At Sea World in San Diego a performing pilot whale in a pen connected with Mission Bay died after a red tide outbreak in the area. Red tides may also be harmful to man, although not directly. A man who swam in a red tide area might experience a mild allergic reaction; however, if he ate shellfish that had been taken in an area where a red tide had occurred or was occurring, he could become seriously ill and possibly even die. Much of the reason that taking shellfish such as clams and mussels on the California coast is forbidden by law between May and October (Fig. 7–7) is the possible occurrence of red tides and the concentration in shellfish of toxins secreted by the dinoflagellates that are the red tide.

Why should red tides be on the increase? One suspected possibility is the effect of man on the oceans. Attempts to correlate rainfall and runoff with red tides tend to indicate a pattern, though not conclusively so.

In the Gulf of Mexico, years of high rainfall seem to correspond roughly with the occurrence of red tides. If this is the case, what might be in the water runoff that is not normally there? There are several possible answers. Phosphate is normally a minor or trace element in the sea. Mining and to some extent agriculture are adding greatly to the phosphate content of runoff, which flows into rivers and then to the ocean. Elements such as iron or cobalt or compounds such as tannic acid might be responsible for the large-scale increase of plankton. There are also a number of other possibilities, and alarmingly most of them are linked to man's changes in his environment.

Many other types of matter have been dumped into the ocean. Industries constantly dump anything from expended rocket engines to large containers of both solid and gaseous industrial chemicals. Nerve gas is now a special worry. Over the years, governmental services commonly have dumped a number of such gases into the ocean in pressure cylinders. Nerve gas has been constantly dumped by the Army into the Gulf of Mexico since World War II. The bulk of captured German nerve gas was dumped into the North Sea. Presumably, these canisters have already rusted through and nerve gas has slowly found its way into and has been absorbed by the environment. But it is also possible that they have not rusted through. While the possibility of extensive damage to the environment is not imminent, if a number of canisters were to rupture at the same time, the highly water-soluble nerve gases could cause large-scale fish and marine life kills over wide areas. This is yet another example of the way the ocean has been used as a dump without thought about its effects. With increased population and industry, dumping cannot go on without more insight toward the preservation of the sea.

7–7 Most of the west coast of the United States and much of the east coast has at least a many-month ban on clamming each year. (Photo by William J. Wallace.)

The amount of lead in the environment is much higher than normal and will probably continue to increase. The lead that people absorb through their lungs or ingest with food may well turn out to be more damaging than the classic example of lead poisoning, the bullet. The tetraethyl lead used to raise the octane rating of automotive and aviation gasolines is now probably the prime source. In 1968 alone, 32 million kg of tetraethyl lead were used in automotive gasolines and an additional 27 million kg in aviation gases. Contrast this with total use up to 1950 of about 136 million kg. One way to get an indication of lead buildup in the environment is to analyze the lead content of the ice cover at the poles. Because the use and subsequent release of lead into the environment has been greatest in the northern hemisphere, any change in the composition of the ice should be greater at the North Pole, and such apparently is the case. While the proportions of other materials over an equivalent time span have shown little variation, the lead content in the Greenland icecap between 1750 and 1940 has increased by about 400 percent. Between 1940 and 1967 this value has increased an additional 300 percent.

Paint, solders used in tin cans, toothpaste tubes, piping, and some kinds of crystal glassware and ceramics also release lead through normal abrasion and wear. The lead content in the blood of an average American is approximately 0.25 parts per million (1968). City dwellers and those who have closer contact with automobiles—such as service station attendants, mechanics, and especially traffic policemen—show an alarmingly higher lead content.

There has been no definitive study of the lead content either in the sea or in marine organisms. It is known, however, that lead is toxic to almost all organisms. Although most of the lead from automotive exhausts is released into the atmosphere, the ultimate reservoir is the sea. Previously, it had been supposed that the lead washed into the sea would mix thoroughly through the ocean. Very recent data shows that the lead tends to concentrate in the surface waters, where most sea life exists, for much longer periods of time than previously supposed. As lead is a cumulative poison, it is well past time to remove it from fuels. It should be mentioned, however, that any substitute for tetraethyl lead, especially if it contains heavy metals, might also become a serious problem.

Much of the hazard of chemical poisons in the environment comes not only from their properties but also from concentration. As has been mentioned, toxins from red tides are concentrated by organisms such as mussels. It has been generally assumed that the slop released in the water and air by man and his industry would dissipate. Unfortunately, nature cannot always conform to this role, and lethal concentrations of substances such as DDT have often been found in dead birds, some fish, shellfish, and even dolphins. The concentration problem becomes twofold—in a region near the source of the contaminant the concentration is high, and organisms also naturally concentrate certain substances.

No place on earth is richer in life than the complex salt marsh region associated with many estuaries, especially in those of the geologically older east coast of the United States. Of the approximately 900 estuarine systems in this country, most are close to man and are affected by this proximity. Either in the number of species present or in the total biomass, the marsh virtually teems with life. But the rivers that are a part of every estuarine system constantly bring chemical pollutants, insecticides, and herbicides to the estuaries, where they do untold damage. Not just the life that lives constantly in the marsh is endangered; a number of commercially important deep sea fish species such as flounder and bass are spawned in these salt water marshes.

The pesticide problem is an especially difficult one. Modern agriculture is dependent on these compounds and will become more dependent as the population, and, therefore, the stresses on cultivated land increase. But their damage to the environment is massive. DDT is such a stable compound that it remains in the environment for long periods of time. Some animals, such as oysters and worms, concentrate it enough to kill the organism that consumes them. The ingestion of DDT by higher organisms can damage vital organs, especially the reproductive organs. Birds are extremely susceptible. The decrease in the number of eagles, pelicans, and many shore birds is probably in large part a result of increased amounts of insecticides. One known effect of DDT and associated chemicals is that of decreasing the thickness of eggshells, making them so fragile that the female cannot even incubate the egg.

Of greater magnitude is the effect of DDT and other such compounds on planktonic creatures. It has recently been determined that a DDT concentration of only 2 to 3 parts per billion in the water, which is less than the present concentration in most estuarine waters, decreases photosynthesis in these plants. Even with the greater use of oxygen in the air by man and his machines over the past 50 years, the percentage level of this vital gas in the atmosphere apparently has remained unchanged. Recall that photosynthesis supplies most of the oxygen to the air, and that about half of this photosynthesis is carried on by phytoplankton. All life in the sea and perhaps on earth is ultimately based on phytoplankton and the photosynthesis they carry on. The implications are obvious.

Less obvious is the fact that DDT affects various planktonic organisms differently. Insecticides could change the structure of the food chain by wiping out some phytoplankton and not others. At least potentially, this might drastically change the upper levels of the food chain that comprise large fish and mammals of the sea. There is little question that man now needs insecticides and herbicides. The ones that will hopefully soon replace DDT should have a transient lifespan, or in other words, should chemically degrade in a short enough period of time that they never reach the sea.

Oil. Anyone who looked carefully at the water anywhere in the world's northern shipping lanes would probably see some small black spheres; this is crude oil. In more southern waters, oil often forms a heavy black matting film. The colder temperature of the northern waters tends to make the oil form into tar balls, ranging from the size of a marble to a little less than that of an orange. The sea does not absorb oil well. Water and oil, because they are chemically and structurally different, have little affinity for one another. The colder the water, the slower will be the rate of absorption. In the warmer waters, tar balls generally will also form because of conditions such as a choppy sea, unless the oil is washed onto beaches. The fabled purity of the open ocean is no longer a fact now. Many fish taken in the North Atlantic have an oily taste from having ingested these tar balls. Scallops and oysters in some areas also have the same oil-like taste.

Each year the world uses 2 billion tons of crude oil. Half of this oil is shipped across the oceans. In the course of this transfer, 1 million tons are lost into the environment each year, primarily through handling in harbor and secondly by seepage at sea. The ever-present oil slick plagues virtually all the world's harbors. The at-sea seepage results from two things: the normal leaking of some of the oil and the fact that many tankers clean their oil tanks by flushing them with seawater. This latter practice is illegal but very difficult to prevent and is still a very common phenomenon. Small oil slicks, not large enough to be a real problem ecologically but large enough to bother the bathers at many of the world's beaches, probably come from the flushing of tanks just off the coast. Normal seepage depends on the quality of the particular tanker. The ships of most nations and oil companies are conscientious about this seepage at sea. Many of the oil tankers throughout the world are large and new, and the seepage is very small. Some countries and individuals, however, choose to register their ships under the flag of Panama. Ships carrying the Panamanian flag are, for all practical purposes, the dregs of the world's ships. They are seldom in good shape and even a freighter would probably leak a fair amount of oil through its fuel tanks. Many large shipping magnates register their vessels in Panama to avoid the often much more stringent regulations and taxes of their own governments.

The figure of 1 million tons of oil lost per year does not include calamities such as the wreck of the tanker *Torrey Canyon* on the English coast of Cornwall, which spread 100,000 tons of crude oil over the beaches, killing a tremendous amount of marine life, especially birds. Nor does it include the spillage in the Santa Barbara Channel when a Union Oil drilling rig, in January 1969, allowed 100,000 tons of crude oil into the channel and beaches, which became choked with oil. It smothered and killed many birds and seals. This Union Oil well, by the way, is 6 miles offshore. The area was leased to Union Oil by the federal government, yet the federal government claims full jurisdiction only out to the 3 mile limit. Union Oil was hardly a purposeful polluter in either of these fiascos. From their standpoint it was just bad luck. These disasters, however, helped focus the attention of the world on the problem. It became immediately evident that the ocean was not equipped to rectify these massive oil slicks when they occurred, and also that tremendous ecological damage was done to wildlife in the areas afflicted. This has long been a problem. With the exception of the diminishing sand on the beaches, oil has been the biggest problem to beaches and associated wildlife since World War II (Fig. 7–8). Many shorelines along the United States, especially those adjacent to harbors, have a constant oil film along the frontage. The Santa Barbara Channel and *Torrey Canyon* incidents were not unusual; large oil spills happen almost every day. Newspapers all over the country report local spills commonly, but they simply do not reach the general public's attention. There is speculation now that some of the oil leaking onto the beaches in Florida may come from the fuel tanks of vessels that were sunk during World War II, which have finally rusted through, allowing oil slowly to seep out. There are also, of course, occasional sinkings of antiquated barges, offshore collisions in harbors and rivers, and other such events that dump thousands of barrels of crude oil annually into shoreline waters (Fig. 7–9).

7–8 The result of a damaging oil spill in Tampa Bay, south of St. Petersburg, Florida. Black and sticky as molasses, crude oil splotches made beach-walking unpleasant for people and fatal for birds. Thousands of ducks, loons, and other shore and water birds were found dead along a ten-mile stretch of shoreline when a Greek tanker, Delian Apollon, *ran aground and spilled 21,000 gallons of oil into the bay on February 13, 1970. (Photo by Carson Baldwin, Jr., 3325 Bradenton Road, Sarasota, Florida. Courtesy of the Outdoor Photographers League, San Diego, California.)*

The damage that a single oil spill can wreak on the waterfowl of an area must almost be seen to be believed. Birds are unable to tell as they land on or plunge into the water that it is covered with oil. Their feathers, which not only insulate from the cold but buoy them up in the water, become penetrated with the oil. They may freeze to death or drown. Because their feathers are matted together, they are no longer capable of flying. Any bird that has become coated with oil will die (Fig. 7–10). Even with individual care, of the thousands of birds saved by volunteers after the San Francisco Bay oil spill, only a few percent lived. Some of these birds were nursed for more than a year before they could be released. Each year thousands of shore birds ranging from swans to gulls, pelicans, and loons die as a result of oil spillage. The Atlantic auk is in danger of extinction partly because of oil spills. One colony on the Newfoundland coast has been virtually wiped out because of oil pollution in that region between 1965 and 1966.

But the damage here is greater than the loss of the birds. In the past, oil companies have quickly pointed out that after the initial kill-off of birds and other forms of wildlife, an oil slick soon dissipates, either naturally or with help, and the problem is solved. After the *Torrey Canyon* and Santa Barbara disasters, however, no one really studied the ocean bottom. It was falsely assumed that all oil and its components remained floating and that little biological damage was done to the ocean. But neither fish catches nor life on the bottom around Cornwall and Santa Barbara had been extensively studied before the spills.

A recent high-caliber report at the famed Woods Hole Oceanographic Institution in Massachusetts shows clearly that the oil spillage problem goes much further. The Woods Hole study was triggered by the grounding of a barge on September 16, 1969, and the subsequent leakage of only 650 to 700 tons of #2 fuel oil—a light, supposedly readily dissipated oil—in an area that had been studied by the institution for years. Within a few weeks the oil appeared gone. Controlled bottom surveys showed, however, that the bottom as deep as 14 m from the surface was contaminated. Eight months later the bottom oil contamination had spread over an area many times greater than that of the initial spill. Once the oil, a complex mixture of chemicals, had settled to the bottom the life there was essentially destroyed. Once the marine life was gone, the cohesion of the bottom sediments was reduced, and the sediments with the oil deposits moved along the bottom to areas not previously polluted. By May 1970, bacterial breakdown of the oil, usually cited by oil companies, had not progressed extensively.

An area inundated by oil can recover only very slowly. There is long-term damage to marsh and marine organisms. Creatures such as clams have been shown to concentrate poisons from the oil that are dangerous for human consumption even months after the oily taste is gone. Obviously the sinking agents proposed

7–9 The oil spill caused by the collision of two tankers on January 18, 1971, in the San Francisco Bay. The foreground is outside the bay. Alcatraz is at the top right. (Courtesy of Western Aerial Photos, Inc., Redwood City, California.)

and used by some oil companies to get rid of oil spills are not a true remedy. They simply remove the problem from sight—and presumably from mind.

In short, many people now believe that while lead and mercury contamination are the more publicized, oil pollution is the biggest threat to the ocean. Only a small percentage of the ocean—generally, the coastal regions—is truly productive of life. It is primarily in these coastal waters that the world's total oil spillage, probably between 5 and 10 million tons annually, occurs. As the world and its people become more industrialized, the need for fuel, especially crude oil, increases. Not only is more of it drilled but more of it is transported, presenting an ever greater danger of leakage into the environment. The new supertankers capable of carrying tremendous volumes of oil are an ecologist's nightmare (Fig. 7–11). If one of these tankers should pile up on the rocks off Alaska, for example, the resulting oil slick could wipe out wildlife along the coast of most of Alaska and the Pacific Northwest and make the *Torrey Canyon* fiasco look like a puddle. The thousands of miles of pipelines that exist along the shores of the United States and other countries are extremely susceptible to earthquakes. These pipeline systems

7–10 A few of the many thousands of ducks and other water birds that succumbed when the Greek tanker Delian Apollon *ran aground (see also 7–8). (Photo by Carson Baldwin, Jr., 3325 Bradenton Road, Sarasota, Florida. Courtesy of the Outdoor Photographers League, San Diego, California.)*

16,419 deadweight tons	523 ft — Oregon Standard (cause of San Francisco Bay disaster)
117,000 deadweight tons	974 ft. — Torrey Canyon (cause of the famous spill off England, 1967)
250,000 deadweight tons	1,100 ft. — ships now being built to carry oil from Valdez, Alaska, to Seattle and San Francisco
500,000 deadweight tons	1,280 ft. — Japanese tankers now under construction
1,472 ft.	Empire State Building
1,000,000 deadweight tons	1,600 ft. — proposed ship now on English drawing boards

7–11 The relative sizes of presently in-use tankers such as the Oregon Standard *(which crashed in San Francisco Bay in 1970) and the* Torrey Canyon, *which crashed off England in 1967. Note that the* Oregon Standard *is only one-fifteenth the size (in carrying capacity) of the ships that would bring oil down the coast from Valdez, Alaska, to San Francisco, and a sixtieth the size of a ship being proposed by the English. In case you think something about their being bigger makes them safer, consider this: If the captain of one of those 250,000 tonners from Alaska sees trouble ahead while going full speed, it will take him a half hour to stop. (Courtesy of Friends of the Earth.)*

inevitably will become much larger and more complex, as in the case of the proposed massive trans-Alaska pipeline, and the possibility of spillage through ruptures increases.

As there is little question that the world's industries will continue to need oil for the time being, the oil must be managed more carefully. Transfer procedures in the harbor could be vastly improved, minimizing leakage in these important closed areas often adjacent to or part of an estuarine system. International pressure on the ships of those countries who have lax shipping regulations can be increased. Methods to detect oil film caused by the flushing of tanks at sea are being developed, and the U.S. Coast Guard is working toward this end. Just stopping this all too common practice of flushing ships' tanks at sea would probably cut down a fair amount of the oil film on beaches (Figs. 7–12 and 7–13).

Oil companies are extremely susceptible to public economic pressure. Had a very large number of Californians turned in their Union Oil credit cards along with a letter of explanation when the Santa Barbara disaster first happened, the manner in which Union Oil attacked the problem might have been a bit more

7–12 Crude oil splashed onto cliffs at Point Loma, near San Diego. The oil stain is ten feet above the average high water mark. (Photo by William J. Wallace.)

7–13 Crude oil covered rock and creatures in a tide pool at Point Loma. Most rocky areas now show this oil coverage to some degree. (Photo by William J. Wallace.)

rigorous—especially since Union Oil sells most of its oil to Californians. While several examples of environmental contamination and destruction have been given, it should be made quite clear that not only at home but throughout the world in, for example, lead emission, DDT manufacture and use, and oil spillage the United States is the world's foremost polluter.

Solid Refuse. The variety of solid refuse dumped in the ocean is almost amazing. As a matter of course, many chemical companies regularly send barges out to sea to drop drums of waste, spent, or dangerous chemicals. The Defense Department and a number of related industries have over the years dumped tremendous amounts of obsolete hardware into the ocean. It is fairly common for large cities like Los Angeles to dump confiscated weapons into the ocean. Narcotics hauls by federal and state agents around the country have sometimes been disposed of in the sea. A number of California motion picture studios have even thrown in old movie sets. A look at a Coast and Geodetic Survey chart for almost any offshore area in the country, especially in some of the larger harbors, will indicate areas where caution is advised because those regions are used normally for bomb disposal. The Navy, for example, constantly uses some areas as gunnery ranges, and it is impossible to say that all the ammunition has detonated. Old bombs and shells are also dumped. Radioactive material has often been disposed of in the ocean, but more on this later. In some areas of the country where the landscape is choked with automobiles, many people want to throw these carcasses into the ocean. While in some areas this might be beneficial because the wrecked auto bodies could serve as a shelter for a number of small fish and could artificially create a sport fishing industry, the reckless dumping of all this material into the ocean is less than desirable for fishermen, who could wreck their nets by catching them on debris.

Land fill. By far the most damaging solid refuse dumped into the nearshore environment is fill. In the name of progress, bays, marshes, and estuaries have been and are being filled to provide more land area near the water for marinas and housing developments. In this, real estate agents are among the primary villains. Comparing charts of most U.S. harbors such as San Francisco or San Diego over the past 50 years shows that the amount of square mileage of harbor or bay has been decreasing, especially since World War II, as large areas are filled in. If filling continues at the present rate even realtors and city fathers should realize that soon there will be no waterfront left for the marinas. There originally were 1131 km^2 of San Francisco Bay; now slightly over 463 km^2 remain after 668 km^2 have been filled in. In San Francisco, as in most places, filling proceeds for the standard reasons: graft, greed, and stupidity, as well as an ill-informed apathetic public. It would be unfortunate aesthetically to lose those regions of the cities that often are the most attractive. It has been truly said that San Francisco is the bay and, without it, would be just another city. But it is more than a question of aesthetics. Unfortunately, a number of the cities around San Francisco Bay—including the cities of San Francisco and Oakland—a large number of companies (such as Dow Chemical), and the U.S. Navy all pump large amounts of sewage and industrial slop into the bay. As fill increases in the bay, the total tidal prism, the amount of tide water moving in and out, decreases. Flushing of the bay by the tides therefore decreases, and pollution increases. Added to this, in the case of San Francisco, is the fact that the Federal Bureau of Reclamation is planning to pump the water effluent from 10,000 acres of farmland into the bay. This effluent water will contain large amounts of pesticide, sulfate, nitrate, and so on.

But from the standpoint of the ocean and the ocean system, filling presents a much greater problem. As mentioned, bays and estuaries are a region of biological productivity second to none. An estuary is over 20 times more productive than the same area of open ocean and easily several times that of the same acreage of filled land. As man continues to encroach on the estuary, not only by pollution but by physically filling it in and destroying it, that tremendous cradle of nearshore and offshore life is cut down.

At present, no federal law controls the filling of tidal and estuarine waters. Either state or local governments have had to see to it themselves. Unfortunately, as one can see from the fill, little attempt at preservation has been made at the local level. With some luck, San Francisco may well be able to save what is left of the bay. A number of other cities are not nearly so fortunate. City and county fathers have in many areas been the primary reason for the filling in of estuaries, perpetrating the myth that marinas bring progress, progress brings people, and people bring greater prosperity. Even the fill used may be damaging. It is common to fill with anything available and often, as with Fiesta Island in San Diego's Mission Bay, dry sewage is used. The tremendous amounts of nutrients in this material may for a time upset the biological balance of the region. A few states have shown themselves singularly responsible in their treatment of estuarine regions. In Massachusetts it is against state law to fill any marsh or estuarine region, even if the marsh is privately owned. Recently in that state, to increase the number of units in his housing development, a contractor filled in some of the marsh land around the edges of the development. A wealthy matron, while touring the area in her chauffeur-driven car, happened to see the filling in progress. She contacted her state congressman, and within two days the contractor had most of his work force armed with shovels laboriously but carefully excavating the fill they had put into the marsh. Examples like this are heartening, though hardly common.

Radioactivity. With the advent of a new power source, the atom, came yet another form of environmental contamination—radioactivity. Natural radioactivity occurs in the physical world. On this planet a small percentage of the atoms of each element are radioactive and, although constant radioactivity is being introduced into the environment by high energy electromagnetic radiation from outer space, the level of radiation was slowly decreasing naturally. The large-scale testing of fission and fusion weapons for a decade or more after World War II noticeably increased the background radiation counts. Although the nations of the world, with the exception of China, have agreed to cease atmospheric testing, there is every reason to believe that the radiation level is still increasing somewhat. The building of nuclear reactors and the large-scale building of nuclear power plants especially are rapidly increasing. These generate very large amounts of radioactive waste products. Before 1961 it was the usual procedure for the Atomic Energy Commission to dispose of radioactive wastes in the sea.

The waters of the deep sea are not as still as the AEC once thought. Fifty-five gallon drums of radioactive wastes have washed ashore in Oregon. North Atlantic fishermen have brought up similar drums in trawl nets in areas well away from restricted dumping regions. Other occurrences such as radioactive crabs in the Rongelap Islands of the western Pacific, or radioactive tuna in the same area, or hot clams in the Marshall Islands have caused a hard second look at the situation.

Apparently all large-scale dumping of radioactive wastes into the ocean by the U.S. has ceased. European countries, however, have made little effort to dispose of their radioactive wastes in any other manner than by dumping it into the sea. The Wind-scale Nuclear Works in England pumps about 50,000 Curies per year into the waters of the Irish Sea. As more and more nuclear reactors are built around the world, it is safe to say that within the next decade 100,000 tons or more of radioactive wastes will be produced per year. Currently, no satisfactory method for the treatment of these waste products is available. Deep sea dumping is not the answer, and large-scale storage in remote (sometimes underground) tanks also leaves something to be desired, although this is the method now used in the United States. The danger of introducing radioactive materials into the environment (especially the marine environment) and its subsequent ingestion and perhaps selective concentration by a number of organisms remains more than a potential hazard.

Heat. One of the most subtle contaminants or pollutants in the environment is heat. In a modern society such as the United States today, the amount of energy that each person uses is tremendous, approximately equal to that of 7.5 tons of coal, oil, or natural gas per individual per year. Within the next ten years, this figure will probably increase to 10 tons per capita. With the expenditure of all this energy comes the accompanying release of large amounts of heat. While much heat is vented into the atmosphere, most of it probably finds its way into water, initially fresh water that flows into salt water or in a number of cases into the salt water directly. The steel and aluminum industries, for instance, not only use a tremendous amount of water but add a large amount of heat to the water. Rivers in heavy industrial areas are extremely polluted and much warmer than they would normally be. One difficulty with both nuclear reactors and conventional electrical power generating stations is the large amounts of hot water they expend in cooling. These generally are pumped into the waters of the environment, although some plants do have cooling towers. While large-scale releases of energy may in some areas ultimately affect the climate of those areas, the harmful effect on the environment of heating the water in rivers, streams, and the ocean is not usually noticed in an area. Often there is no direct harmful effect, although many fish like salmon become almost drugged in warmer water, and they become much more susceptible to injury by disease and predation. In the case of salmon and most species, there is probably little outright fish kill. The temperature of the Columbia River, for example, where salmon are fighting for survival, is increasing, due to the effect of industry, dams, and nuclear reactors such as those at Hanford, Washington. Salmon are a cold water fish. In seeking out the river or stream in which they were spawned, if they come across warmer water than is natural, instead of surging upstream anyway they will mill around at the mouth of that stream or river waiting until the temperature drops. If the temperature does not drop, then the salmon will not make a spawning run. Quite simply, thermal pollution forms a barrier to their spawning grounds for the salmon and other anadromous fish. The increased temperature may also affect fish metabolism, feeding habits, and growth rate, as well as reproductive patterns. As the temperature of a river or stream is changed by the heat that man adds, the total local ecological cycle —microscopic and plant life and ultimately animal life —may be permanently altered.

Although a number of states, such as Washington and Oregon, that border the Columbia River are greatly concerned about the possibility of the loss of salmon

(a)

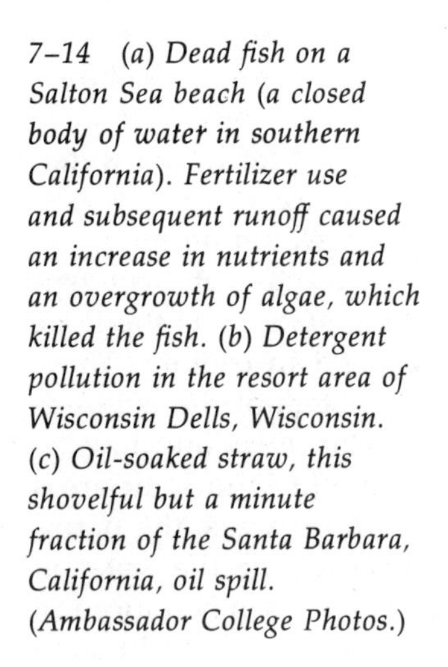

7–14 (a) Dead fish on a Salton Sea beach (a closed body of water in southern California). Fertilizer use and subsequent runoff caused an increase in nutrients and an overgrowth of algae, which killed the fish. (b) Detergent pollution in the resort area of Wisconsin Dells, Wisconsin. (c) Oil-soaked straw, this shovelful but a minute fraction of the Santa Barbara, California, oil spill. (Ambassador College Photos.)

runs and although Congress since 1965 seems to have embarked on a program to conserve anadromous fish populations, the regulatory agencies of the federal government that could enforce thermal controls have not taken a firm stand. The AEC, for example, which is responsible for the licensing of nuclear power plants, has conveniently ignored the strong recommendations of such government agencies as the Fish and Wildlife Service for some form of thermal control in at least ten recent nuclear power plants. Private power utilities, both nuclear and conventional, have been extremely powerful opponents to conservationists. There is little question that in a number of regions the warm water effluents of private plants have caused extremely large numbers of fish kills. Not too long after the opening of a Consolidated Edison Plant at Indian Point on New York's Hudson River, for example, at least one concerned conservationist noted that large numbers of bass, a semi-anadromous fish, were dying in the warm waters of Consolidated Edison's thermal effluent. Tons of dead fish were furtively removed by truck to nearby dumps, where the fish were treated with lime to more quickly remove the evidence.

The Future

To predict the future or attempt to do so is probably man's most difficult and least productive activity. Very often this is so because he does not understand the present. The logical conclusion to a number of the situations discussed above is not too difficult. Man obviously can alter the composition of large communities of organisms. He can also alter the nature of his entire environment and now can even change the climate. A number of calamities are building up in the environment because man hasn't known enough about his environment to predict how it might be affected as a result of his actions (Fig. 7–14). It seems that man usually has to be confronted with a catastrophe before he begins to take action and work on the problem. In the past, this reaction time often has been a number of years. Just how much time does the environment have before serious change occurs in it?

Each problem has a number of possible solutions. In the fall of 1969, when mackerel taken by Los Angeles region fishermen were found to contain more than the

(*b*)

(*c*)

federally acceptable 10 parts per million of DDT, the fish were put into frozen storage while lobbyists in Congress attempted to coerce the federal government to change its DDT content levels to 15 or 20 parts per million so the mackerel could be sold. Solutions like this have occurred most often. Although the number of concerned people in the United States and the world is growing, they are often almost offset by the fact that industry generally at best gives only token answers to the problems. These have taken the form of subterfuge or advertising double-talk that only seems to solve some aspect of a problem. The price of a clean environment is money, perhaps as little as 10 percent of the GNP. The question is "Are we willing to pay for this clean-up?"

The above is but a glance at the bleak picture man has painted for himself on the globe. Only rationality, a commodity historically absent in the affairs of man, can save this earth from destruction, whether by bomb or by pollution. Whatever is put into the ocean will remain there for a long, long time, and the catastrophe will endure.

Further Reading

Bardach, John, *Harvest of the Sea*. New York: Harper & Row, 1968.

Blumer, Max, Howard L. Sanders, J. Fred Grassle, and George R. Hampson, "A Small Oil Spill." *Environment*, vol. 13, no. 2 (March 1971), pp. 2–12.

Buell, R. K., and C. N. Sklada, *Sea Otters and the China Trade*. New York: David McKay, 1968.

Carson, Rachel, *The Sea around Us*. New York: Oxford University Press, 1951.

Coker, R. E., *This Great and Wide Sea*. Chapel Hill: University of North Carolina Press, 1947.

Commoner, Barry, Michael Corr, and Paul J. Stamler, "The Causes of Pollution." *Environment*, vol. 13, no. 3 (April 1971), pp. 2–19.

Ehrlich, Paul R., "Ecocatastrophe." *Ramparts Magazine*, September 1969, pp. 23–32.

Fonselius, Stig H., "Stagnant Sea." *Environment,* vol. 12, no. 6 (July/August 1970), pp. 2–11.

Gammon, C., "Danes Scourge the Seas: Decimating Migratory Atlantic Salmon." *Sports Illustrated,* December 15, 1969, pp. 28–30.

Gullion, Edmund A., ed., *Uses of the Seas.* Englewood Cliffs, N.J.: Prentice-Hall, 1968.

Hedgpeth, Joel, "The Oceans: World Sump." *Environment,* vol. 12, no. 3 (April 1970), pp. 40–46.

Hickel, W. J., "When a Race Breathes No More: Extinction of Forty-Seven Species of U.S. Wildlife." *Sports Illustrated,* December 14, 1970, pp. 70–71.

Hirdman, Sven, "Weapons in the Deep Sea." *Environment,* vol. 13, no. 3 (April 1971), pp. 28–42.

"Horizon to Horizon: A Review of Oil Spills." *Environment,* vol. 13, no. 2 (March 1971), pp. 13–21.

Iversen, E. S., *Farming the Edge of the Sea.* London: Fishing News Books, 1968.

Kormondy, E. J., *Readings in Ecology.* Englewood Cliffs, N.J.: Prentice-Hall, 1965.

Loftas, Tony, *The Last Resource.* Chicago: Henry Regnery, 1970.

Marx, Wesley, *The Frail Ocean.* New York: Coward-McCann, 1967.

McDougal, M. S., and W. T. Burke, *The Public Order of the Oceans.* New London, Conn.: Yale University Press, 1962.

Moxness, Ron, "The Long Pipe." *Environment,* vol. 12, no. 7 (September 1970), pp. 12–23.

Olson, T. A., and F. J. Burgess, *Pollution and Marine Ecology.* New York: Interscience, 1967.

Petrow, Richard, *In the Wake of Torrey Canyon.* New York: David McKay, 1968.

Ricketts, E. F., Jack Calvin, and J. Hedgpeth, *Between Pacific Tides.* Stanford, Calif.: Stanford University Press, 1968.

Sammartino, P., ed., *Conference on Oceanology.* Rutherford, N.J.: Fairleigh Dickinson University Press, 1968.

Scheffer, V. B., "Can We Save the Great Whales?" *McCall's,* May 1970, p. 54.

Scheffer, V. B., "Cliché of the Killer." *Natural History,* October 1970, pp. 26–28.

Scheffer, V. B., *Seals, Sea Lions, and Walruses: Review of the Pinnipedia.* Stanford, Calif.: Stanford University Press, 1958.

Scheffer, V. B., *Year of the Seal.* New York: Charles Scribner's Sons, 1969.

Scheffer, V. B., *Year of the Whale.* New York: Charles Scribner's Sons, 1969.

Steinbeck, John, *The Log from the Sea of Cortez.* New York: Viking Press, 1951.

"Whale of a Failure." *Time Magazine,* July 13, 1970, p. 44.

28P1307

8 Man and Technology

Sank through easeful
azure. Flower
creatures flashed and
shimmered there—
lost images
fadingly remembered.
Swiftly descended
into canyon of cold
nightgreen emptiness.
Freefalling, weightless
as in dreams of
wingless flight,
plunged through infra-
space and came to
the dead ship,
carcass that swarmed with
voracious life.
Angelfish, their
lively blue and
yellow prised from
darkness by the
flashlight's beam,
thronged her portholes.
Moss of bryozoans
blurred, obscured her
metal. Snappers,
gold groupers explored her,
fearless of bubbling
manfish. I entered
the wreck, awed by her silence,
feeling more keenly
the iron cold.

Robert Hayden

I. *Transportation*

A. *Oil tankers*

B. *Submarines*

C. *Hydrofoils*

II. *Submergence*

A. *Scuba*

B. *Deep sea submersible vehicles*

C. *Projects*

III. *Mineral Exploitation*

A. *Oil*

B. *Manganese*

C. *Phosphate*

D. *Gold*

E. *Diamonds*

F. *Desalination*

IV. *Information: H.M.S.* Challenger *vs.* Glomar Challenger

A. *Purpose*

B. *People involved*

C. *Description of vessels*

D. *Voyages*

E. *Results*

V. *Use of the Sea Bed*

A. *UN proposals*

B. *Military use*

In four major areas, ocean technology has expanded significantly: (1) transportation, (2) submergence, (3) resource exploitation, and (4) information acquisition. As a result, international political problems and worldwide conservation problems have come into sharp focus recently.

Transportation

Although newspapers have reported the demise of trans-Atlantic tourist ships from lack of public interest, the broad field of marine transportation has made astounding technological advances. These advances have been mainly in the transportation of commercial commodities and in military vessels. Lesser advances have been made in short-haul passenger services.

A number of oil tankers now operating on the ocean are each capable of carrying more than 300,000 deadweight tons of oil (Fig. 8–1). Plans are being made for oil tankers that carry 1,000,000 tons. The largest oil tankers of 15 years ago carried only 30,000 deadweight tons. This development is rapidly eliminating the need for canals such as the Suez Canal in transporting oil. At the same time, however, potential pollution dangers are created by these "supertankers." Oil spills in the ocean have increased during the past 10 years, with some turning into marine life disasters either by direct toxicity or by the unwise use of toxic dispersants such as detergents. Oil is broken down slowly by bacteria, mussels, oysters, snails, and many other organisms. Tar clumps even provide surfaces on which barnacles, bryozoans, and other sessile organisms can attach and live successfully. Many diving birds, including bald eagles, ospreys, cormorants, and pelicans die rapidly when their feathers become oil-soaked. The question is not "Should oil tankers be allowed to operate?" but rather "Why can't oil tankers be operated without a standard 1 percent spillage and with more precautionary devices?"

Military vessel technology has advanced sharply in the past 15 years, partly because of other advancements in such areas as missile technology. Three major factors establishing naval superiority 20 years ago were battleships, aircraft carriers, and diesel-powered submarines. All three are now largely obsolete because of the accuracy of guided missiles and detection devices. Rapid achievements have been made in nuclear submarine, rapid hydrofoil, and hovercraft technology. The key strategic research and development pertains to nuclear submarines. Most important recent advances are covetously guarded. Some that are well known, however, include: deep water operation, especially below water mass density layers; long periods of continuous submergence, now up to at least 60 days; and missile delivery capability from below the ocean surface.

Hydrofoils that move at speeds over 50 knots can deliver extensive fire power and can operate in shallow water. They are not significant in a major military sense, although they have proved their usefulness in such places as the Mekong Delta of South Vietnam. Hydrofoils are also now being used to carry passengers rapidly over short distances. They are presently in operation between England and France, between Miami and the Bahamas, and in many other places around the world.

Submergence

Man has developed the technology to submerge himself into the marine environment for many different purposes, under a variety of conditions, and for varying periods of time. Self-contained underwater breathing apparatus (scuba) can be purchased at sporting goods stores in almost any town in the United States. A well-equipped diver has a tank filled with compressed air, an air regulator that is attached to the tank on one end and has a mouthpiece on the other end, a face plate, rubber flippers, a rubber insulated wet suit,

8–1 *The supertanker* Universe Ireland. *(Courtesy of Gulf Oil Corporation.)*

8–2 *Diver among fish examining a sand channel off the north coast of Jamaica. (Official United States Navy Photograph.)*

a weight belt, and an inflatable life preserver. An experienced diver with this equipment can explore the nearshore areas of the ocean to a depth of 40 m. Reliable equipment can be purchased for less than $500. Recent developments such as recycled air units and special gas mixtures have enabled professional divers to reach depths of more than 300 m on test dives and work for as long as an hour at about 200 m. These activities are still in an experimental stage, and several experienced divers have died testing apparatus. Amateur divers using scuba are now a common sight from southern California waters to the Mediterranean Sea. It has become a rewarding hobby for those swimmers who are willing to keep physically fit and who enjoy the thrill of exploring a hostile environment (Fig. 8–2).

Underwater habitats for man are also being used in research. The U.S. Navy sponsored the Sea Lab Project (Fig. 8–3). The first two phases of the project were conducted off La Jolla, California. An underwater domicile was placed at a depth of approximately 65 m. Two crews each spent two weeks living inside and working for various periods of time on scientific and technological problems. Ex-astronaut Scott Carpenter spent the entire test period of four weeks underwater. Sea Lab III was scheduled for the winter of 1969 near San Clemente Island off southern California. In a preparatory dive to check the domicile at a depth of 200 m, one of the divers died. Sea Lab III has since been discontinued.

The oldest oceanographic institute in the world, located at Monaco in the Mediterranean Sea, sponsored the Conshelf Project under the direction of the inventor of scuba, Jacques Cousteau. The project's major purpose was to observe and test man's manipulative skill and physical endurance at a depth of about 65 m. While ashore, the divers learned how to assemble an oil well valve system called a "Christmas tree" and then assembled such a device successfully for the first time while on the bottom.

The first Tektite Project dive was made in 1969 and the second in 1970 (Fig. 8–4). This ongoing project is a joint effort involving private business, several government agencies, and several academic institutions. Its main purpose is scientific, with *in situ* observations of and experiments on marine life being the major activities. The dive domicile is located at a depth of approxi-

8–3 SeaLab II *habitat, which was used by the U.S. Navy off La Jolla, California. (Official United States Navy Photograph.)*

8–4 A cutaway model of the Tektite I *habitat. (Courtesy United States Navy.)*

mately 20 m in Lameshur Bay at St. John's, one of the U.S. Virgin Islands in the Caribbean Sea. Tektite II in 1970 had two phases, the first of which involved an all-female diving team.

From a scientific standpoint, being able to enter the marine environment to observe and measure variables without appreciably disturbing any part of the environment has been an important technological advancement. As scuba opened the nearshore and shallow ocean environment to exploration, so the recent generation of deep diving submersible vehicles has made every part of the ocean environment potentially accessible to observation.

The submersible vehicles presently in operation fall into four classes: bathyscaphes, submarines, rescue vehicles, and mid-water drifters (see Table 8–1). Bathyscaphes are deep-diving chambers with little or no horizontal movement capabilities. They operate in a vertical plane, using the buoyancy principle of adding and dropping ballast. The *Trieste,* presently stored at San Diego, California, is a bathyscaphe (Fig. 8–5). It set the world record for manned deep dives when it carried two men to a depth of about 12,000 m in the Challenger Deep southwest of Guam. Most submersible vehicles are manned submarines capable of vertical and horizontal propulsion by self-contained power. They generally have observation portals, externally mounted mechanical arms, and various attachments for collecting plankton, water, or sediment, and for television or photographic equipment (Fig. 8–6). The Deep Sea Rescue Vehicle (DSRV) is designed to attach to U.S. Navy nuclear submarines to rescue crew members. It is being tested at San Diego, California. One of the most unique submersibles is the mid-water drifting vessel, *Ben Franklin.* It is the brain child of Jacques Piccard, who also designed the *Trieste.* The *Ben Franklin* was designed to drift at relatively shallow depths. Its maiden voyage took place in the fall of 1969 when Dr. Piccard and his crew drifted in the Gulf Stream from Florida to the region off Long Island, New York. The vessel was named after Benjamin Franklin, who made several trans-Atlantic trips as emissary to the king of England and later as ambassador to France. During these trips, he measured the temperature, density, and current direction of the water, and produced the first map of the Gulf Stream and North Atlantic drift current.

Table 8–1 Some Deep Submersible Vehicles

Vessel	Operator	Depth limit (m)
DSRV (rescue)	U.S. Navy	1,600
Alvin (submarine)	Woods Hole Oceanographic Institution	2,000
Deepstar 4,000 (submarine)	Westinghouse	1,200
Aluminaut (submarine)	Reynolds Metals	5,000
Star I, II, III (submarines)	General Dynamics	2,000
Beaver (submarine)	North American Rockwell	1,800
Deepquest (submarine)	Lockheed Aircraft	2,000
Trieste (bathyscaphe)	U.S. Navy	12,000 +
Ben Franklin (drifter)	Grumman Industries	~750

Mineral Exploitation

Oil. Although many minerals and other natural resources are known to exist in ocean bottom sediments few of these are being mined for commercial purposes. Oil and natural gas are the main commercially productive ocean bottom resources. Oil and gas wells are located on the continental shelves in many areas of the world. Large fields are located off northern Australia, in the Persian Gulf, and in the North Sea near Norway. Bordering the United States are three major offshore oil and gas producing areas: the Gulf of Mexico coast, especially near the Mississippi River delta; the southern California coast; and the southern Alaska coast (Fig. 8–7). By far, the largest area of producing wells is in the Gulf coast region.

A recent oil field discovery of major proportions has been found at Prudhoe Bay on the Arctic Ocean coast of Alaska. Because northern Alaska is virgin wilderness, many conservationists feel that a last and firm stand to preserve primeval nature from the polluting onslaught of industrial mankind should be made. Oil companies have maintained, at the same time, that oil and gas consumption increases plus lack of security in South American and Middle Eastern oil fields make development of the Prudhoe Bay field essential. The specific worries of conservationists concern the potential ecological effects of a trans-Alaska pipeline and the potential effects of offshore oil spills on the Arctic Ocean and the continental and island coasts.

In 1969 a drill core taken in the middle of the Gulf of Mexico, where the water is more than 2,000 m deep, contained oil-soaked sediment. It was the first time an oil reservoir of potential commercial significance had been located in deep ocean sediment. How much oil is located in such sediments throughout the ocean is unknown.

Manganese. Much has been written about the presence of manganese nodules on the ocean floor at depths greater than 2,000 m. During the summer of 1970 several companies tested recovery processes that allow nodules to be recovered at costs low enough to return a profit. Deposits on land, however, have higher manganese and iron concentrations. Photographs of the nodules have been taken, samples for geochemical analysis have been made, and theories of origin have been published by E. D. Goldberg, G. Arrhenius, J. L. Mero, and others. Manganese nodule concentrations are known to exist off southern California, around the Blake Plateau (off Georgia), and off southwest Africa on the Angolan abyssal plain.

Phosphate. Phosphate compounds have accumulated on submarine terraces on many continental shelves. These terraces were formed by erosion when sea level was lower than at present, as during the periods of continental glaciation in the past one million years. The two largest known reserves are off the south-

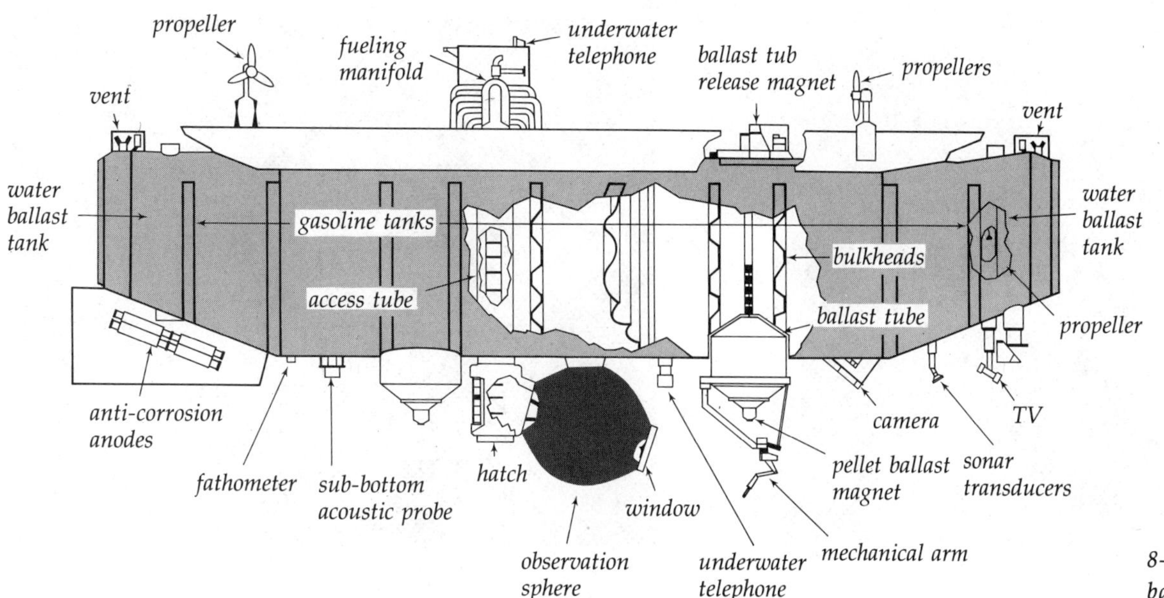

8–5 *A diagram of the bathyscaphe* Trieste.

west coast of Florida and off the southwest coast of California and Baja California.

Gold. Since the world monetary system is based on gold, every country in the world would like to find more. The U.S. Geological Survey searches continuously for additional gold in all parts of the country. One of their main prospecting regions is in the river delta sediments on continental shelves downstream from known gold-producing mountains and rivers. As a result of this exploration, gold has been found off the Columbia River near Washington and in many delta regions along the western coast of the continental United States and Alaska. Because the price of gold is rising, many of these deposits can be profitably exploited.

Diamonds. Diamonds, like gold, are formed in crystalline rocks and are transported to deltas by streams after the host rocks are eroded. Along the coast of South Africa, diamonds are mined in delta sediments on the continental shelf. Potentially rich deposits may occur near Brazil and Venezuela, but they have not yet been explored.

Desalination. One of the most significant technological developments of the past ten years has been the emergence of methods to remove salts from saline water. Because industrial nations use and pollute fresh water at a rapid rate, desalination is important for them. In addition, underdeveloped nations commonly are unable to progress because they have insufficient water to grow enough crops to feed their people. Many of these nations, such as Ethiopia, Tanzania, and Peru, border oceans and could benefit greatly from desalination technology.

Six methods of desalination are presently known: (1) distillation, (2) humidification, (3) freezing, (4) reverse osmosis, (5) electrodialysis, and (6) absorption. A number of different processes using the principle of distillation have been developed. Simple distillation involves boiling saline water, channeling the water vapor through tubes, cooling the water vapor, and collecting the fresh water. The salt in this process precipitates onto the original container. Variations on this process, long-tube vertical distillation and multistage flash distillation (Fig. 8–8), are designed to minimize energy requirements and maximize fresh water production. In the long-tube process, some fresh water is produced in each evaporator. In the flash process, fresh water is also produced at each stage.

8–6 *The* Aluminaut, *a research submarine.* *(Courtesy of Reynolds Metals Company.)*

8–7 *Shell Oil Company Platform, Cook Inlet, Alaska. (Courtesy of Shell Oil Company.)*

Solar humidification is based on evaporation and precipitation under controlled conditions. Salt water is placed inside the apparatus. Solar radiation penetrates the plastic, glass, or canvas cover heating the salt water. The water evaporates and rises. As the water vapor comes in contact with the cover, condensation occurs and the water trickles down to a fresh water catching trough. Large-scale modification of this process has been used in Israel, West Africa, and Peru for watering crops with ocean water under experimental conditions. The advantage of this process is that it uses only solar energy.

When seawater freezes under natural conditions, most of the salt remains in solution while the water changes to ice. The remaining solution has an increased salinity, which causes its density to be increased. The dense ocean water formed in this way sinks. The ice is even less dense than pure water and floats on the surface. Because of this natural separation of fresh water from salts under cold conditions, freezing seawater is a practical way of producing fresh water.

The reverse osmosis process of desalination is the most promising for large-scale use. Osmosis is the process of migration of less salty water to more salty water until equilibrium is reached. Reverse osmosis is the process of forcing water to flow from a saltier solution to a less salty solution (Fig. 8–9). Reverse osmosis is possible when pressure is applied to ocean water that is separated from fresh water by a semipermeable membrane. The pressure forces water out of the ocean water solution into the fresh water solution. The salts in ocean water, however, cannot pass through the semipermeable membrane. This process has also been effective in recovering fresh water from city sewage water and industrial waste water. The major problem with the process is inventing a semipermeable membrane that is sufficiently long lasting and does not have to be replaced often.

A fifth method of desalination is electrodialysis. This process uses two semipermeable membranes that allow salt ions to pass through but prevent water flow. Seawater is separated by these membranes from fresh water placed on either side of it. Each fresh water compartment contains an electrode. When a current passes through the solutions, positive ions such as sodium are attracted to the negative electrode and negative ions such as chloride are attracted to the positive electrode (Fig. 8–10). The seawater becomes progressively less salty until all the salt ions have been removed. This process is not currently used for large-scale desalination.

The U.S. Navy uses still another method in its survival kits. This method is absorption of salts by activated charcoal and resin. When a small cake of charcoal and resin is placed in a plastic bag containing ocean water, the salts in the ocean water are absorbed.

Humans are using fresh water at a rate increasing faster than population. Fresh water pollution is also increasing. Fortunately, desalination technology is developing rapidly. Its costs have dropped enough to make its use worthwhile in many parts of the world from Key West, Florida, to Israel.

8–8 Scripps Institution of Oceanography's former seawater conversion study used this 90-foot tower in a "multiple effect flash" evaporation system that produced 3,000 gallons of fresh water from seawater daily. (Courtesy of Scripps Institution of Oceanography, University of California, San Diego.)

Information: H.M.S. *Challenger* vs. *Glomar Challenger*

One way to appreciate the technological advancements in oceanology during the past one hundred years is to compare two oceanographic vessels and their respective four-year historic cruises, which have been referred to in this book and which have played important roles in the evolution of the science of oceanology.

The purposes of the most famous voyage of H.M.S. *Challenger* and of the JOIDES Project of the *Glomar Challenger* were different. H.M.S. *Challenger* sought to investigate the deep sea in all the major ocean basins. As *Punch* said before the voyage:

The Challenger Her Challenge

I'm a spar-decked corvette, built of wood not of iron,
I am good under steam, under sail:
No Sheffield-plate dead-weights my topsides environ,
So I ride like a duck through a gale.
By my Lords I'm about to be in commission,
For a cruise of three years, if not four;
And for all I'm short-handed, I carry provision
Such as corvette ne'er victualled before.

Mine's no cruise to train officers, boys or blue-jackets,
Or BRITANNIA'S old flag to display;
To observe and report South American rackets,
Or enjoy life in Naples' blue bay:
To practise manœuvres, or study steam-tactics,
Hunt down pirate-junk or slave-dhow;
The Challenger now aims at higher didactics,
And on different quests sets her prow!

Her task's to sound Ocean, smooth humours or rough in,
To examine old NEP'S deep-sea bed;
Dredge up samples precise of his mattress's stuffing,
And the bolsters that pillow his head:
To study the dip and the dance of the needle;
Test the currents of ocean and air—
In a word, all her secrets from Nature to wheedle,
And the great freight of facts homeward bear.

And by way of a treat—when the Fauna and Flora
Of all lands and all seas I've run through,
And learnt if the Austral Antarctic Aurora

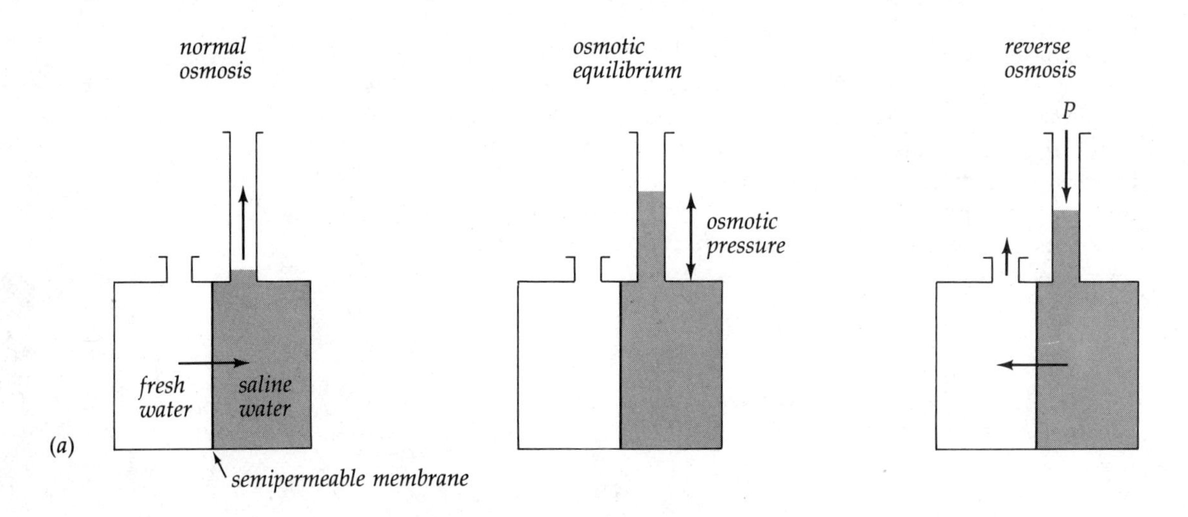

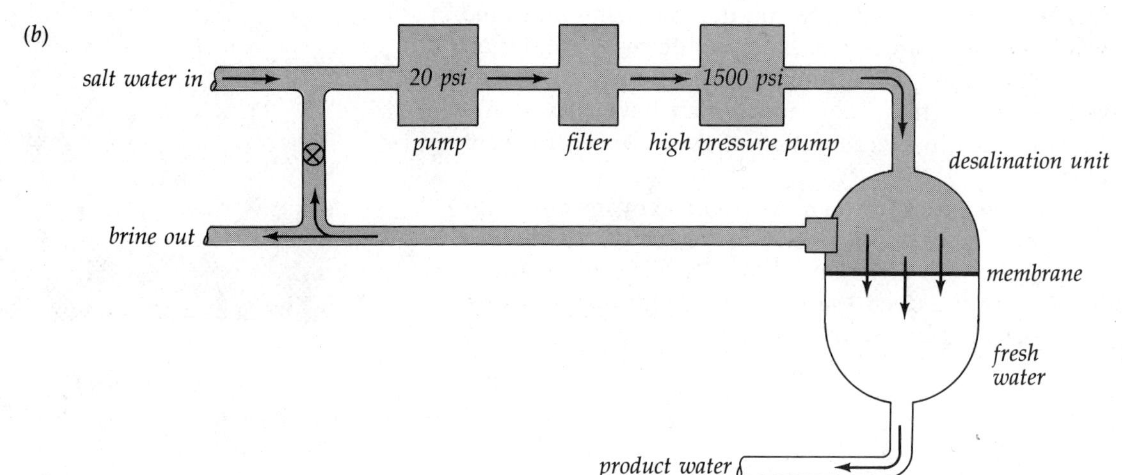

8–9 (a) Principles of reverse osmosis. (b) Schematic of the reverse osmosis process.

Our Boreal in beauty outdo—
In the Isle of Kerguelen, with nothing between us
But the thinnest of clouds—O what fun!—
I'm to lurk and look on at the transit of Venus,
Across the broad blush of the sun!

For this I bear science to seamanship plighted,
In THOMPSON and NARE and MACLEAR,
While from highest to lowest aboard all united,
To serve both alike volunteer.
Broadside guns have made room to ship batteries magnetic,
Apparatus turns out ammunition,
From main-deck to ground-tier I'm a peripatetic
Polytechnic marine exhibition.

"Mighty fine!" says JOHN BULL. "But, pray, how about cost?
Cash soon makes ducks and drakes in the Ocean."
Treasury leave was asked first: prayer, of course, aside tost,
Till LOWE went to figures with GÖSCHEN.
When they found that the outlay for all this provision,
To question the land, and the sea,
Would be no more than keeping my hull in commission,
With nothing to show for't, would be!

Said LOWE, laughing, "To pay by results is my plan;
For results here'll be nothing to pay.
Let the Challenger go: and I'll challenge the man,
Be it RYLANDS himself, who'll gainsay;

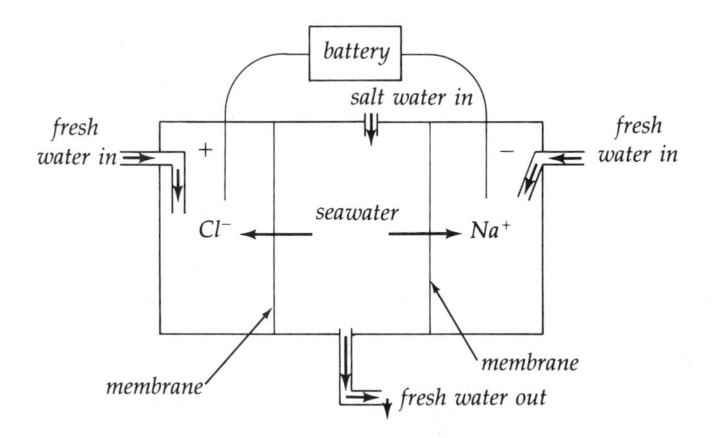

8–10 Schematic of the electrodialysis process.

For he, like myself, though he's not been to college,
And's a shallowish sort of a snob,
Has, at bottom, I'm sure, no objection to knowledge,
So long as it don't cost a bob."

And so I'm to sail on my grand cruise of science,
And a prouder ship ne'er put to sea;
In the good of my mission high souls have reliance,
Whatever the LOWE view may be.
Of the axiom that "nothing of nothing can come,"
I'm the Challenger. How is it true?
When 'tis clear to BOB LOWE, as a rule-of-three sum,
Good for nothing I'm not, 'cause I do.[1]

On the other hand, the *Glomar Challenger*'s purposes were more specific and mainly geological. Its purposes were: (1) to establish a national program for deep sea sediment coring, (2) to determine the age and process of development of the ocean basins, (3) to develop deep sea drilling technology, and (4) to test the global plate tectonics model.

The people most responsible for each voyage included some of the well-known scientists of the respective times. Sir Charles Wyville Thomson was the principal investigator in 1872. After Thomson's death (soon after the end of the voyage), Sir John Murray became its leader. The crew and the group of participating scientists were mostly British. For the JOIDES Project, Dr. W. A. Nierenberg, present Director of Scripps Institution of Oceanography, is the principal investigator. Dr. M. N. A. Peterson, a scientist at Scripps, was the project's chief scientist at its inception. The crew and scientists in this case are mainly from the United States.

A comparison of the vessels reveals little in common other than the fact they both float (Fig. 8–11). H.M.S. *Challenger* was a corvette. It was sail-rigged with an auxiliary steam engine and was designed as a warship. It was converted for oceanographic use especially for the voyage, although two guns remained in position during the four years. Among the equipment were 260 km of sounding rope, 22.5 km of sounding wire, a donkey engine to operate winches, sinkers, winches, dredges, various samplers, plankton nets, thermometers, hydrometers, sextants, a library, and "spirits of wine" for specimen preservation. The *Glomar Challenger* was built by Global Marine, Inc., especially for deep sea drilling (Fig. 8–12). The ship is 135 m long and has a beam of 22 m. Its displacement when loaded is 10,500 tons. A large drill rig is mounted just forward of amidships with more than 10,000 m of 30 m pipe lengths stacked on the forward deck. Auxiliary engines mounted on the hull are able to drive the ship sideways or rotate it. Navigation is accomplished by a new satellite navigation system operated by a shipboard computer.

A general outline of the voyage of H.M.S. *Challenger* shows the broad extent of its exploration (Fig. 8–13a). Only the tropical Indian Ocean and the Arctic Ocean were not investigated. This voyage lasted four years (1872–1876) from home port to home port. General outlines of the routes for the 25 legs of the JOIDES Project (1968–1972) are even more inclusive geographically than that of her sister ship (Fig. 8–13b).

It may be unfair, yet interesting, to compare the results of the two voyages. At this writing, 22 legs of the JOIDES Project have been completed. The results of H.M.S. *Challenger*'s voyage were truly broad in scope. The information gathered was significant for every field of the natural sciences. As an example, 715 genera and 4,417 species of previously unknown organisms were collected and described. Think of the effect on the field of zoology. A sounding was taken in the Mariana Trench to a depth of almost 9,000 m proving that the

[1] *Punch,* December 14, 1872, p. 245.

(*a*)

(*b*)

8–11 (a) Dredging and sounding arrangements aboard H.M.S. Challenger. *(b) The Deep Sea Drilling Project vessel* Glomar Challenger, *which is drilling and coring for ocean sediment in all the oceans of the world. Scripps Institution of Oceanography, of the University of California at San Diego, is the managing institution for DSDP. (Courtesy of Scripps Institution of Oceanography, University of California, San Diego.)*

ocean was much deeper than previously believed. The final report, entitled *The Challenger Report,* fills 50 large volumes, reprints of which can be found in most good university libraries. This voyage was responsible for the emergence of descriptive oceanology as a unique interdisciplinary science. In addition, "swinging the compass" to acquire accurate magnetic compass readings enabled the most precise location fixing to be achieved at sea up to that time.

The JOIDES Project has made significant contributions to the science of geology and to deep sea drilling technology. So far, the *Glomar Challenger* has traveled 207,000 km. Cores have been drilled in water more than 6,800 m deep, and one core penetrated almost 1,200 m below the ocean floor. More than 17 km of core sediments have been recovered. These accomplishments are far less broad than those of the *Challenger* expedition. The most important scientific aspect of the project has been the acquisition of deposition dates for oceanic sediments and igneous rocks, of rates of spreading for global plates, and, in general, the substantiation of the global plate tectonics model. The most important tech-

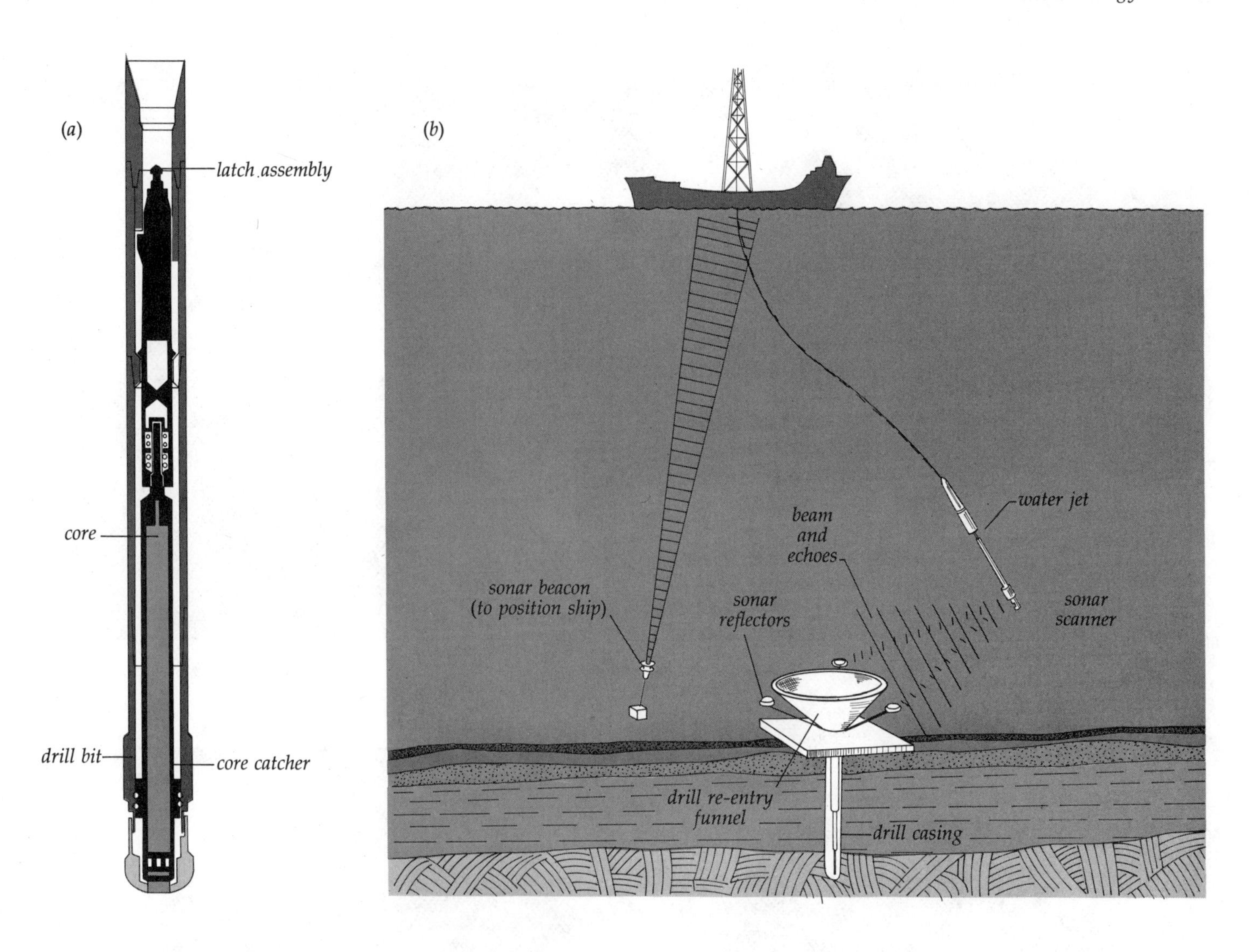

8–12 (a) Drilling pipe and core barrel. (b) Reentering a drill hole in the deep ocean floor. A newly developed mechanism enables the drill bit to be guided into a previously drilled hole. The reentry funnel is attached to the drill string when it is first lowered to the bottom and remains on the ocean floor when the string is withdrawn. At the time reentry is attempted, the drill string is relowered with a sonar scanner on the bit assembly that emits sound signals. These signals are echoed back from three reflectors spaced around the funnel. Position information is relayed to the ship, and the water jet is used to steer the bit directly over the funnel.

nical aspects have been on station positioning and core-hole reentry capability. In addition, the satellite navigation system enables precise location of the ship within a range of 100 m.

Oceanologists also use many other research vessels. They can be conventional-looking craft with highly specialized equipment for a particular field of study (Figs. 8–14 and 8–15) or they can be specially constructed, unique vessels such as *Flip* (Fig. 8–16), which is used to study subsurface sound waves and their movement along and across subsurface density layers.

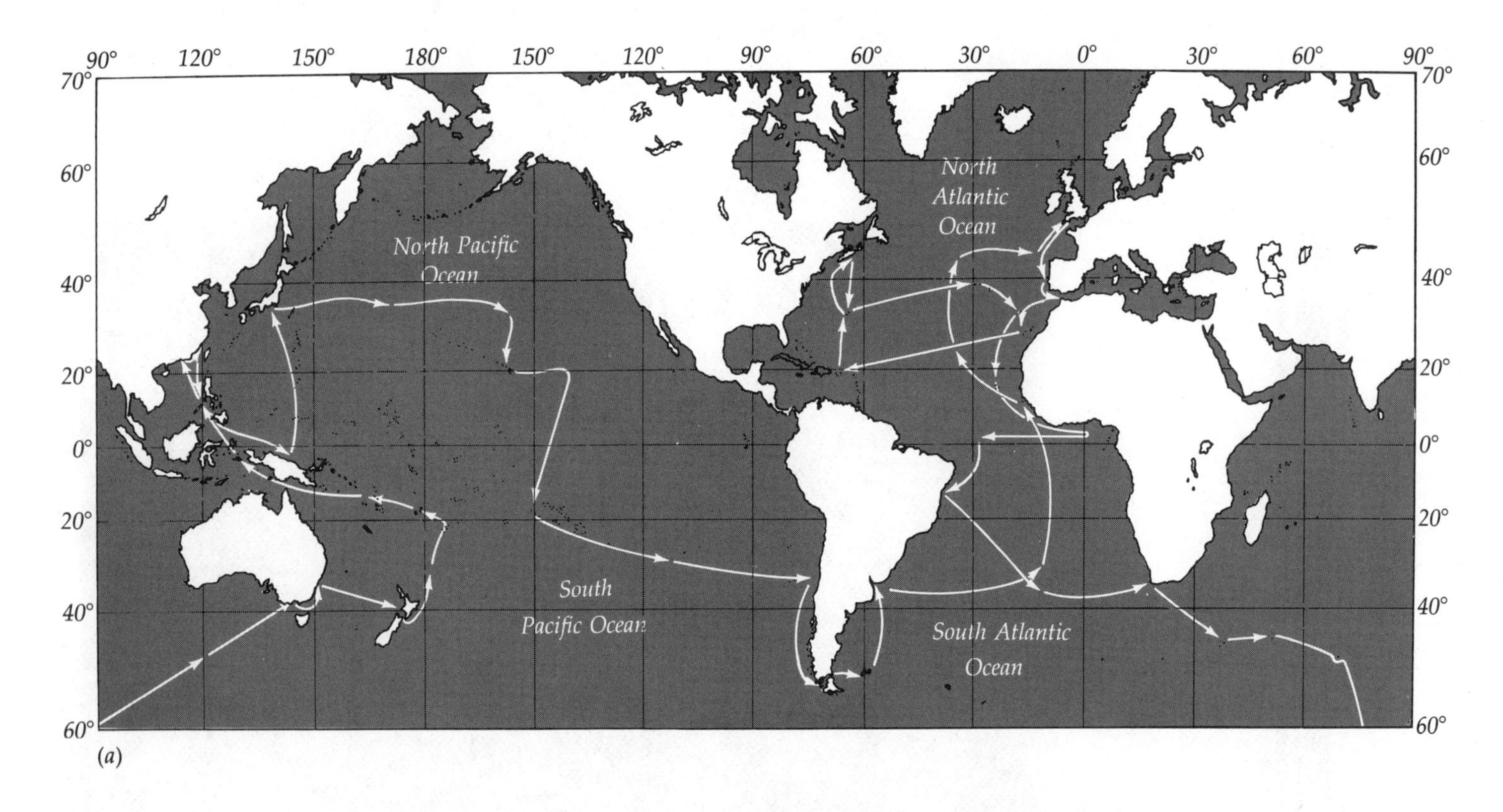

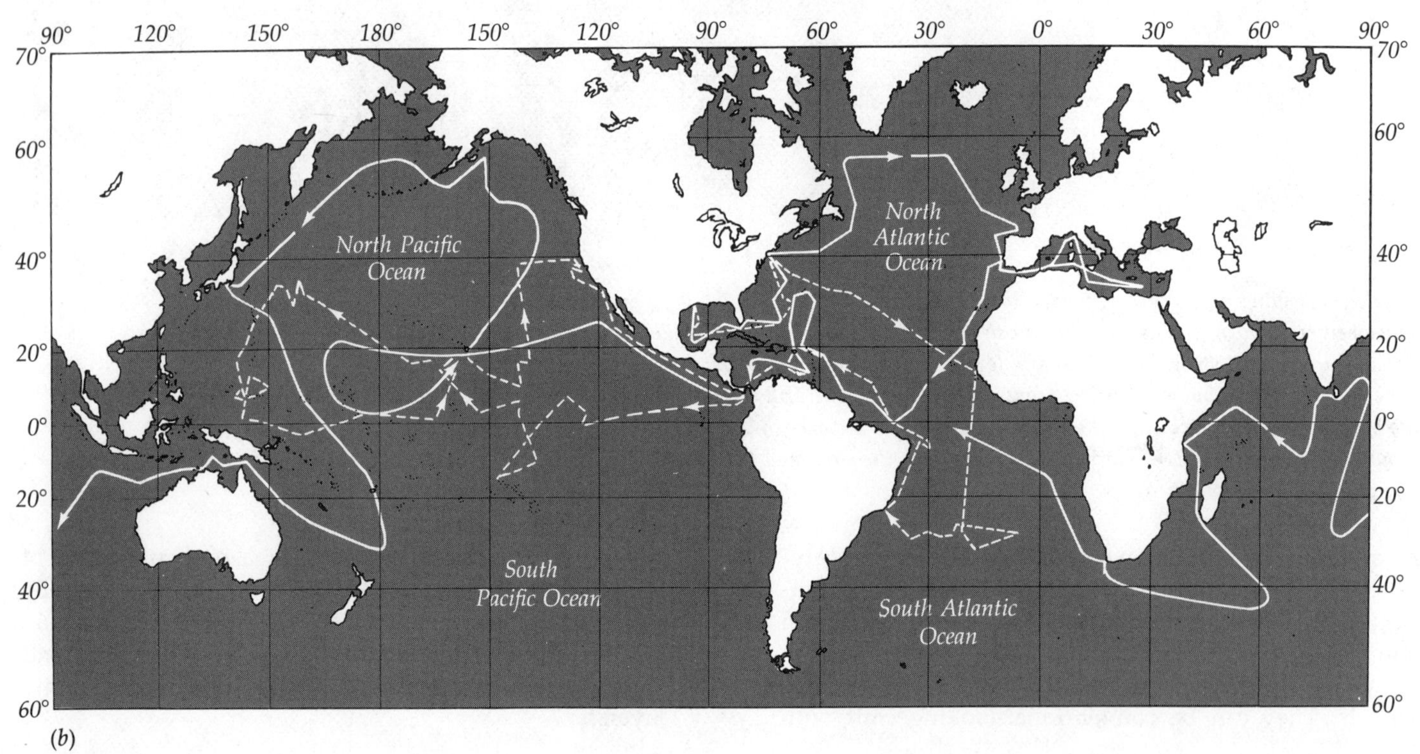

8–13 *(a) The route of H.M.S.* Challenger. *(b) The drilling route of the* Glomar Challenger *for the Deep Sea Drilling Project.*

8–14 The oceanographic research vessel Melville, *built for Scripps Institution of Oceanography. The 245-foot, 2,075-ton vessel was built at an estimated cost of $7 million, including equipment, and has a maximum capacity of 62 scientists, technicians, and crew members. The ship's unique propulsion system uses vertically mounted, multi-bladed, cycloidal propellers, one at the bow and one at the stern. This enables her to proceed forward, backward, or sideways, or remain stationary over a fixed point in 35 knot winds and heavy seas. (Courtesy of Scripps Institution of Oceanography, University of California, San Diego.)*

8–15 The research vessel Alpha Helix, *133 feet long and operated by Scripps Institution of Oceanography, is essentially an ocean-going biological laboratory. She contains an analytical laboratory; a "wet" laboratory that can be maintained at any temperature down to 5°C; an electrophysiological, or optical, laboratory; and a special machine shop, completely equipped. (Courtesy of Scripps Institution of Oceanography, University of California, San Diego.)*

Use of the Sea Bed

Role of the United Nations. On December 2, 1969, the Malta Resolution was passed in the First Committee, United Nations General Assembly. The United States supported this resolution. Two major points in the resolution asked for a poll of UN member nations on the desirability of reconvening the 1958 Convention on the Continental Shelf and for further discussion on present laws pertaining to oceanic provinces. On December 15, 1970, an amendment to this resolution passed, which asked to review the historical systems and codes of operation on the high seas. In addition, another resolution, passed on the same day, placed a moratorium on exploitation by any nation of the sea bed beyond national jurisdiction.

These resolutions are part of a continuing controversy regarding a previous proposal by Malta to turn the administration of ocean bed exploitation over to the UN with the profits being distributed by the UN to presently underdeveloped countries according to their need. The technical capabilities for potential exploitation have been largely developed by the United States, France, Japan, and the Soviet Union, with the vast majority by the United States. In the United States, private corporations are unhappy with the prospect of turning profits over to the UN. The U.S. oil industry is particularly concerned because their technology is relatively advanced even to the capability of deep sea drilling.

In June 1970, President Nixon proposed a Seabed Treaty. Some major points in this proposal are: (1)

(a)

(b)

national jurisdiction would extend from the shoreline to a depth of 220 m; (2) the balance of the oceans would be controlled by an international governmental system; (3) profits from deep sea exploitation would be used for international community benefits; and (4) coastal nations would act as trustees for the international community, receiving a share of the profits of exploitation. The issues are complex and nationalistic interests are generally opposed to most of the proposals. The exploitation of the oceanic province for commercial purposes promises to be a hotly debated issue during the next decade.

Military Use. The oceans have been used for military purposes for several thousand years. Today, the military use of the oceans, especially by the United States and the Soviet Union, is far more extensive than at any previous time in history. Both countries have warships or other facilities operating at all times in every ocean of the world. A wide variety of ships and facilities are used, including nuclear-powered submarines, aircraft carriers, seek-and-destroy vessels, and numerous classified listening devices. Efforts are being made to limit the military use of the oceans to its present extent and especially to prevent the implantation of submarine bases and fixed nuclear weapons on the ocean floor. Such a treaty is under diplomatic discussion now at Geneva, Switzerland, and at other diplomatic negotiations between the involved nations.

Epilogue

The survival of man and, in fact, of all life on earth is at the crossroads today. What man does during the next ten years will decide the future existence of life on earth. Information currently available indicates that the oceans are finite. Man should not continue to pollute the oceans, exterminate life in them, and plan to rape the seas as he has the land. If man does not soon establish a value system that will enable all people to live in a way that benefits the earth as a whole, the extinction of man and other species is imminent and inevitable.

Where in this picture can hope be found? First, man must learn to limit his own population. Young people are becoming especially aware of this necessity. Second,

(c)

8–16 Flip (*Floating Instrument Platform*) *was developed by the Marine Physical Laboratory at Scripps Institution of Oceanography.* Flip *has no motive power and must be towed to the research position* (*a*)*. Ballast tanks are flooded to "flip" her* (*b*)*. When in vertical position* (*c*)*, she affords scientists an extremely stable platform from which to carry out oceanographic research.* Flip *is 355 feet long, 300 feet of which are submerged in vertical position.* (*Photographs courtesy of Scripps Institution of Oceanography, University of California, San Diego.*)

man must learn to preserve as well as exploit. The United Nations and many private organizations are exerting pressure on behalf of the preservation of species and environment. Supporting environmental improvement has become a necessary political stand for a growing number of politicians. Third, man must learn to farm the soil and the sea while putting less emphasis on hunting, fishing, and bulldozing. This is probably the least understood change which, we believe, must occur. Fourth, man must improve and expand education to all people of the world. In each area, progress is accelerating. The big question is "How can the definition of progress eliminate its present connotations of expansion, exploitation, extermination, and maintenance of a poverty class?"

Further Reading

Bascom, Willard, "Technology and the Ocean." *Scientific American*, September 1969, pp. 198–217.

Bibliography on Marine Law. Sea Grant 70's, vol. 1, no. 6, February 1971.

Gullion, E. A., ed., *Uses of the Seas*. Englewood Cliffs, N.J.: Prentice-Hall, 1968.

Kane, T. E., *Aquaculture and the Law*, Sea Grant Technical Bulletin, No. 2. Miami, Fla.: University of Miami Sea Grant Program, November 1970.

Levine, S. N., *Desalination and Ocean Technology*. New York: Dover Publications, 1968.

Padelford, N. J., *Public Policy and the Use of the Seas*, 2nd ed. Cambridge, Mass.: MIT Press, 1970.

Wenk, Edward, Jr., "The Physical Resources of the Ocean." *Scientific American*, September 1969, pp. 166–177.

Wolff, Thomas, "Peruvian–U.S. Relations Over Maritime Fishing 1945–1969." Kingston: University of Rhode Island Law of the Sea Institute, 1970 (occasional paper).

Appendixes

Latitude and Longitude

I

Like streets in a city set up perpendicularly to one another (that is, streets named A, B, C, . . . at right angles to streets numbered 1, 2, 3, . . .) the earth's surface is divided, for the same reasons, by parallels of latitude perpendicular to meridians of longitude. Because the earth's surface is essentially two-dimensional it might seem, like the street grid, that the latitude–longitude system is simply the result of linear measurements of the exterior. Such is not the case. Actually the system is produced by angular measurements from the center of the earth.

The rim of the equatorial plane that divides the earth into two hemispheres is the equator, 0° N and S latitude. In degrees of 1 to 90 north (the North Pole) and 1 to 90 south measured from the equatorial plane and the earth's center are the degrees of north and south latitude (Fig. I–1). Salem, Oregon, for example, is on the circle formed by the angle of 45° N—its latitude is 45° N.

Longitude is expressed in degrees at the north–south axis as the angle (θ) formed by the intersection of two lines drawn from two points (one a reference) to the earth's center (Fig. I–2). Greenwich, England, is the reference point and is assigned the meridian of 0°, running from pole to pole. A 360° reference could have been used, as the reference system for satellites does, but it has been common to express longitude east and west of Greenwich, up to, of course, 180° E and W.

Each degree of latitude and longitude is divided into 60 minutes, each minute in turn divided into 60 seconds. A nautical mile (6076 ft.) is approximately equal to one minute of latitude. As the earth is not a perfect sphere, this varies somewhat. One degree of latitude at 0° measures 59.701 nautical miles, at 30° latitude one degree is 59.853 nautical miles, and at 60° latitude one degree is 60.159 nautical miles. The latitude and longitude of Washington, D.C., is 38° 53′ N, 77° 0′ W; of San Francisco 37° 47′ N, 122° 26′ W.

A great circle (or geodesic) is a plane that passes through the earth's center. On the earth's surface a great circle is the shortest distance between any two points. Each meridian of longitude is a great circle, but for the circles of latitude only the equator is a great circle.

The meridian on the opposite side of the planet from Greenwich, corresponding to 180° E and 180° W, is also called the international date line.

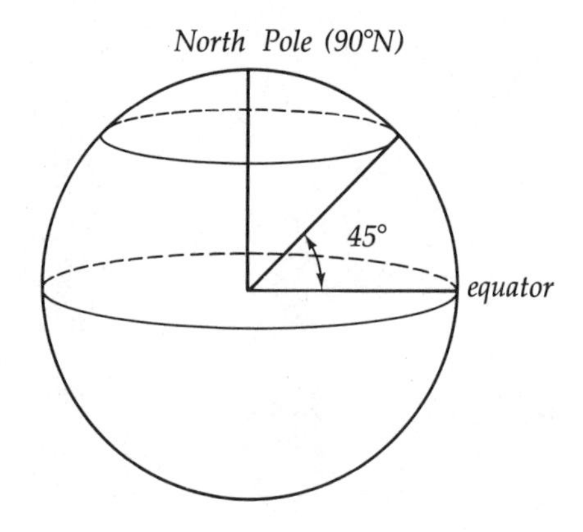

I–1 *Determining latitude.*

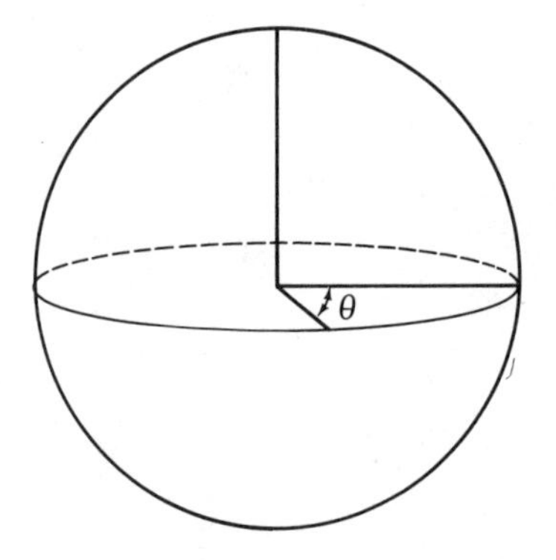

I–2 *Determining longitude.*

Time and the Date Line

II

Most people know that the time varies from place to place. When it is noon in San Diego, for example, it is 3 P.M. in Boston. It might seem simpler and more desirable to have the same time prevailing throughout the world, so that when it was 2 A.M. at one point it would be 2 A.M. everywhere. But the time zones, though arbitrarily determined, are almost a necessary consequence of the earth's rotation. Noon at any place occurs when the sun is directly overhead. Only one place (one meridian of longitude) can then have noon at any one time. Indeed because of the earth's rotation, if there were not different time zones concepts like noon and relative time in general would have little meaning, especially for a person traveling from place to place. The earth rotates 360° about once every 24 hours relative to the sun—or 15° each hour. Consequently the

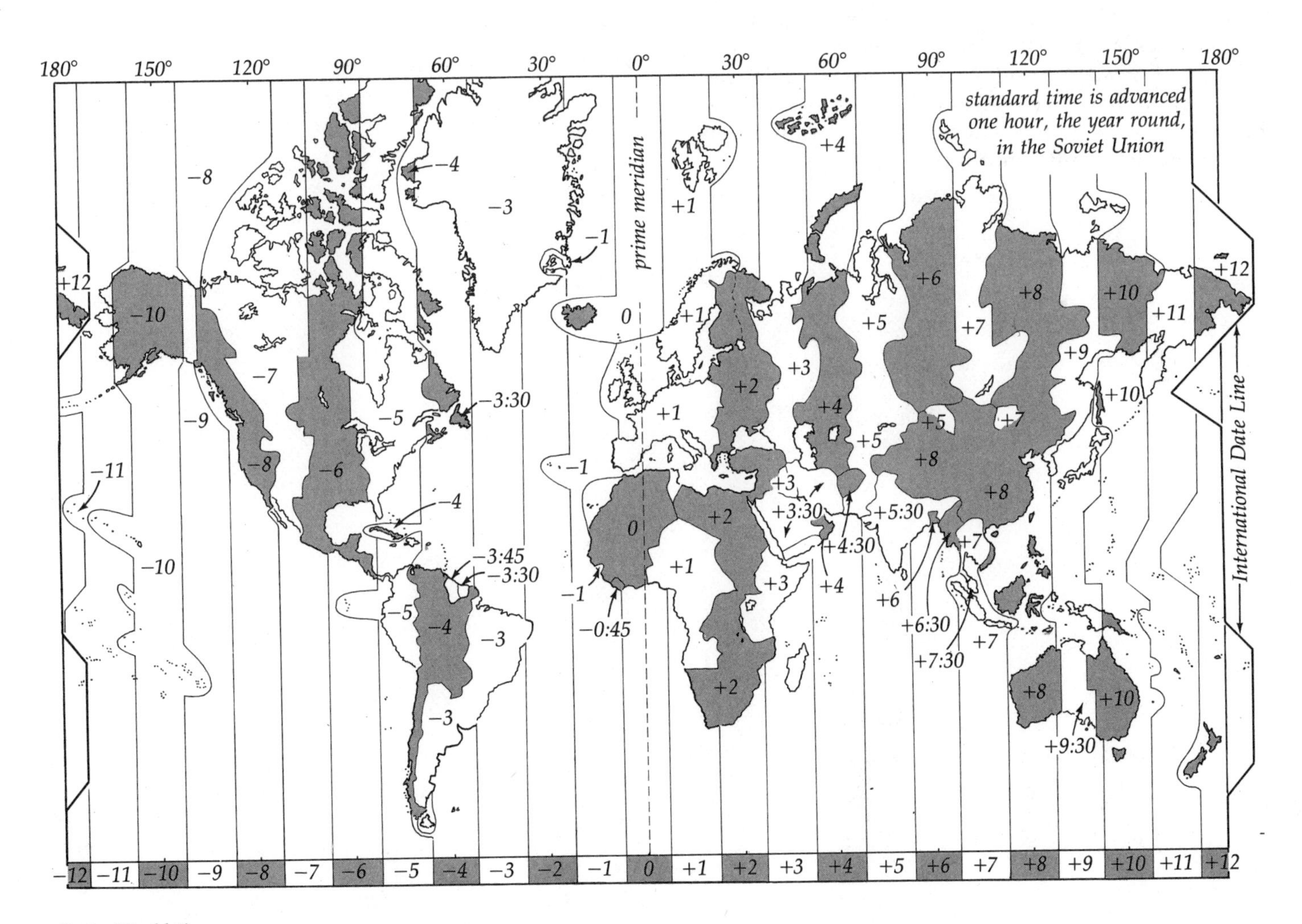

II–1 World time zones.

earth is divided into 24 time zones according to each 15° of longitude. The prime meridian, passing through Greenwich, is the reference point for time and is called Greenwich Mean Time (GMT). Moving from zone to zone westward, since the earth rotates counterclockwise, time gets earlier (Fig. II–1). Because any one zone has the same time throughout, and the earth doesn't rotate in 24 equal clicks per day, noon will be off by 30 minutes, earlier or later, at regions on the western or eastern edge of a zone.

When it is noon at Greenwich it is 00:00 on the west longitude side of the international date line and 24:00 on the east side. Here at longitude 180° E and W, where the day begins, is the international date line. East of the line it is one day earlier than it is to the west.

As the charts of world and U.S. time zones (Fig. II–2) indicate, the 15° time meridians are not perfectly straight lines. So that time zones bisect as few populated regions, states, and countries as possible, the dividing lines have been reworked slightly to avoid the confusion that would be caused if the ends of Main Street were using different times.

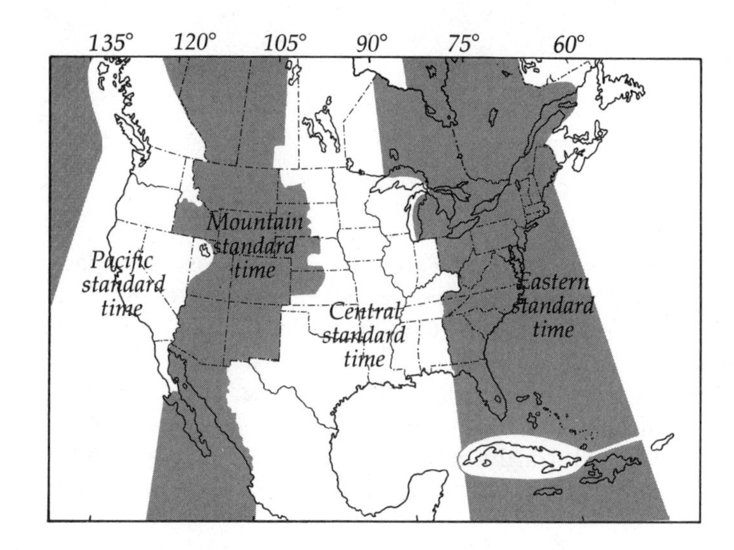

II–2 U.S. time zones.

Scientific Notation III

In the study of the physical world one encounters many large and small numbers determined by scientists to help describe the world. For example, the average distance from the sun to the earth is 93,000,000 miles, the speed of light is 297,800,000 m/sec, the age of the earth is 4,700,000,000 years, the number of atoms or molecules in a molecular weight—Avogadro's number—is 602,300,000,000,000,000,000,000/mole, the rest mass of an electron is 0.000000000000000000000000000000-0009109 kg, and the distance light travels in a year is 9,460,550,000,000,000 m/yr. Obviously these numbers are cumbersome to work with. To shorten the numbers and therefore make them easier to use, a type of shorthand notation is generally used. Avogadro's number becomes 6.023×10^{23}/mole; the rest mass of an electron 9.109×10^{-31} kg.

The number 1000 can be written 1×10^3, or 10×10^2, or 100×10^1 or 100×10. By convention only one integer is placed to the left of the decimal. So

$$1.4 \times 10^4 = 14{,}000$$

Scientific notation takes the general form $M \times 10^n$, M equaling the number with the decimal placed so that one integer is to the left of the decimal point and the exponent ^{n} equaling the number of places the decimal must be moved (the power to which the number is raised) to yield M.

If the decimal is moved to the left the exponent is positive.

$$274{,}000 = 2.74 \times 10^5$$

If the decimal is moved to the right the exponent is negative.

$$0.00000926 = 9.26 \times 10^{-6}$$

Thus, the expression 2.74×10^5 means that the number 2.74 is to be multiplied by 10 five times, to give 274,000. Numbers expressed in scientific notation are especially valuable because they can easily be used in computations without rechanging their forms.

Multiplication and division are performed as follows. To multiply $(2 \times 10^4) \times (3 \times 10^3)$ multiply the M factors and add the exponents:

$$2 \times 3 = 6 \qquad 10^4 + 10^3 = 10^7$$

The answer is 6×10^7. As a second example, multiply $(4.2 \times 10^5) \times (6.0 \times 10^{-2})$.

$$4.2 \times 6.0 = 25.2 \qquad 10^5 + 10^{-2} = 10^3$$

The answer is 25.2×10^3 or preferably 2.52×10^4. For division such as $(8 \times 10^3) \div (2 \times 10^{-5})$ divide the M factors and subtract the exponents:

$$8 \div 2 = 4 \qquad 10^3 - 10^{-5} = 10^8$$

So the answer is 4×10^8.

Addition and subtraction generally are not often used, but to add or subtract numbers in scientific notation the exponents must be the same.

$$\begin{array}{r} 4.5 \times 10^6 \\ + \underline{5.0 \times 10^6} \\ 9.5 \times 10^6 \end{array}$$

4.5×10^4 cannot be added to (or subtracted from) 5×10^3 unless the exponents agree:

$$\begin{array}{rl} 4.5 \times 10^4 & \\ + \underline{0.5 \times 10^4} & (= 5 \times 10^3) \\ 5.0 \times 10^4 & \end{array}$$

The table below lists the more common prefixes.

Table III–1 Prefixes

Multiples and submultiples	Prefixes	Symbols
10^{12}	tera	T
10^9	giga	G
10^6	mega	M
10^3	kilo	k
10^2	hecto	h
10	deka	da
10^{-1}	deci	d
10^{-2}	centi	c
10^{-3}	milli	m
10^{-6}	micro	μ
10^{-9}	nano	n
10^{-12}	pico	p
10^{-15}	femto	f
10^{-18}	atto	a

Constants and Equations

IV

Length

1 micron = 0.0000394 inches
1 millimeter (mm) = 0.0394 inches
1 centimeter (cm) = 0.394 inches
1 meter (m) = 39.4 inches = 3.28 feet
1 kilometer (km) = 0.621 mile = 0.540 nautical mile
1 fathom = 6 feet
1 nautical mile = 1.85 kilometers ≈ 1 minute of latitude

Volume

1 milliliter (ml) = 1 cubic centimeter (cm^3) = 0.061 cubic inch
1 liter = 1,000 milliliters = 1.0567 US quarts

Mass

1 kilogram (kg) = 2.2 pounds = 1,000 grams

Speed

1 kilometer per hour = 27.8 centimeters per second
1 knot = 51.5 centimeters per second = 1 nautical mile per hour

Pressure

1 atmosphere = 1.013 bar = 14.7 pounds per square inch = 760 millimeters of mercury at 0°C

Temperature

°C	0	10	20	30	40	100
°F	32	50	68	86	104	212

Composition of dry air (that is, without water vapor) at sea level

Nitrogen	N_2	78.084 ± .004%
Oxygen	O_2	20.946 ± .002
Argon	Ar	.934 ± .001
Carbon dioxide	CO_2	.033 ± .001

The remaining gases include neon, krypton, and nowadays nitrogen oxides and other pollutants.

Partial table of chlorinity, salinity (34.33 to 37.94‰), and σ_0 (density)

If σ_0 is 27.58, the density is 1.02758 g/ml.

A sample of seawater with a chlorinity (Cl‰) of 19.00 has a salinity of 34.33‰ (using the equation S‰ = 1.8050 × Cl‰ + .030) and a density (σ_0) of 1.02758 g/ml.

The directions below and the table on p. 293 are from the original *Hydrographical Tables* of 1901 by Martin Knudsen. (The centigrade temperature scale is used in the table.)

"*Cl* means the amount of chlorine in ‰, that is to say, the weight of chlorine in grams, found in 1000 grams of seawater.

S means the salinity in ‰, that is to say, the total weight of salt in grams, found in 1000 grams of seawater. S = 0.030 + 1.8050 Cl.

$\sigma_0 = (s_0 - 1)\,1000$, where s_0 means the specific gravity of seawater at 0° referred to distilled water at 4°.

$\sigma_0 = -0.069 + 1.4708\ \text{Cl} - 0.001570\ \text{Cl}^2 + 0.0000398\ \text{Cl}^3$.

$\rho_{17.5} = \left(\frac{S_{17.5}}{S_{17.5}} - 1\right) 1000$, where $S_{17.5}$ means the specific gravity of seawater at 17.5° referred to distilled water at 4° and $S_{17.5}$ means the specific gravity of distilled water at 17.5° in proportion to distilled water at 4°. Thus $\left(\frac{S_{17.5}}{S_{17.5}}\right) = S\left(\frac{17.5}{17.5}\right)$ means the specific gravity of seawater at 17.5° referred to distilled water of the same temperature.

$\rho_{17.5} = (0.1245 + \sigma_0 - 0.0595\,\sigma_0 + 0.000155\,\sigma_0^2) \times 1.00129$."

Cl	S	σ_0	$\rho_{17.5}$	Cl	S	σ_0	$\rho_{17.5}$	Cl	S	σ_0	$\rho_{17.5}$	Cl	S	σ_0	$\rho_{17.5}$
19.00	34.33	27.58	26.22	19.50	35.23	28.31	26.91	20.00	36.13	29.04	27.60	20.50	37.03	29.77	28.29
.01	.34	.60	.23	.51	.25	.32	.92	.01	.15	.05	.61	.51	.05	.78	.31
.02	.36	.61	.24	.52	.26	.34	.94	.02	.17	.07	.63	.52	.07	.79	.32
.03	.38	.63	.26	.53	.28	.35	.95	.03	.18	.08	.64	.53	.09	.81	.33
.04	.40	.64	.27	.54	.30	.37	.96	.04	.20	.10	.66	.54	.10	.82	.35
.05	.42	.65	.29	.55	.32	.38	.98	.05	.22	.11	.67	.55	.12	.84	.36
.06	.43	.67	.30	.56	.34	.40	.99	.06	.24	.12	.68	.56	.14	.85	.38
.07	.45	.68	.31	.57	.35	.41	27.01	.07	.26	.14	.70	.57	.16	.87	.39
.08	.47	.70	.33	.58	.37	.43	.02	.08	.27	.15	.71	.58	.18	.88	.40
.09	.49	.71	.34	.59	.39	.44	.03	.09	.29	.17	.73	.59	.19	.90	.42
19.10	34.51	27.73	26.36	19.60	35.41	28.46	27.05	20.10	36.31	29.18	27.74	20.60	37.21	29.91	28.43
.11	.52	.74	.37	.61	.43	.47	.06	.11	.33	.20	.75	.61	.23	.93	.44
.12	.54	.76	.38	.62	.44	.48	.07	.12	.35	.21	.77	.62	.25	.94	.46
.13	.56	.77	.40	.63	.46	.50	.09	.13	.36	.23	.78	.63	.27	.95	.47
.14	.58	.79	.41	.64	.48	.51	.10	.14	.38	.24	.79	.64	.29	.97	.49
.15	.60	.80	.42	.65	.50	.53	.12	.15	.40	.26	.81	.65	.30	.98	.50
.16	.61	.82	.44	.66	.52	.54	.13	.16	.42	.27	.82	.66	.32	30.00	.51
.17	.63	.83	.45	.67	.53	.56	.14	.17	.44	.29	.84	.67	.34	.01	.53
.18	.65	.84	.47	.68	.55	.57	.16	.18	.45	.30	.85	.68	.36	.03	.54
.19	.67	.86	.48	.69	.57	.59	.17	.19	.47	.31	.86	.69	.38	.04	.56
19.20	34.69	27.87	26.49	19.70	35.59	28.60	27.19	20.20	36.49	29.33	27.88	20.70	37.39	30.06	28.57
.21	.70	.89	.51	.71	.61	.62	.20	.21	.51	.34	.89	.71	.41	.07	.58
.22	.72	.90	.52	.72	.62	.63	.21	.22	.53	.36	.91	.72	.43	.09	.60
.23	.74	.92	.54	.73	.64	.64	.23	.23	.55	.37	.92	.73	.45	.10	.61
.24	.76	.93	.55	.74	.66	.66	.24	.24	.56	.39	.93	.74	.47	.12	.63
.25	.78	.95	.56	.75	.68	.67	.25	.25	.58	.40	.95	.75	.48	.13	.64
.26	.79	.96	.58	.76	.70	.69	.27	.26	.60	.42	.96	.76	.50	.14	.65
.27	.81	.98	.59	.77	.71	.70	.28	.27	.62	.43	.97	.77	.52	.16	.67
.28	.83	.99	.60	.78	.73	.72	.30	.28	.64	.45	.99	.78	.54	.17	.68
.29	.85	28.00	.62	.79	.75	.73	.31	.29	.65	.46	28.00	.79	.56	.19	.69
19.30	34.87	28.02	26.63	19.80	35.77	28.75	27.32	20.30	36.67	29.47	28.02	20.80	37.57	30.20	28.71
.31	.88	.03	.65	.81	.79	.76	.34	.31	.69	.49	.03	.81	.59	.22	.72
.32	.90	.05	.66	.82	.81	.78	.35	.32	.71	.50	.04	.82	.61	.23	.74
.33	.92	.06	.67	.83	.82	.79	.36	.33	.73	.52	.06	.83	.63	.25	.75
.34	.94	.08	.69	.84	.84	.80	.38	.34	.74	.53	.07	.84	.65	.26	.76
.35	.96	.09	.70	.85	.86	.82	.39	.35	.76	.55	.08	.85	.66	.28	.78
.36	.97	.11	.72	.86	.88	.83	.41	.36	.78	.56	.10	.86	.68	.29	.79
.37	.99	.12	.73	.87	.90	.85	.42	.37	.80	.58	.11	.87	.70	.30	.81
.38	35.01	.14	.74	.88	.91	.86	.43	.38	.82	.59	.13	.88	.72	.32	.82
.39	.03	.15	.76	.89	.93	.88	.45	.39	.83	.61	.14	.89	.74	.33	.83
19.40	35.05	28.16	26.77	19.90	35.95	28.89	27.46	20.40	36.85	29.62	28,15	20.90	37.75	30.35	28.85
.41	.07	.18	.78	.91	.97	.91	.48	.41	.87	.63	.17	.91	.77	.36	.86
.42	.08	.19	.80	.92	.99	.92	.49	.42	.89	.65	.18	.92	.79	.38	.87
.43	.10	.21	.81	.93	36.00	.94	.50	.43	.91	.66	.20	.93	.81	.39	.89
.44	.12	.22	.83	.94	.02	.95	.52	.44	.92	.68	.21	.94	.83	.41	.90
.45	.14	.24	.84	.95	.04	.96	.53	.45	.94	.69	.22	.95	.84	.42	.92
.46	.16	.25	.85	.96	.06	.98	.54	.46	.96	.71	.24	.96	.86	.44	.93
.47	.17	.27	.87	.97	.08	.99	.56	.47	.98	.72	.25	.97	.88	.45	.94
.48	.19	.28	.88	.98	.09	29.01	.57	.48	37.00	.74	.26	.98	.90	.46	.96
.49	.21	.30	.90	.99	.11	.02	.59	.49	.01	.75	.28	.99	.92	.48	.97
19.50	35.23	28.31	26.91	20.00	36.13	29.04	27.60	20.50	37.03	29.77	28.29	21.00	37.94	30.49	28.99

Heats of fusion and vaporization for water

Heat of fusion (0°C) = 79.71 cal/g

Heat of vaporization (100°C) = 539.55 cal/g

Pressure at sea level

Pressure = force ÷ area

$$P = F/A$$

1 atmosphere = the pressure at the bottom of a column of mercury 760 mm or 29.921 in. high at 32°F = 14.696 lb/in.^{2} = 33.899 ft of water at 39.1°F = 2116.2 lb/ft^2 = 1033.2 g/cm^2 = 1.0332×10^4 kg/m^2 = 1.01325×10^6 dynes/cm^2.

1 dyne = the force to give a 1 gram mass an acceleration of 1 sec/sec = 1 g/sec^2.

Archimedes' principle

Buoyant force = density × acceleration of gravity × volume of submerged body

$$F = dgV$$

d expressed in g/cm^3, g = 980 cm/sec^2, V expressed in cm^3

Acceleration of gravity at different locales

Place	Elevation (m)	g (cm/sec^2)
New Orleans	2	979.324
Colorado Springs	1841	979.490
Denver	1638	979.609
San Francisco	114	979.965
St. Louis	154	980.001
New York	38	980.267
Chicago	182	980.278
Portland, Oregon	8	980.646

Newton's law of gravitation

Attractive force = acceleration of gravity × (mass of object$_1$ × mass of object$_2$) ÷ distance between objects^2

$$F = G\frac{M_1M_2}{R^2}$$

In the earth–sun system:

$F = 5.876 \times 10^{-2}$ dynes/g
$M_{sun} = 1.971 \times 10^{32}$ g
$R_{earth\ to\ sun} = 1.495 \times 10^{13}$ cm

In the earth–moon system:

$F = 3.317 \times 10^{-3}$ dynes/g
$M_{moon} = 7.347 \times 10^{25}$ g
$R_{earth\ to\ moon} = 3.844 \times 10^{10}$ cm

Force

Force = mass × acceleration

$$F = ma$$

Since objects tend to move in a straight line, an equation for an object moving in a curved path (circle) must include a force directed toward the center of the circle. In this case

Acceleration = velocity^2 ÷ distance to center

$$a = \frac{v^2}{R}$$

therefore

$$F = \frac{mv^2}{R}$$

Centrifugal force at any latitude per unit mass

Force = mass × velocity^2 × radius at that latitude × cosine of latitude

$$F = Mv^2R\phi(\cos\phi)$$

Centrifugal acceleration is $FM = v^2R\phi(\cos\phi)$. Centrifugal acceleration is 0 at the North and South poles and a maximum at the equator.

Angular radius of halos and rainbows

Corona (caused by small water droplets in air)	1–10°
Small halo (caused by 60 angles of ice crystals)	22°
Large halo (caused by 90 angles of ice crystals)	46°
Rainbow, primary	41°20′
Rainbow, secondary	52°15′

(from *CRC Handbook of Chemistry and Physics*, 34th ed.)

Seismic waves (km/sec)

Depth	Longitudinal	Transverse
0–20	5.4–5.6	3.2
20–45	6.25–6.75	3.5
1300	12.5	6.9
1400	13.5	7.5

(from *CRC Handbook of Chemistry and Physics*, 34th ed.)

Geologic Time Scale V

Time Unit Began (years ago)	Era or Eon	System or Period		Epoch	Apex Marine Predators
0.01×10^6	Cenozoic	Quaternary		Holocene	whales, sharks
10^6				Pleistocene	
10×10^6		Tertiary	Neogene	Pliocene	
25×10^6				Miocene	
40×10^6			Paleogene	Oligocene	
60×10^6				Eocene	
70×10^6				Paleocene	
130×10^6	Mesozoic	Cretaceous			reptiles
180×10^6		Jurassic			
230×10^6		Triassic			
270×10^6	Paleozoic	Permian			
310×10^6		Pennsylvanian			
350×10^6		Mississippian			
400×10^6		Devonian			placoderms
440×10^6		Silurian			eurypterids, cephalopods
500×10^6		Ordovician			trilobites
600×10^6		Cambrian			
3.5×10^9	Cryptozoic				dominant form unknown
$\sim 4.5 \times 10^9$	Azoic				none

Taxonomy VI

Classification of common dolphin:

Kingdom: Metazoa
Subkingdom: Eumetazoa
Phylum: Chordata
Subphylum: Vertebrata
Class: Mammalia
Subclass: Theria
Infraclass: Eutheria
Order: Cetacea
Suborder: Odontocetia
Family: Dolphinidae
Genus: *Delphinus*
Species: *delphis*
Common Name: common dolphin

Index of genera of some common marine organisms

1. Algae:
   Halimeda: calcareous, tropical lagoons, banks, green algae
   Lithothamnion: calcareous, rim of coral reef, red algae
   Macrocystis: giant kelp, west coast N. America, brown algae
   Penicillus: calcareous, tropical lagoons, banks, green algae
   Trichodesmium: green algae, namesake of Red Sea
   Ulva: sea lettuce, green algae, intertidal
   Zooanthellae: green algae, symbiotic with reef coral

2. Barnacles:
   Balanus: acorn barnacle, intertidal
   Conchoderma: whale barnacle
   Pollicipes: Pacific goose-necked barnacle, intertidal

3. Bivalves:
   Crassostrea: Atlantic oyster
   Donax: little bean clam or coquina
   Mya: "steamer" clam
   Mytilus: edible mussel, intertidal
   Tivela: Pismo clam
   Tridacna: giant clam, tropical Pacific
   Venus: quahog clam

4. Coral:
   Acropora: tropical, reef slope
   Favia: brain coral, back reef
   Fungia: reef front
   Millepora: reef edge

5. Crabs:
   Callinectes: blue crab
   Emerita: sand crab
   Uca: Fiddler crab

6. Dinoflagellates:
   Ceratium: bioluminescent
   Gymnodinium: causes red tide in Gulf of Mexico
   Prorocentrum: causes red tide in southern California

7. Fish:
   Anguilla: eel
   Cynoscion: sea trout
   Mugil: mullet
   Salmo: salmon

8. Grasses:
   Phyllospadix: surf grass
   Spartina: marsh grass
   Thalassia: mud flat grass
   Zostera: mud flat grass

9. Mammals:
   Balaenoptera: blue whale
   Callorhinus: Pacific fur seal
   Dugong: dugong
   Enhydra: sea otter
   Odobenus: walrus
   Zalophus: California sea lion

10. Snails:
    Acmaea: limpet
    Haliotes: abalone
    Littorina: periwinkle
    Stenoplax: chiton

11. Starfish:
    Acanthaster: crown of thorns starfish
    Pisaster: common starfish, intertidal

12. Worms:
    Lineus: more than 25 m long, Indian Ocean
    Thoracophelia: beach bloodworm

Cycles in the Ocean VII

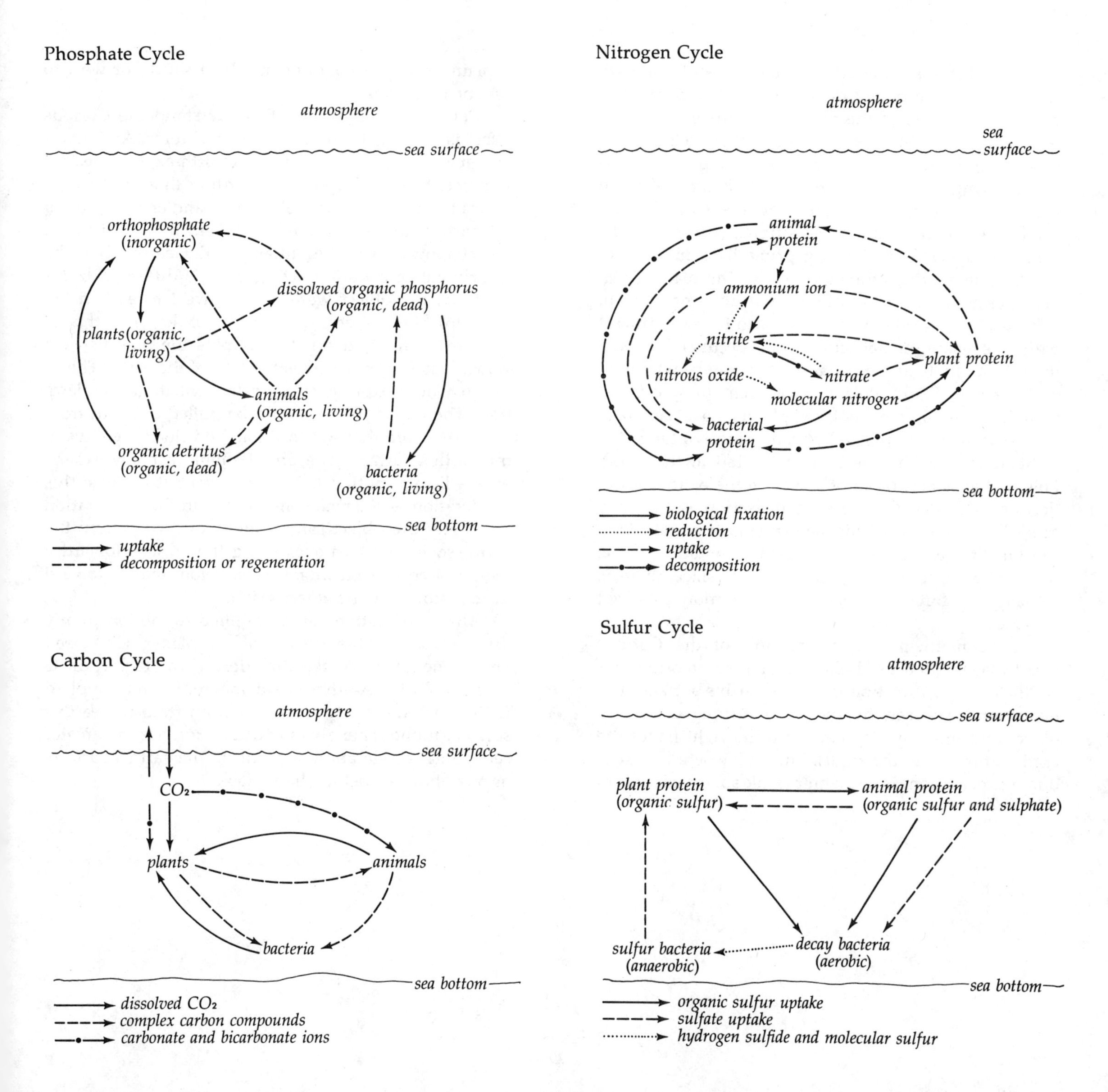

The Coriolis Effect VIII

Any object moving across the surface of the earth experiences an apparent deflection—to the right of its direction of travel in the northern hemisphere, to the left in the southern. The term *apparent* is used because the change in direction of travel is relative. If a missile were moving west to east in the northern hemisphere, it would, relative to an observer on the earth, deflect somewhat to the right of its original direction—the extent of the deflection being primarily a function of the latitude at which the motion is taking place. To an observer not on the earth or within its rotating frame of reference, the missile would appear to travel a straight path. The maximum deflection would occur at the poles; the minimum, 0, at the equator. This may be why penguins at the South Pole do not seem to waddle in straight lines, or why polar explorers have generally tended to miss the poles, by wandering too far to the right at the North Pole and to the left at the South Pole. For an object or an observer on the earth, the deflection is a real one. Because deflection is normally caused by some force, this deflection is often called the Coriolis "force." If the term is used it should be borne in mind that it is an apparent force, since there is actually no force or push or acceleration involved ($F = ma$).

A nonmathematical explanation of the Coriolis effect that is often used cites an object in motion in a south to north direction over the earth's surface. The deflection is caused by the earth turning under the object and moving it somewhat to its right (since the earth is turning to the right). This is a woefully inadequate explanation. If you don't think so, try using it to explain an object's motion north to south, or west to east or vice versa.

Perhaps the best way to understand the Coriolis effect is to consider it as a centrifugal force. Any object at rest at a particular point on the earth experiences the same centrifugal force as the earth at that locale. This varies from place to place obviously, and is a function of the earth's rate of rotation at different places. But if an object moves over the surface of the earth, then the centrifugal acceleration acting on it is different. If, for example, the object were moving west to east in the northern hemisphere, its velocity would equal that of the earth's rotation on its axis plus the object's own velocity as it moves across the face of the earth. Therefore it would experience a greater centrifugal acceleration. The object, then, should be pulled outward from the earth more than when it had no velocity relative to the earth's surface. The object should actually experience a lifting motion from the earth. But because this acceleration is so small compared with the acceleration of gravity the object cannot move off or up from the earth, so it simply moves along the surface toward a region of equal centrifugal acceleration, that is, toward the equator, or to the object's right.

This explanation can be applied to motion in any direction across the surface of the planet. One need only remember that the direction of the deflection is reversed for the southern hemisphere. It can be applied with equal success to vertical motion from the earth's surface, if one remembers which region has the greater centrifugal acceleration, a point on the earth's surface, or one above or below the surface.

Glossary

A

*_abyssal knoll_—An elevation rising less than 1,000 m from the sea floor and of limited extent across the summit.

abyssal plain—A flat, gently sloping, or nearly level region of the sea floor.

abyssopelagic—(or abyssal). Pertaining to the portion of the ocean below 3,700 m.

acceleration—The rate of change with time of the velocity of a particle. In the cgs system it is expressed as centimeters per second per second.

acceleration of gravity—The acceleration of a freely falling body because of the gravitational attraction of the earth.

*_advection_—The horizontal or vertical flow of seawater as a current. In meteorology, the predominantly horizontal, large-scale motions of the atmosphere.

aerobe—An organism that can live and grow only in the presence of oxygen.

age of water—The time elapsed since a water mass was last at the surface and in contact with the atmosphere. The water's age helps indicate its rate of overturn, an important factor in the use of the oceans for dumping radioactive wastes and determining the rate of replenishment of nutrients.

Agulhas Current—A fast current flowing southwestward along the southeast coast of Africa.

Alaska Current—A current that flows northwestward and westward along the coasts of Canada and Alaska to the Aleutian Islands. It contains water from the North Pacific Current and has the character of a warm current, so it influences climatic conditions.

*_alga(e)_—A simple plant, without a true stem, leaves, or roots, having a one-celled sex organ and possessing chlorophyll; includes almost all seaweeds. _See_ red alga, blue-green alga, brown alga, green alga, golden alga.

algal ridge—The elevated margin of a windward reef built by actively growing calcareous algae.

algal rim—A low rim built by actively growing calcareous algae on the lagoonal side of a leeward reef, or on the windward side of a reef patch in a lagoon.

alluvium—The detrital deposits eroded, transported, and deposited by streams.

amensalism—Interaction of two populations in which one is inhibited, but the other is not affected.

amphipod—One of an order (Amphipoda) of elongate and usually laterally compressed crustaceans. The species live in a variety of habitats from the parasitic state to the deep pelagic.

amplitude—For an ocean surface wave, the vertical distance from still water level to wave crest, that is, one-half the wave height.

anaerobe—An organism for whose life processes a complete or nearly complete absence of oxygen is essential.

anaerobic sediment—A highly organic sediment rich in H_2S formed in the absence of free oxygen, where little or no circulation or mixing of the bottom water occurs.

annelid—One of a phylum (Annelida) of segmented worms, with the majority of marine forms possessing a distinct head. Members of the group are free swimming, burrowing, or tube building.

Antarctic Bottom Water—_See_ water mass.

Antarctic Circumpolar Current—_See_ West Wind Drift.

*_Antarctic Convergence_—The convergence line circling the South Pole. It is the best defined convergence line in the oceans, being recognized by a relatively rapid northward increase in the surface temperature. _See_ convergence.

Antarctic Ocean—The name commonly applied to those portions of the Atlantic, Pacific, and Indian oceans that reach Antarctica on the south and are bounded on the north by the Subtropical Convergence. It is not a recognized ocean body.

anthozoan—One of a class (Anthozoa) of coelenterates in which the medusa stage is absent and the polyp stage is better developed than in the other coelenterates. Examples are sea anemones, sea pens, and corals.

*_aphotic zone_—That portion of the ocean waters where light is insufficient for plants to carry on photosynthesis. _See_ euphotic zone.

Archimedes's principle—The statement that a new upward force (buoyancy), equal in magnitude to the weight of the displaced fluid, acts on a body either partly or wholly submerged in any fluid under the influence of gravity.

archipelago—A sea or part of a sea studded with islands or island groups; often synonymous with island group.

Most of the definitions in this glossary are quoted or adapted from B. B. Baker, Jr., W. R. Deebel, and R. D. Geisenderfer, eds., _Glossary of Oceanographic Terms_ (Washington, D.C.: U.S. Naval Oceanographic Office, 1966).

*Starred glossary terms are especially important.

Arctic Convergence—The northern hemisphere polar convergence. Because of the configuration of the oceans in the northern latitudes, this convergence zone is poorly defined. *See* convergence.

arthropod—One of a phylum (Arthropoda) of animals with segmented external skeletons and jointed appendages; for example, the crustaceans, spiders, and insects.

Atlantic Ocean—The Atlantic Ocean extends from Antarctica northward to the southern limits of the Greenland and Norwegian seas. It is separated from the Pacific Ocean by the meridian of Cape Horn and from the Indian Ocean by the meridian of Cape Agulhas.

atmosphere—The pressure exerted per square centimeter by a column of mercury 760 mm high at a temperature of 0°C where the acceleration of gravity is 980.665 cm/sec/sec. One atmosphere of pressure equals 1.0133×10^6 dynes/cm^2. When considering pressures within the ocean, surface atmospheric pressure is considered to be zero.

atoll—A ring-shaped organic reef that encloses a lagoon in which there is no preexisting land, and which is surrounded by the open sea.

atoll reef—A ring-shaped coral and limestone reef often carrying low sand islands, enclosing a body of water.

azoic—Without life; but most ocean areas described as azoic contain at least a bacterial flora.

B

backshore—The part of a beach that is usually dry, being reached only by the highest tides. *See* foreshore.

baleen whale—A member of the cetacean suborder Mysticeti, which comprises the right whales, gray whales, and rorquals. Baleen, the horny material growing down from the upper jaw of these plankton-feeding whales, forms a strainer or filtering organ.

bank—An elevation of the sea floor located on a continental (or island) shelf and over which the depth of water is relatively shallow but sufficient for safe surface navigation. Shoals or bars on its surface may be dangerous to navigation.

**bar*—A submerged or emerged embankment of sand, gravel, mud, *or* mollusk shells built on the sea floor in shallow water by waves and currents. *See* longshore bar, barrier beach, baymouth bar.

barnacle—One of an order (Cirripedia) of crustaceans that are enclosed in a calcareous shell and sessile during their adult life. They are of two general types, acorn barnacles and stalked (or gooseneck) barnacles.

barrier beach—A bar essentially parallel to the shore, the crest of which is above high water.

barrier reef—A coral reef parallel to and separated from the coast by a lagoon that is too deep for coral growth.

basalt—(or sima). An igneous rock that is low in silica and that forms a worldwide layer under the oceans and the granitic continents.

bathyal—Pertaining to ocean depths between 180 and 3,700 m; sometimes identical with the continental slope environment.

bathymetry—The science of measuring ocean depths to determine the sea floor topography.

bathypelagic—A depth zone of the ocean that lies between 900 and 3,700 m.

bathythermograph—(abbreviated BT). A device for obtaining a record of temperature against depth (strictly speaking, pressure) in the ocean, from a ship underway.

**bay*—A recess in the shore or an inlet of a sea between two capes or headlands; not as large as a gulf but larger than a cove.

baymouth bar—A bar extending partially or entirely across the mouth of a bay.

**beach*—The zone of unconsolidated material that extends landward from the low water line to the line of permanent vegetation (usually the effective limit of storm waves). A beach includes foreshore and backshore. *See also* shoreline, coast.

beach face—The section of the beach normally exposed to the action of the wave uprush. The foreshore zone of a beach. (Not synonymous with shoreface.)

beach scarp—An almost vertical slope along the beach caused by wave erosion. It may vary in height from a few inches to several feet, depending on wave action and the nature and composition of the beach.

bends—(decompression sickness). A condition resulting from the formation of gas bubbles in the blood or tissues of a diver during ascent. Depending on their number, size, and location, these bubbles may cause pain, paralysis, unconsciousness, and occasionally death. The blocking of an artery by a bubble is called an air embolism.

Benguela Current—A strong current flowing northward along the southwest coast of Africa; it is formed by the West Wind Drift and the Agulhas Current.

benthic—(or benthonic). 1. The portion of the marine environment inhabited by marine organisms that live permanently in or on the bottom. 2. Pertaining to all submarine bottom terrain regardless of water depth.

berm—The nearly horizontal portion of a beach or backshore having an abrupt fall and formed by deposition of material by wave action. It marks the limit of ordinary high tides.

bilaterally symmetrical—Capable of being halved in one plane so that the halves are mirror images of each other.

biological oceanography—(or marine biology). The study of the ocean's plant and animal life in relation to the marine environment, including the effects of habitat, sedimentation, physical and chemical changes in the environment, and other factors on the distribution of marine organisms, as well as the action of organisms on the environment.

bioluminescence—(also called phosphorescence). The production of light without sensible heat by living organisms as a result of a chemical reaction either within certain cells or organs or outside the cells in some form of secretion.

**biomass*—The amount of living matter per unit of water surface or volume expressed in weight units.

**biosphere*—The transition zone between earth and atmosphere within which most terrestrial life is found; the outer portion of the geosphere and inner or lower portion of the atmosphere.

biotic factors—Biological factors such as availability of food, competition between species, and predator-prey relationships, which affect the distribution and abundance of a given species of plant or animal.

blue-green alga—Very simple single-celled or filamentous plants (Monera) in which the blue color is imparted by a water-soluble accessory pigment.

**bottom water*—The water mass at the deepest part of the water column. It is the densest water that is permitted to occupy that position by the regional topography. *See* water mass.

brachiopod—(or lamp shell). One of a phylum (Brachiopoda) of sessile, marine, mollusc-like animals in which the body, whose construction differs considerably from that of the molluscs, is enclosed in calcareous or horny bivalve shell.

brackish water—Water in which salinity values range from approximately 0.50 to 17.00‰.

Brazil Current—The warm ocean current that flows southward along the Brazilian coast below Natal.

breaker—A wave breaking on the shore, over a reef, etc. They are roughly classified into three kinds: spilling breakers break gradually over a considerable distance; plunging breakers tend to curl over and break with a crash; surging breakers peak up, but then instead of spilling or plunging they surge up on the beach face.

breccia—(or agglomerate). A rock composed of angular (not water-worn) fragments, as distinguished from conglomerates.

brine—Seawater containing a higher concentration of dissolved salt than the ordinary ocean. Brine is produced by the evaporation or freezing of seawater.

brittle star—One of a class (Ophiuroidea) of echinoderms having five, sometimes six, rarely seven or eight, elongate, slender, cylindrical arms radiating from a flat central disc.

brown alga—One of a phylum (Phaeophyta) of greenish yellow to deep brown, filamentous to massively complex plants, in which the color is imparted by the predominance of orangish pigments over the chlorophylls. This group includes the rockweeds, gulfweeds, and the large kelp. Brown algae are most abundant in the cooler waters of the world.

bryozoan—One of a phylum (Bryozoa) of minute, mostly colonial, aquatic animals with body walls often hardened by calcium carbonate and growing attached to aquatic plants, rocks, and other firm surfaces.

buoyancy—1. The property of an object that enables it to float on the surface of a liquid, or ascend through and remain freely suspended in a compressible fluid such as the atmosphere. 2. (or buoyant force). The upward force exerted on a parcel of fluid (or an object within the fluid) in a gravitational field by virtue of the density difference between the parcel (or object) and that of the surrounding fluid.

C

caballing—The mixing of two water masses having identical *in situ* densities but different *in situ* temperatures and salinities, such that the resulting mixture is denser than its components.

calcareous algae—Marine plants that form a hard external covering of calcium compounds.

calf—A piece of floating ice that has broken away from a larger piece of sea ice or land ice.

California Current—The wide, sluggish ocean current that flows southward along the west coast of the United States to northern Baja California.

Callao Painter—(or *El Pintor*). Mariners' reference to the catastrophic destruction of marine life that causes the blackening of paint on ships within the harbor of Callao, Peru. Hydrogen sulfide released during the decomposition of organisms that die when the water temperature increases as warmer oceanic currents turn inshore.

calm—The apparent absence of motion of the water surface when there is no wind or swell; water is generally considered calm if the current speed is less than 0.1 knot.

calorie—(abbreviated cal). A unit of heat, often defined as the amount of heat required to raise the temperature of 1 g of water through 1°C.

Canary Current—The prevailing southward flow along the northwest coast of Africa; it helps to form the North Equatorial Current.

capillary wave—A wave whose velocity of propagation is controlled primarily by the surface tension of the liquid in which the wave is traveling. Water waves less than 1.73 cm long are considered capillary waves.

carapace—A chitinous or bony shield covering the whole or part of the back of certain animals, such as crustaceans.

carbon dioxide—A heavy, colorless gas of chemical formula CO_2. It is the fourth most abundant constituent of dry air. Over 99 percent of the terrestrial CO_2 is found in the oceans.

cellular convection—An organized, convective, fluid motion characterized by convection cells, usually with upward motion (away from the heat source) in the central portions of the cell and sinking or downward flow in the cell's outer regions.

Celsius temperature scale—(abbreviated C). Same as centigrade temperature scale, by recent convention. 0°C = 273.16°K. See Kelvin temperature scale.

center of gravity—A point at which the mass of the entire body may be regarded as being concentrated.

**centrifugal force*—The force with which a body moving under constraint along a curved path reacts to the constraint. Centrifugal force acts away from the center of curvature of the path of the moving body. As a force caused by the rotation of the earth on its axis, centrifugal force is opposed to gravitation and combines with it to form gravity.

cephalopod—One of a class (Cephalopoda) of free-swimming molluscs with a large head, large eyes, and a circle of arms or tentacles around the mouth; the shell is external, internal, or absent, and an ink sac usually is present. Examples are squid, octopus, and nautilus.

cetacean—A marine mammal of the order Cetacea, which includes the whales, dophins, and porpoises.

**cgs system*—The system of physical measurements in which the fundamental units of length, mass, and time are the centimeter, gram, and second, respectively.

chaetognath—One of a phylum (Chaetognatha) of small, elongate, transparent, worm-like animals pelagic in all seas from the surface to great depths. Also called arrow worm.

chemical oceanography—The study of the chemical composition of dissolved solids and gases, material in suspension, and acidity of ocean waters and their variability both geographically and temporally in relationship to the atmosphere and the ocean bottom.

chiton—One of a class (Amphineura) of flattened molluscs protected either by calcareous spicules or plates.

chlorinity—(Symbol Cl). A measure of the chloride content, by mass, of seawater (grams per kilogram of seawater, or per mille). Because the proportion of chloride to sodium is reasonably constant the amount of chlorinity in a seawater sample is generally used to establish the sample's salinity.

chlorophyll—A group of green pigments active in photosynthesis.

chordate—One of a phylum (Chordata) of animals that possess a notochord (a middorsal cylindrical rod), a series of paired gill slits, both of which features are present only in the embryo of air-breathing members, and a dorsal central nervous system. Representative chordates are the tunicates, fishes, and mammals.

circulation—A general term describing a water current flow within a large area; usually a closed circular pattern such as the North Atlantic.

clapotis—(French). A type of standing wave. In American usage it is usually associated with the standing wave phenomenon caused by the reflection of a wave train from a breakwater, bulkhead, or steep beach.

clastic—A rock composed principally of detritus transported mechanically into its place of final deposition. Sandstones and shales are the commonest clastics.

**clay*—As a size term, refers to sediment particles smaller than silt. Mineralogically, clay is a hydrous aluminum silicate material with plastic properties and a crystal structure.

**coast*—The general region of indefinite width that extends from the sea inland to the first major change in terrain features.

coccolithophore—One of a family (Coccolithophoridaceae) of microscopic, often abundant planktonic algae, the cells of which are surrounded by an evelope on which numerous small calcareous discs or rings (coccoliths) are embedded.

coelenterate—One of a phylum (Coelenterata) of two-staged (sessile and free-floating) organisms. The sessile stage basically is cylindrical and is called a polyp; the free-swimming stage is disc- or bell-shaped and is called a medusa or jellyfish. Many coelenterates, particularly the hydrozoans and corals, are colonial, consisting of many polyps united in complex or massive structures. All contain stinging cells, and many are bioluminescent. *See* hydrozoan, scyphozoan, anthozoan.

**colligative property*—One of four characteristic properties of solutions, namely the interdependent changes in vapor pressure, freezing point, boiling point, and osmotic pressure, with a change in amount of dissolved matter. If, under a given set of conditions, the value for any one property is known, the others may be computed. In general, with an increase in dissolved matter (for example, salt in seawater) freezing point and vapor pressure decrease, and boiling point and osmotic pressure increase.

colloid—As a size term, refers to particles smaller than clay size.

colonial organism—Organism in which the individuals are attached together as units and do not exist as separate animals. Groups of individuals specialize in function.

comb jelly—One of a phylum (Ctenophora) of spherical, pear-shaped, or cylindrical animals of jelly-like consistency ranging from less than 3 cm to about 1 m in length. The outer surface of the body bears eight rows of comb-like structures. Many species are luminescent.

commensalism—A symbiotic relationship between two species in which one species is benefited and the other is not harmed.

community—An integrated, mutually adjusted assemblage of plants and animals inhabiting a natural area. The assemblage may or may not be self-sufficient and is considered to be in a state of dynamic equilibrium. The community usually is characterized as having a more or less definite species composition and may be defined by the habitat it occupies or by the species present.

competition—Interaction of two populations in which each inhibits the other.

**condensation*—The physical process by which a vapor becomes a liquid or solid; the opposite of evaporation. When water vapor condenses, heat is released, and the surrounding temperature is raised.

**conservative property*—A property whose values do not change in the course of a particular series of events.

consolidated sediment—Sediment that has been converted into rocks by compaction, cementation, other physical, or chemical changes.

continent—A large landmass rising abruptly from the deep ocean floor, including marginal regions that are shallowly submerged. Continents constitute about one-third of the earth's surface.

continental borderland—A region adjacent to a continent, normally occupied by or bordering a continental shelf, that is highly irregular with depths well in excess of those typical of a continental shelf.

continental drift—The concept that the continents can drift on the surface of the earth because of the weakness of the suboceanic crust, such as ice can drift through water.

continental rise—A gentle slope with a generally smooth surface, rising toward the foot of the continental slope.

continental shelf—A zone adjacent to a continent or around an island, and extending from the low water line to the depth at which there is usually a marked increase of slope to greater depth.

continental slope—A declivity seaward from a shelf edge into greater depth.

**convection*—In general, mass motions within a fluid resulting in transport and mixing of the properties of that fluid. Convection, along with conduction and radiation, is a principal means of energy transfer.

convection cell—*See* cellular convection.

**convergence*—Situation in which waters of different origins come together at a point or, more commonly, along a line known as a convergence line. Along such a line the denser water from one side sinks under the lighter water from the other side. The recognized convergence lines in the oceans are the polar, subtropical, tropical, and equatorial convergence lines.

copepod—One of a subclass (Copepoda) of minute shrimp-like crustaceans, most species of which range between about 0.5 and 10.0 mm in length. Copepods occur in the surface layers of temperate and subpolar waters in large concentrations.

coral—1. The hard calcareous skeleton of various anthozoans and a few hydrozoans, or the stony solidified mass of a number of such skeletons. In warm waters colonial coral forms extensive reefs of limestone. In cool or cold water coral usually appears in the form of isolated solitary individuals. 2. The entire animal; a compound polyp that produces the skeleton.

coral reef—A ridge or mass of limestone built up of detritus deposited around the skeletal remains of mollusks, colonial coral, and massive calcareous algae. Coral may constitute less than half of the reef material.

cordillera—An entire mountain system, including all the subordinate ranges, interior plateaus, and basins.

core—1. A vertical, cylindrical sample of the bottom sediments, from which the nature and stratification of the bottom may be determined. 2. The central zone of the earth.

**Coriolis effect*—An apparent force on moving particles resulting from the earth's rotation. It causes the moving particles to be deflected to the right of motion in the northern hemisphere and to the left in the southern hemisphere; the force is proportional to the speed and latitude of the moving particle and cannot change the speed of the particle.

crinoid—(or *sea lily*). One of a class (Crinoidea) of echinoderms most of which are permanently attached by a long stalk to the bottom; species without stalks either creep slowly about or swim. About 2,000 fossil species are known.

Cromwell Current—An eastward-flowing subsurface current that extends about $1^1/_2$ degrees north and south of the equator, and from about 150°E to 92°W. It is 300 km wide and 200 m thick; at its core the speed is 100 to 150 cm/sec.

crust—The outer shell of the solid earth. The crust varies in thickness from approximately 5 to 7 km under the ocean basins to 35 km under the continents.

crustacean—One of a class (Crustacea) of arthropods that breathe through gills or branchiae and have a body commonly covered by a hard shell or crust. The group includes barnacles, crabs, shrimps, and lobsters.

crystalline—The term applied to rocks containing grains of regular polyhedral form bounded by plain surfaces and having an orderly molecular structure. Usually applied to igneous and metamorphic rocks but not to sedimentary rocks.

current—A horizontal movement of water.

D

Davidson Current—A coastal countercurrent flowing north inshore of the California Current from northern California to Washington to at least 48°N during the winter months.

dead water—The phenomenon that occurs when a ship of low propulsive power negotiates water in which thin layer of fresher water is over a deeper layer of more saline water. As the ship moves, part of its energy generates an internal wave, which causes a noticeable drop in efficiency of propulsion.

decapod—One of an order (Decapoda) of crustaceans that includes the shrimps, lobsters, and crabs.

decay of waves—The change that waves undergo after they leave a generating area (fetch) and pass through a calm or region of lighter winds. In the process of decay, the significant wave height decreases and the significant wavelength increases, principally because of angular spreading, dispersion, and opposing winds.

deep scattering layer—(DSL). The stratified population(s) of organisms in most oceanic waters that scatter sound. Such layers generally are found during the day at depths from 200 to 800 m and may be from 50 to 200 m thick.

deepwater wave—A surface wave the length of which is less than twice the depth of the water; its velocity is independent of the depth of the water. *See* shallow water wave.

delta—An alluvial deposit, roughly triangular or digitate in shape, formed at the mouth of a stream or tidal inlet.

**density*—1. The ratio of the mass of any substance to the volume occupied by it; the reciprocal of specific volume. 2. In oceanology, density is equivalent to specific gravity and represents the ratio, at atmospheric pressure, of the weight of a given volume of seawater to that of an equal volume of distilled water at 4.0°C (39.2°F). It is thus dimensionless.

density layer—A layer of water in which density increases with depth enough to increase the buoyancy of a submarine. (Submariner's term for pycnocline.)

detritus—(or debris). Any loose material produced directly from rock disintegration.

detritus feeder—An animal that feeds on bacteria and dead or dying algae.

diatom—One of a class (Bacillariophyceae) of microscopic plankton organisms, possessing a wall of overlapping halves impregnated with silica. Diatoms are one of the most abundant groups of organisms in the sea and the most important primary food source of marine animals.

diffracted wave—A wave whose front has been bent by an obstacle in the medium other than by reflection or refraction. For example, when a portion of a train of waves is interrupted by a barrier such as a breakwater, diffracted waves are propagated in the sheltered region within the barrier's geometric shadow.

dinoflagellate—One of a class (Dinophyceae) of single-celled microscopic or minute organisms. Dinoflagellates may possess both plant (chlorophyll and cellulose plates) and animal (ingestion of food) characteristics.

discontinuity—The abrupt change or jump of a variable at a line or surface. *See* interface.

**diurnal*—1. Daily, especially pertaining to actions completed within 24 hours and that recur every 24 hours. 2. Having a period or cycle of approximately one lunar day (24.84 solar hours).

**divergence*—A horizontal flow of water, in different directions, from a common center or zone; often associated with upwelling.

doldrums—A nautical term for the equatorial trough, with special reference to the light and variable nature of the winds.

dolphin—1. A member of the cetacean suborder Odontoceti. The name is used interchangeably with porpoise by some. More properly it is given generally to the beaked members of the family Delphinidae, except the larger members which have been given the name "whale," such as the killer whale and pilot whale. 2. A pelagic fish of the genus *Coryphaena* noted for its brilliant colors.

dredge—A simple cylindrical or rectangular device for collecting samples of bottom sediment and life. It is generally made of heavy gauge steel plate or pipe and depends on a scooping action to obtain the sample.

drown—To submerge land beneath water either through a rise in the level of the water or by sinking of the land.

dyne—A force which, acting on a mass of 1 g, imparts to that mass an acceleration of 1 cm/sec^2. Since 1930, gravity has been reported in terms of the gal rather than the dyne.

E

earthquake—A sudden, transient motion or trembling of the earth's crust resulting from waves in the earth caused by faulting of the rocks or by volcanic activity.

East Greenland Current—A current flowing south along the east coast of Greenland and carrying water of low salinity and low temperature. It flows through the Denmark Strait between Iceland and Greenland to join the Irminger Current.

East Wind Drift—A west-setting current close to Antarctica caused by the polar easterlies.

ebb current—The tidal current associated with the decrease in the height of a tide. Ebb currents generally flow seaward, or in an opposite direction to the tide progression. Erroneously called ebb tide.

echinoderm—One of a phylum (Echinodermata) of principally benthic marine animals having calcareous plates with projecting spines forming a rigid or articulated skeleton or plates and spicules embedded in the skin; the animals have radial symmetry, usually with a five-rayed body. Some echinoderms are the sea stars, sea urchins, crinoids, and sea cucumbers.

echo sounding—Determining the depth of water by measuring the time interval between emission of a sonic or ultrasonic signal and the return of its echo from the bottom.

**eddy*—A circular movement of water usually formed, where currents pass obstructions, between two adjacent currents flowing counter to each other, or along the edge of a permanent current.

Ekman spiral—A theoretical representation of the effect that a wind blowing steadily over an ocean of unlimited depth and extent and of uniform viscosity would cause the surface layer to drift at an angle of 45° to the right of the wind direction in the northern hemisphere. Water at successive depths would drift in directions more to the right until at some depth it would move in the direction opposite to the wind. Velocity decreases with depth throughout the spiral. The depth at which this reversal occurs is of the order of 100 m. The net water transport is 90° to the right of the direction of the wind in the northern hemisphere.

electrical conductivity—The facility with which a substance conducts electricity; it is an intrinsic property of seawater and varies with temperature, salinity, and pressure.

El Niño—A warm current flowing south along the coast of Ecuador. It generally develops just after Christmas concurrently with a southerly shift in the tropical rain belt. In exceptional years, plankton and fish are killed in the coastal waters and a phenomenon somewhat like the red tide of Florida results.

**environment*—The sum total of all the external conditions that effect an organism, community, material, or energy.

epicenter—The point on the earth's surface directly above the focus of an earthquake.

epipelagic—The upper portion of the oceanic province, extending from the surface to a depth of about 200 m.

Equatorial Convergence—The zone along which waters from the northern and southern hemispheres converge. This zone generally lies in the northern hemisphere, except in the Indian Ocean.

Equatorial Countercurrent—An ocean current flowing eastward near the equator.

**estuary*—A tidal bay formed by submerging the lower part of a nonglaciated river valley and containing a measurable quantity of salt.

euphausiid—One of an order (Euphausiacea) of shrimp-like, planktonic crustaceans, widely distributed in oceanic and coastal waters, and especially abundant in colder waters. Euphausiids grow to 8 to 10 cm in length, and nearly all possess luminous organs. Some form the principal food for many of the whalebone whales. *See* krill.

**euphotic zone*—The layer of a body of water that receives ample sunlight for the photosynthetic processes of the plants. The depth of this layer varies with the angle of incidence of the sunlight, length of day, and cloudiness, but it is usually 80 m or more.

eutrophic—Pertaining to bodies of water containing abundant nutrient matter.

evaporation—The physical process by which a liquid or solid is transformed to a gas; the opposite of condensation.

F

Fahrenheit temperature scale—(abbreviated F). A temperature scale with the freezing point of water at 32° and the boiling point at 212° at standard atmospheric pressure.

Falkland Current—A current flowing northward along the Argentine coast and originating from part of the West Wind Drift.

fan—A gently sloping, fan-shaped feature normally located near the lower end of a canyon.

fast ice—Sea ice that generally remains in the position where originally formed and may attain a considerable thickness. It is formed along coasts where it is attached to the shore or over shoals.

fathom—The common unit of depth in the ocean for countries using the English system of units, equal to 6 ft (1.83 m). It is also sometimes used in expressing horizontal distances, in which case 120 fathoms make one cable or very nearly one-tenth nautical mile.

fault—A fracture or fracture zone in rock along which one side has been displaced relative to the other side. The intersection of the fault surface with a surface, such as the sea bottom, is called a fault line. If a fault is not a single clean fracture but a wide zone with small interlacing faults and filled with breccia, it is called a fault zone.

fault block—A rock body bounded on at least two opposite sides by faults. It usually is longer than it is wide; when it is depressed relative to the adjacent regions it is called a graben or rift valley, and when it is elevated it is called a horst.

fauna—The animal population of a particular location, region, or period.

**fetch*—1. (also called generating area). An area of the sea surface over which seas are generated by a wind having a constant direction and speed. 2. The length of the fetch area, measured in the direction of the wind in which the seas are generated.

filter feeder—An animal that strains small particles of food from water.

fish—A member of the class Pisces, which includes the true fishes that have a bony endoskeleton, paired fins, and an operculum covering the gills.

**fjord*—(also spelled fiord). A narrow, deep, steep-walled inlet of the sea, formed either by the submergence of a mountainous coast or by an entrance of the sea into a deeply excavated glacial trough after the melting away of the glacier. A fjord may be several hundred fathoms deep and often has a relatively shallow entrance sill of rock or gravel.

Florida Current—A fast current with speeds of 2 to 5 knots that flows through the Straits of Florida to a point north of Grand Bahama Island where it joins the Antilles Current to form the Gulf Stream.

fog—A visible aggregate of minute water droplets suspended in the atmosphere near the earth's surface, which reduces visibility below 1 km.

food chain—The sequence of organisms in which each is food for a higher member of the sequence.

food cycle—The production, consumption, and decomposition of food in the sea, and the energy relationships involved in this cycle. Decomposition products are transformed by bacteria into inorganic nutrients suitable for use by the producers (marine plants) which, directly or indirectly, are the food source for all animals in the sea.

food web—A group of interrelated food chains.

foraminifer—(or foraminiferan, foram). One of an order (Foraminifera) of benthic and planktonic protozoa possessing variously formed shells of calcium carbonate, silica, chitin, or an agglomerate of materials.

foreshore—The zone that lies between the ordinary high and low water marks; the tides rise and fall daily across this area.

fracture zone—An extensive linear zone of unusually irregular topography of the sea floor characterized by large seamounts, steep-sided or asymmetrical ridges, troughs, or scraps.

freezing point—The temperature at which a liquid solidifies under any given set of conditions. Pure water under atmospheric pressure freezes at 32°F (0°C). The freezing point of water is depressed with increasing salinity.

fringing reef—A reef attached directly to the shore of an island or continental landmass. Its outer margin is submerged and often consists of algal limestone, coral rock, and living coral.

G

gal—A unit of acceleration equal to 1 cm/sec^2. The term was invented to honor the memory of Galileo.

gastropod—(or snail). One of a class (Gastropoda) of molluscs in which the animals possess a distinct head, generally with eyes and tentacles, and a broad flat foot and usually are enclosed in a spiral shell.

geological oceanography—The study of the floors and margins of the oceans, including description of submarine relief features, chemical and physical composition of bottom materials, interaction of sediments and rocks with air and seawater, and action of various forms of wave energy in the submarine crust of the earth.

geophysics—The physics or nature of the earth. It deals with the composition and physical phenomena of the earth and its liquid and gaseous envelopes; it embraces the study of terrestrial magnetism, atmospheric electricity, and gravity and includes seismology, volcanology, oceanology, meteorology, and related sciences.

geosphere—The "solid" portion of the earth, including water masses (hydrosphere). Above the geosphere lies the atmosphere and at the interface between these two regions is found almost all of the biosphere, or zone of life.

geosyncline—A large generally linear subsident trough in which many thousands of feet of sediments are accumulating or have accumulated. Deep oceanic trenches paralleling island arcs are considered to be developing geosynclines.

giant kelp—One of a genus (*Macrocystis*) of large vine-like brown algae, which grow attached to the sea bottom by a massive holdfast and reach lengths to 50 m. Members of this genus are the largest algae in existence.

glacial epoch—The Pleistocene epoch, the earlier of the two divisions of geologic time included in the Quaternary period; characterized by continental glaciers that covered extensive land areas now free from ice.

glacial trough—A U-shaped valley, excavated by a glacier either on land or sea bottom.

golden alga—Any member of the class Chrysophyceae, a usually one-celled mobile photosynthetic alga. Some forms are planktonic; others occur in tidepools or marshes.

graben—See *fault block.*

grab sampler—An instrument with jaws that enclose a part of the bottom for retrieval and study. The sample may be unrepresentative in coarse sediments where the jaws may be propped open by gravel and stones, permitting part of the sample to wash out.

graded bedding—A type of stratification in which each stratum displays a gradation in grain size from coarse below to fine above.

**gram*—A cgs unit of mass; originally defined as the mass of 1 cm^3 of water at 4°C; now one-thousandth of the standard kilogram.

granite—(or sial). A crystalline rock consisting essentially of alkali feldspar and quartz. Granitic is a textural term applied to coarse and medium-grained granular igneous rocks.

gravimeter—A weighing instrument of sufficient sensitivity to register variations in the weight of a constant mass when the mass is moved from place to place on the earth and thereby subjected to the influence of gravity at those places.

gravitation—In general, the mutual attraction between masses of matter (bodies). Gravitation is the component of gravity that acts toward the earth.

**gravity wave*—A wave whose velocity of propagation is controlled primarily by gravity. Water waves more than 5 cm long are considered gravity waves.

green alga—One of a phylum (Chlorophyta) of grass-green, single-celled, filamentous, membranous, or branching plants in which the color, imparted by chlorophylls *a* and *b*, is not masked by accessory pigments.

**greenhouse effect*—In the ocean where a layer of low salinity water overlies a layer of more dense water the short wavelength radiation of the sun is absorbed in the deeper layers. The radiation given off by the water is in the far infrared, and since this cannot radiate through the low salinity layer, a temperature rise results in the deeper layers. In the atmosphere the same effect is produced by a layer of clouds and the long wave radiation is trapped between the clouds and the earth.

**groin*—A low artificial wall of durable material extending from the land to seaward to protect the coast or to force a current to scour a channel.

Gulf Stream—A warm, well-defined, swift, and relatively narrow ocean current that originates north of Grand Bahama Island where the Florida Current and the Antilles Current meet. The Gulf Stream extends to the Grand Banks at about 40°N 50°W. The Florida Current, Gulf Stream, and North Atlantic Current together form the Gulf Stream system. Sometimes the entire system is referred to as the Gulf Stream.

guyot—(or tablemount). A seamount having a relatively smooth, flat top.

gyre—A closed circulatory system, larger than a whirlpool or eddy.

H

habitat—Place with a particular kind of environment inhabited by organisms.

hadal—Pertaining to the greatest depths of the ocean.

harbor—An area of water affording natural or artificial protection for ships.

**heat*—A form of energy transferred between systems through a difference in temperature, and existing only in the process of energy transformation. By the first law of thermodynamics, the heat absorbed by a system may be used by the system to do work or to raise its internal energy.

**heat budget*—Accounting for the total amount of the sun's heat received on the earth during any one year as being exactly equal to the total amount lost from the earth by reflection and radiation into space. The portion reflected by the atmospheres does not affect the earth's heat budget. The portion absorbed must balance the long-range radiation into space from the earth's entire system. The portion absorbed into the oceans causes the surface warming critical to the phenomenon of layer depth. Transport by currents further extends the distribution of heat.

heat capacity—The ratio of the heat absorbed (or released) by a system to the corresponding temperature rise (or fall).

heat conduction—The transfer of heat from one part of a body to another, or from one body to another in physical contact with it without displacement of the particles of the body.

**heat transport*—The process by which heat is carried past a fixed point or across a fixed plane; thus, a warm current such as the Gulf Stream represents a poleward flux of heat.

high energy environment—A region with considerable wave and current action that prevents the settling and accumulation of fine-grained sediment smaller than sand size.

higher high water—(abbreviated HHW). The higher of two high waters occurring during a tidal day where the tide exhibits mixed characteristics.

high water—(abbreviated HW; also called high tide). The highest limit of the surface water level reached by the rising tide. High water is caused by the astronomic tide-producing forces and the effects of meteorological conditions.

hook—A spit or narrow cape of sand or gravel whose outer end bends sharply landward.

horse latitudes—The belts of latitude over the oceans at approximately 30° to 35°N and S where winds are predominantly calm or very light and weather is hot and dry. In the North Atlantic Ocean these are the latitudes of the Sargasso Sea.

Humboldt Current—The cold ocean current flowing north along the coasts of Chile and Peru. Its northern limit can be placed a little south of the equator.

hummocked ice—Pressure ice, characterized by haphazardly arranged mounds or hillocks. This has less definite form and shows the effects of greater pressure than either rafted or tented ice, but in fact may develop from either of these.

hurricane—Severe tropical cyclone in the North Atlantic Ocean, Caribbean Sea, Gulf of Mexico, or the eastern North Pacific off the west coast of Mexico.

hydrography—The science dealing with the measurement and description of the physical features of the oceans, seas, lakes, rivers, and their adjoining coastal areas, with particular reference to their use for navigational purposes.

hydroid—The polyp form of a hydrozoan, as distinguished from the medusa or jellyfish form.

hydrology—The scientific study of the waters of the earth, especially the effects of precipitation and evaporation on the water in streams, lakes, or below the land surface.

hydrozoan—One of a class (Hydrozoa) of coelenterates. The highly branched polyp or hydroid stage of many members is an important component of fouling.

hypsographic chart—A chart or part of a chart showing land or submarine bottom relief in terms of height above datum. Hypsography is the science of measuring or describing elevations above datum.

I

iceberg—A large mass of detached land ice floating in the sea or stranded in shallow water. Irregular icebergs generally calve from glaciers, whereas tabular icebergs and ice islands are usually formed from shelf ice. An iceberg is usually defined as being the size of a ship or larger, although any piece of glacier ice more than 5 m high is often called an iceberg.

ice cap—A perennial cover of ice and snow over an extensive portion of the earth's land surface. The most important of the existing ice caps are those on Antarctica and Greenland (the latter often called inland ice).

ice floe—A single piece of sea ice, other than fast ice, large or small, described if possible as "light" or "heavy" according to thickness. Designations are: vast, over 10 km across; big, 1 to 10 km across; medium, 200 to 1,000 m across; small, 10 to 200 m across.

ice island—A large tabular fragment of shelf ice found in the Arctic Ocean. Most appear to have calved from the Ward Hunt ice shelf off the northern coast of Ellesmere Island. Ice islands are smaller than the largest tabular icebergs of the Antarctic, the largest one known being about 300 square miles in area and about 50 m thick.

ice rind—A thin, elastic, shining crust of ice formed by the freezing of slush on a quiet sea surface. It is easily broken by wind or swell and is generally less than 5 cm thick.

igneous rock—Rock formed by solidification of molten material or magma.

Indian Ocean—That ocean bounded on the north by the southern limits of the Arabian Sea, Laccadive Sea, Bay of Bengal, the limits of the East Indian archipelago and the Great Australian bight; on the east from the meridian of Southeast Cape, Tasmania; and on the west from Cape Agulhas southward. The limits of the Indian Ocean exclude the seas lying within it.

infrared radiation—(abbreviated IR). Electromagnetic radiation lying in the wavelength interval from about 0.8 micron to an indefinite upper boundary sometimes arbitrarily set at 1,000 microns (0.01 cm). At the lower limit of this interval, the infrared radiation spectrum is bounded by visible radiation and on its upper limit by microwave radiation.

inlet—A short, narrow waterway connecting a bay or lagoon with the sea.

inshore—In beach terminology, the zone of variable width between the shoreface and the seaward limit of the breaker zone.

**in situ*—In place; in the natural or original position.

interface—A surface separating two media, across which there is a discontinuity of some property, such as density, velocity, etc.

**internal wave*—A wave that occurs within a fluid whose density changes with depth, either abruptly at an interface or gradually. Its amplitude is greatest at the density discontinuity or, in the case of a gradual density change, somewhere inside the fluid and not at the free upper surface where the surface waves have their maximum amplitude. A relatively small amount of energy is required to set up and maintain an internal wave. Wave heights, periods, and lengths are usually large as compared to surface waves.

intertidal zone—(or littoral zone). Generally considered to be the zone between mean high water and mean low water levels.

invertebrate—Any animal without a backbone or spinal column.

**ion*—An electrically charged negative or positive atom or group of atoms. The dissolved salts in seawater dissociate into ions.

island—A body of land surrounded by water; relatively smaller than a continent.

island arc—A term used for a group of islands usually having a curving arch-like pattern, generally convex toward the open ocean, with a deep trench or trough on the convex side and usually enclosing a deep sea basin on the concave side.

isopod—One of an order (Isopoda) of generally flattened crustaceans. They are mostly scavengers.

J

jellyfish—(or medusa). 1. Any of various free-swimming coelenterates having a disc- or bell-shaped body of jelly-like consistency. Many have long tentacles with stinging cells. Some are luminescent. The term is often applied to the comb jellies.

**jetty*—A structure, such as a wharf, pier, or breakwater, located to influence current or protect the entrance to a harbor or river. *See also* groin.

K

**kelp*—One of an order (Laminariales) of usually large, blade-shaped, or vine-like brown algae. *See* giant kelp.

Kelvin temperature scale—(abbreviated K). An absolute temperature scale independent of the thermometric properties of the working substance. For convenience the Kelvin degree is identified with the Celsius degree (0°K = −273.16°C).

key—A low, flat island or mound of sand built up on a reef flat slightly above high tide; the sand may be mixed with coral or shell fragments.

kilometer—(abbreviated km). The unit of distance measurement in the metric system equal to 0.62 statute mile or 0.54 nautical mile. A statute mile equals 1.61 km; a nautical mile equals 1.85 km.

**kinetic energy*—The energy that a body possesses as a consequence of its motion, defined as one-half the product of its mass and the square of its speed, $\frac{1}{2} mv^2$.

knot—A speed unit of one nautical mile (6,076.12 ft) per hour. It is equivalent to a speed of 1.688 ft/sec or 51.4 cm/sec.

krill—(Norwegian kril). A term used by whalers and fishermen for euphausiids.

Kuroshio—(or Japan Current). A fast ocean current (2 to 4 knots) flowing northeastward from Taiwan to the Ryukyu Islands and close to the coast of Japan to about 150°E.

L

Labrador Current—A current that flows southward from Baffin Bay, through the Davis Strait, and southeastward along the Labrador and Newfoundland coasts.

lagoon—A shallow sound, pond, or lake generally separated from the open sea.

landlocked—A body of water enclosed or nearly enclosed by land, thus protected from the sea. San Francisco Bay is a classic example.

larva—An embryo that becomes self-sustaining and independent before it has assumed the characteristic features of its parents.

lee—Shelter, or the part or side sheltered or turned away from the wind or waves.

leeward—The direction toward which the wind is blowing; the direction toward which waves are traveling.

limestone—A general term for a class of rocks that are at least 80 percent carbonates of calcium or magnesium. Varieties of limestone take their names from the source material, for example, algal limestone.

**limnology*—The physics and chemistry of fresh water bodies and the classification, biology, and ecology of the organisms living in them.

liquid—A state of matter in which the molecules are relatively free to change their positions with respect to each other but restricted by cohesive forces so as to maintain a relatively fixed volume.

lithology—The study and description of rocks based on macroscopic and microscopic examination of samples.

**littoral*—(or intertidal). The benthic zone between high and low water marks; according to some authorities, between the shore and water depths of approximately 200 m.

load—The quantity of sediment transported by a current. It includes the suspended load of small particles, which float in suspension distributed through the whole body of the current, and the traction load, bottom load, or bed load of large particles which move along the bottom by traction, that is, saltation, rolling, and sliding.

longshore bar—A bar generally parallel to the shore and submerged at high tide.

longshore current—A current running roughly parallel to the shoreline and produced by waves being deflected at an angle by the shore. The amount of material a longshore current is capable of carrying depends on its velocity and the particle size of the material; however, any obstruction cutting across the path of the current will cause loss of velocity and consequently loss of carrying power.

low energy environment—A region characterized by a general lack of wave or current motion, permitting the settling and accumulation of very fine-grained sediment (silt and clay).

lower low water—(abbreviated LLW). The lower of two low waters of any tidal day where the tide exhibits mixed characteristics. *See* mixed tide.

low water—(abbreviated LW; or low tide). The lowest limit of the surface water level reached by the lowering tide. Low water is caused by the astronomic tide-producing forces and the effects of meteorological conditions.

lunar day—(or tidal day). The interval between two successive upper transits of the moon over a local meridian. The period of the mean lunar day, approximately 24.84 solar hours, is derived from the rotation of the earth on its axis relative to the movement of the moon about the earth.

M

maelstrom—A confused and often destructive current usually caused by the combined effects of high, wind-generated waves and a strong opposing tidal current; the rapid flows may follow eddying patterns or circular paths with whirlpool characteristics. Named after the frequently cited phenomenon along the south shore of Moskenesoy Island in the Lofoten Islands off the Norway coast; here, the maelstrom reaches its strength when the tidal current ebbs westward with speeds up to 9 knots at springs during a strong opposing westerly wind. Similar phenomena occur in Pentland Firth, Scotland, and off Cape de la Hague, Normandy.

**magma*—Mobile rock material generated within the earth from which igneous rock is derived by solidification. When extruded it is called lava.

magnetic anomaly—A distortion of the regular pattern of the earth's magnetic field due to local concentrations of ferromagnetic minerals.

magnetometer—An instrument for measuring the intensity and direction of the earth's magnetic field.

mangrove—One of several genera of tropical trees or shrubs that produce many prop roots and grow along low-lying coasts into shallow water.

mantle—The relatively plastic region between the crust and core of the earth.

marine biology—*See* biological oceanography.

**marsh*—An area of soft wet land. Flat land periodically flooded by salt water is called a salt marsh. Sometimes called a slough.

mass transport—The transfer of water from one region to another originating from the orbital motion of waves.

mean high water—(abbreviated MHW). The average height of all the high waters recorded over a 19-year period, or a computed equivalent period.

mean low water—(abbreviated MLW). The average height of all the low waters recorded over a 19-year period, or a computed equivalent period.

mean sea level—(abbreviated MSL). The mean surface water level determined by averaging heights at all stages of the tide over a 19-year period.

Mean sea level is usually determined from hourly height readings measured from a fixed predetermined reference level.

mean sphere depth—The uniform depth to which the water would cover the earth if the solid surface were smoothed off. This depth would be about 2,440 m.

mean water level—(abbreviated MWL). The mean surface level determined by averaging the height of the water at equal intervals of time, usually at hourly intervals, over a considerable period of time.

medusa—*See* jellyfish.

mesopelagic—That portion of the oceanic province extending from about 200 to 1,000 m.

messenger—A cylindrical metal weight approximately 7.5 cm long and 2.5 cm in diameter; it is attached around an oceanographic wire and sent down to trip devices such as Nansen bottles and current meters after they have been lowered to the desired depth.

metamorphic rock—Rock that has undergone structural and mineralogical changes, such as recrystallization, in response to marked changes of temperature, pressure, and chemical environment.

**meter*—The basic unit of length of the metric system, equal to 1,650,763.73 wavelengths of Kr^{86} orange-red radiation.

mid-ocean ridge—A great median arch or sea bottom swell extending the length of an ocean basin and roughly paralleling the continental margins. *See* cordillera.

mid-ocean rift—A deep, narrow-notched cleft valley, or graben, which is reportedly found almost continuously along the crest of a cordillera or ridge. *See* cordillera.

**mixed layer*—The layer of the water that is mixed through wave action or thermohaline convection.

mixed tide—The type of tide in which a diurnal wave produces large inequalities in heights or durations of successive high and law waters. This term applies to the tides between the predominantly semidiurnal and the predominantly diurnal.

mollusc—(also spelled mollusk). One of a phylum (Mollusca) of soft unsegmented animals, most of which are protected by a calcareous shell. The phylum is second only to the insects in number of species. The group includes snails, pelecypods, chitons, squids, and octopuses.

**monsoon*—A name for seasonal winds (from Arabic "mausim," a season). It was first applied to the winds over the Arabian Sea, which flow for six months from northeast and for six months from southwest, but it has been extended to similar winds in other parts of the world.

moraine—Rock debris, deposited chiefly by direct glacial action. Where glaciers float on or discharge into the sea, or glaciated regions are drowned by the sea, moraines form marine deposits.

mussel—One of a family (Mytilidae) of elongate, tapering pelecypods, usually dark colored, growing in masses on floating objects, underwater structures, or rocks and rocky cliffs, covering mud flats in the intertidal zone, and boring into rock.

mutualism—A symbiotic relationship between two species in which both are benefited but neither requires the interaction.

myctophid—One of a family (Myctophidae) of small bioluminescent oceanic fishes which normally live at depths between about 200 and 4,000 m. Many species undergo extensive daily vertical migrations and are thought to contribute to the sound scattering layers in the sea.

N

Nansen bottle—A device used by oceanologists to obtain subsurface samples of seawater.

nautical mile—(abbreviated n. mile). In general a unit used in marine navigation equal to a minute of arc of a great circle on a sphere. In the United States one international nautical mile equals 6,076.11549 ft.

neap tide—Tide of decreased range, which occurs about every two weeks when the moon is in quadrature.

nearshore current system—The current system caused by wave action in and near the surf zone. The nearshore current system consists of four parts: the shoreward mass transport of water, longshore currents, rip currents, and longshore movement of expanding heads of rip currents. Sometimes called inshore currents.

nearshore zone—Pertaining to the zone extending seaward from the shore to an indefinite distance beyond the surf zone.

nekton—Pelagic animals that are active swimmers, such as most of the adult squids, fishes, and marine mammals.

neritic—The portion of the pelagic division extending from low water level to the approximate edge of a continental shelf. Some writers have used this term in describing bottom organisms of a continental shelf, but its recommended usage is restricted to the waters overlying a shelf.

niche—The functional role and position of an organism in the ecosystem.

nitrate nitrogen—The most abundant and readily assimilable form of nitrogen for marine organisms. Like phosphate, it is an essential nutrient. Estimates of primary productivity have been made by determining the concentrations of nitrates in a water sample.

nitrogen cycle—The series of chemical changes that nitrogen undergoes in its use by plants and animals. Inorganic nitrogen compounds (nitrates, nitrites, and ammonium) and, to a small extent, organic nitrogen compounds in the sea are used by marine plants, which form other nitrogen compounds, such as amino acids. More complex amino acids and proteins are synthesized from these by the marine animals, which feed on the plants. Finally, these compounds, in the waste products and the dead bodies of the animals, are broken down by bacteria into inorganic compounds and simple organic compounds, completing the cycle.

North Atlantic Current—A wide slow-moving continuation of the Gulf Stream originating in the region east of the Grand Banks of Newfoundland at about 40°N and 50°W.

North Atlantic Drift—The weak, sluggish, northeast part of the North Atlantic Current that is easily influenced by winds; currents have been observed to change speeds and directions frequently, and at times reverse directions.

North Equatorial Current—Ocean currents driven by the northeast trade winds blowing over the tropical oceans of the northern hemisphere.

nudibranch—(or sea slug). One of the order (Nudibranchia) of gastropods in which the shell is entirely absent in the adult. The body bears projections which vary in color and complexity among the species.

**nutrient*—In the ocean any one of a number of inorganic or organic compounds or ions used primarily in the nutrition of primary producers. Nitrogen and phosphorus compounds are essential nutrients. Silicates are essential to diatoms. Vitamins such as B_{12} are essential to many algae.

O

ocean—(or sea). The intercommunicating body of salt water occupying the depressions of the earth's surface, or one of its major primary subdivisions, bounded by continents, the equator, and other imaginary lines.

ocean basin—That part of the floor of the ocean that is more than about 200 m below sea level.

**ocean current*—A regular movement of ocean water, either cyclic or more commonly a continuous stream flowing along a definable path. Three general classes, by cause, may be distinguished: (1) currents related to seawater density gradients, comprising the various types of gradient currents; (2) wind-driven currents, which are those directly produced by the stress exerted by the wind on the surface and; (3) currents produced by long-wave motions, principally the tidal currents, but including internal wave, tsunami, and seiche currents. The major ocean currents are continuous flowing streams and are of first-order importance in maintaining the earth's thermodynamic balance.

oceanic—The portion of the pelagic division seaward from the approximate edge of a continental shelf.

oceanic crust—A mass of simatic material approximately 5 km thick which lies under the ocean bottom and may be more or less continuous beneath the continental crust.

**oceanology*—The study of the sea, embracing and integrating all knowledge pertaining to the sea's physical boundaries, the chemistry and physics of seawater, and marine biology. In strict usage oceanography is the description of the marine environment, whereas oceanology is the study of the oceans and related sciences.

ocean slick—An area of the sea surface, variable in size and markedly different in color or oiliness; may be caused by internal waves.

octopus—A cephalopod with a round or sac-like body, eight arms, no shell, and generally without fins.

offshore—The comparatively flat zone of variable width that extends from the outer margin of the rather steeply sloping shoreface to the edge of the continental shelf.

ooze—A fine-grained pelagic sediment containing undissolved sand- or silt-sized, calcareous or siliceous skeletal remains of small marine organisms in proportion of 30 percent or more, the remainder being amorphous clay-sized material. There are diatomaceous, foraminiferal, pteropod, and radiolarian oozes.

orbit—In water waves, the path of a water particle affected by the wave motion. In deepwater waves the orbit is nearly circular, and in shallow water waves the orbit is nearly elliptical. In general, the orbits are slightly open in the direction of wave motion, giving rise to mass transport.

organic reef—A sedimentary rock aggregate composed of living and dead colonial organisms such as algae, coral, crinoids, and bryozoa. When it is covered by more than 12 m of water, it is an organic bank.

ostracod—One of a subclass (Ostracoda) of minute crustaceans, the individuals of which are unsegmented, laterally compressed, and enclosed in a bivalve shell. Some members are benthic; others are planktonic. Many species are luminescent.

**oxygen minimum layer*—A subsurface layer in which the dissolved oxygen content is very low or nil.

P

Pacific Ocean—The ocean area bounded on the east by the western limits of the coastal waters of southwest Alaska and British Columbia, the southern limits of the Gulf of California, and from the Atlantic Ocean by the meridian of Cape Horn, on the north by the southern limits of Bering Strait and the Gulf of Alaska; on the west by the eastern limits of the Sea of Okhotsk, Japan Sea, Philippine Sea, the East Indian archipelago from Luzon Island to New Guinea, Bismarck Sea, Solomon Sea, Coral Sea, Tasman Sea, and from the Indian Ocean by the meridian of Southeast Cape, Tasmania.

pack ice—The term used to denote any area of sea ice other than fast ice, no matter what form it takes or how disposed.

pancake ice—Pieces of newly formed ice, usually approximately circular, about 30 cm to 3 m across, and with raised rims caused when the pieces strike together as a result of wind and swell.

parasitism—A relationship between two species in which one lives on or in the body of its host and obtains food from its tissues, often benefiting at the expense of its host.

pelagic division—A primary division of the sea that includes the whole mass of water. The division is made up of the neritic province (water shallower than 200 m) and the oceanic province (water deeper than 200 m).

pelecypod—(or bivalve). One of a class (Pelecypoda) of molluscs generally sessile or burrowing into soft sediment, rock, wood, or other materials. Individuals possess a hatchet-shaped foot, which in some is used for digging. Clams, oysters, and mussels belong to this class.

per mille—(symbol ‰). Per thousand or 10^{-3}; used in the same way as percent (%, per hundred, or 10^{-2}). Per mille by weight is commonly used in oceanology for salinity and chlorinity; for example, a salinity of 0.03452 (or 3.452%) is commonly stated as 34.52‰.

**photosynthesis*—The manufacture of carbohydrate food from carbon dioxide and water in the presence of chlorophyll, by using light energy and releasing oxygen.

physical oceanography—The study of the physical aspects of the ocean, such as its density, temperature, ability to transmit light and sound, and sea ice; the movements of the sea, such as tides, currents, and waves; and the variability of these factors both geographically and temporally in relationship to the adjoining domains, namely, the atmosphere and the ocean bottom.

**physical properties*—The physical characteristics of seawater; for example, temperature, salinity, density, velocity, sound, electrical conductivity, and transparency.

phytoplankton—The plant forms of plankton. They are the basic synthesizers of organic matter (by photosynthesis) in the pelagic division. The most abundant of the phytoplankton are the diatoms.

pinger—A battery powered acoustic divice equipped with a transducer that transmits sound waves. When the pinger is attached to a wire and lowered into the water, the direct and bottom reflected sound can be monitored with a listening device. The difference between the arrival time of the direct and reflected waves is used to compute the distance of the pinger from the ocean bottom.

pinniped—A marine mammal of the order Pinnipedia, which comprises the seals, sea lions, and walruses.

**plankton*—The passively drifting or weakly swimming organisms in marine and fresh waters. Members of this group range in size from microscopic plants to jellyfishes measuring up to 2 m across the bell, and includes the eggs and larval stages of the nekton and benthos. *See* phytoplankton, zooplankton.

**plankton bloom*—An enormous concentration of plankton (usually phytoplankton) in an area, caused either by an explosive or a gradual multiplication of organisms (sometimes of a single species) and usually producing an obvious change in the physical appearance of the sea surface, such as discoloration. Blooms consisting of millions of cells per liter often have been reported. *See* red tide.

plankton net—A net for collecting plankton. A great variety of plankton nets have been constructed in attempts to fulfill specific requirements. Typically, the nets are cone shaped, but several modifications of this shape as well as completely different shapes exist.

polychaete—One of an order (Polychaeta) of annelids, which includes most of the marine segmented worms, some of which are the tubeworms of fouling. Some of these worms are luminescent during spawning.

polynya—A water area enclosed in ice, generally fast; this water area remains constant and usually has an oblong shape; sometimes limited to one side by the coast.

polyp—An individual sessile coelenterate.

population—Group of individuals of one species living in an area.

porpoise—A small to moderate sized member of the cetacean suborder Odontoceti. The name is used interchangeably with dolphin by some.

predation—Interaction in which a population of larger animals benefits by eating the smaller, or prey, population.

**primary production*—The amount of organic matter synthesized by organisms from inorganic substances in unit time in a unit volume of water or in a column of water of unit area cross section and extending from the surface to the bottom.

prime meridian—The meridian of longitude 0°, used as the origin for measurements of longitude. The meridian of Greenwich, England, is the internationally accepted prime meridian.

probe—A measuring device or sensor inserted into the environment to be measured. As applied to oceanology the term is used for devices lowered into the sea for *in situ* measurements.

**progressive wave*—A wave that is manifested by the progressive movement of the wave form.

propagation—The transmission of energy through a medium.

protocooperation—Interaction of two populations in which both benefit and require the interaction for survival.

Protozoa—A kingdom of mostly microscopic, one-celled animals. This group constitutes one of the largest populations in the sea, including some bioluminescent genera.

pteropod—One of an order (Pteropoda) of pelagic, free-swimming gastropods in which the foot is modified into fins; both shelled and nonshelled forms exist.

**pycnocline*—The vertical gradient of density.

pycnogonid—(or sea spider). One of a class (Pycnogonida) of spider-like benthic arthropods which range from shallow water to great depths. The species inhabiting shallow waters range in size from a fraction of an inch to a few inches; the deepwater species may attain a spread of several feet.

R

radially symmetrical—Capable of being halved in either of two or more planes so that the halves are approximately mirror images of each other.

**radiation*—The emission and propagation of energy through space or through a material medium in the form of waves; for instance, the emission and propagation of electromagnetic waves.

radiolarian—One of an order (Radiolaria) of single-celled planktonic protozoa possessing a skeleton of siliceous spicules and radiating thread-like pseudopodia. Most members are pelagic, and many are luminescent.

rafting—The overriding of one ice floe by another under pressure.

ram—An underwater ice projection from an iceberg or a hummocked ice floe. Its formation is usually due to a more intensive melting of the unsubmerged part of the floe.

ray—A cartilege skeletoned fish of the order Batoidei, in which the body generally is compressed dorso-ventrally, the eyes are on the upper surface, the gills on the lower surface, and the tail often reduced to a whip-like appendage. The order includes electric rays, stingrays, and manta rays.

red alga—One of a phylum (Rhodophyta) of reddish, filamentous, membranous, encrusting, or complexly branched plants in which the color is imparted by the predominance of *r*-phycoerythrin over the chlorophylls and other pigments. Red algae are worldwide in their distribution, being more abundant in temperate waters and ranging to greater depths than other algae.

red tide—A red or reddish-brown discoloration of surface waters, most frequently in coastal regions, caused by concentrations of microscopic organisms, particularly dinoflagellates. Toxins produced by the dinoflagellates can cause mass kills of fishes and other marine animals. In some regions, notably off the west coast of Florida, red tide appears to follow increased rainwater runoff from the land; in this way, it is believed, one or more scarce nutrient elements flow into the sea, permitting the dinoflagellates to multiply rapidly.

reef—An offshore consolidated rock hazard to navigation with a least depth of 20 m or less. *See* shoal. (For many years, a depth of 12 m has been considered critical for navigational safety. Because of the increased drafts of modern ships, a depth of 20 m is now considered critical.)

reef flat—A flat expanse of dead reef rock that is partly or entirely dry at low tide. It includes shallow pools, potholes, gullies, and patches of coral debris and sand.

reef front—The upper seaward face of the reef, extending above the dwindle point of abundant living coral and coralline algae to the reef edge. This zone commonly includes a shelf, bench, or terrace that slopes to 16 to 30 m, as well as the living wave-breaking face of the reef.

**reflected wave*—The wave that is returned seaward when a wave impinges on a very steep beach, barrier, or other reflecting surface.

**refraction of water waves*—The process by which the direction of a wave moving in shallow water at an angle to the contours is changed. That part of the wave advancing in shallower water moves more slowly than the other part still advancing in deeper water, causing the wave crest to bend toward alignment with the underwater contours.

**reversing thermometer*—A mercury-in-glass thermometer that records temperature on being inverted and thereafter retains its reading until returned to the first position.

ridge—A long, narrow elevation of the sea floor with steep sides and irregular topography.

rift valley—*See* fault block.

rip current—The return flow of water piled up on shore by incoming waves and wind; a strong narrow surface current flowing away from the shore.

S

**salinity*—A measure of the quantity of dissolved salts in seawater. It is formally defined as the total amount of dissolved solids in seawater in parts per thousand (‰) by weight when all the carbonate has been converted to oxide, the bromide and iodide to chloride, and all organic matter is completely oxidized. In practice, salinity is not determined directly but is computed from chlorinity, electrical conductivity, refractive index, or some other property whose relationship to salinity is well established. The equation presently used for determining salinity from chlorinity is $S = 1.80655\ Cl$.

**salt*—Any substance that yields ions, other than hydrogen or hydroxyl ions. A salt is obtained by displacing the hydrogen of an acid by a metal.

**salt marsh*—Flat, poorly drained coastal swamps that are flooded by most high tides.

**salt water wedge*—An intrusion in a tidal estuary of seawater in the form of a wedge characterized by a pronounced increase in salinity from surface to bottom.

Sargasso Sea—The region of the North Atlantic Ocean to the east and south of the Gulf Stream system. This is a region of convergence of the surface waters and is characterized by clear, warm water, a deep blue color, and large quantities of floating *Sargassum.*

Sargassum—(or sargasso weed). A genus of brown algae, characterized by a bushy form, a substantial holdfast when attached, and a yellowish brown, greenish yellow, or orange color. Two species (*S. fluitans* and *S. natans*) make up 99 percent of the macroscopic vegetation in the Sargasso Sea.

scaphopod—(or tooth shell). One of a class (Scaphopoda) of benthic marine molluscs having tubular, tapering, slightly curved shells, open at both ends; the body has no distinct head but possesses a foot.

scarp—An elongated and comparatively steep slope of the sea floor, separating flat or gently sloping areas.

**scattering*—The random dispersal of sound energy after it is reflected from the sea surface or sea bottom or off the surface of solid, liquid, or gaseous particles suspended in the water.

scavenger—Animal that feeds on dead animals.

scyphozoan—One of a class (Scyphozoa) of coelenterates in which the polyp stage is minimized or insignificant and the medusa stage is well developed. The true jellyfishes belong to this group.

sea—A subdivision of an ocean. All seas except inland seas are physically interconnected parts of the earth's total salt water system. Two types are distinguished, mediterranean and adjacent. Mediterraneans are groups of seas, collectively separated from the major body of water as an individual sea. Adjacent seas are those connected individually to the larger body.

sea anemone—Any of numerous anthozoans whose form, bright colors, and tentacles about the mouth often give them a superficial resemblance to a flower.

sea cow—(or sirenian). An aquatic herbivorous mammal of the order Sirenia which includes the dugong, the manatee, and the extinct Stellar sea cow.

sea cucumber—(or holothurian). One of a class (Holothuroidea) of elongate, usually worm-like echinoderms which have a flexible body wall and creep over the bottom from shallow water to great depths.

sea floor—The bottom of the ocean where there is a generally smooth, gentle gradient. In many uses depth is disregarded and the term may be used to designate areas in basins or plains or on the continental shelf.

sea ice—Ice formed by the freezing of seawater; opposed, principally, to land ice. In brief, it forms first as small crystals, thickens into sludge, and coagulates into sheet ice, pancake ice, or ice floes of various shapes and sizes. Thereafter, sea ice may develop into pack ice or become a form of pressure ice.

sea level—The height of the surface of the sea at any time.

seamount—An elevation rising 1,000 m or more from the sea floor and of limited extent across the summit.

sea smoke—Fog formed when water vapor is added to air that is much colder than the vapor's source; most commonly, when very cold air drifts across relatively warm water.

sea snake—A reptile of the family Hydrophiidae; a group comprising about 50 species of truly marine forms distantly related to the cobras and possessing similar venom. All are inhabitants of warm coastal waters of the Indian Ocean and western Pacific with one exception, the yellowbellied sea snake, which is oceanic.

sea state—The numerical or written description of ocean surface roughness. For more precise usage, sea state may be defined as the average height of the highest one-third of the waves observed in a wave train, referred to a numerical code.

sea turtle—Any of various large marine turtles belonging to the reptilian order Testudinata and having the feet modified into paddle-like appendages; they are widely distributed in warm seas.

sea urchin—One of a class (Echinoidea) of echinoderms in which the body is covered by a hard shell (or test) composed of fitted immovable plates; spines articulated at their bases and of various sizes, often large and sharp, are present in the test. Many species of urchins have venomous spines. Sand dollars are also members of the class.

seawater—The water of the seas, distinguished from fresh water by its appreciable salinity. Commonly, seawater is used as the antithesis of specific types of fresh water, as river water, lake water, or rain water, whereas salt water is merely the antithesis of fresh water in general.

seaweed—Any macroscopic marine alga or seagrass.

sediment—Particles of organic and inorganic matter that accumulate in a loose unconsolidated form.

sedimentary rocks—Rocks formed by the accumulation of sediment in water or from the air. The sediment may consist of rock fragments, the remains or products of animals or plants, the products of chemical action or evaporation, or mixtures of these materials.

**seiche*—A standing wave oscillation of an enclosed or semienclosed water body that continues, pendulum fashion, after the cessation of the originating force, which may have been either seismic, atmospheric, or wave induced.

seismic profile—The data from a single series of observations made at one geographic location with a linear arrangement of seismometers.

semidiurnal tide—The type of tide having two high waters and two low waters each tidal day, with small inequalities between successive high and successive low water heights and durations.

serpulid worm—A polychaete that

builds a calcareous or leathery tube on a submerged surface.

**sessile*—Permanently attached; not free to move about.

shallow water—Commonly, water of such a depth that surface waves are noticeably affected by bottom topography. It is customary to consider water of depths less than half the surface wavelength as shallow water.

**shallow water wave*—A progressive gravity wave that is in water less than ½ the wavelength in depth.

shark—Any of approximately 250 species of cartilege skeletoned fish-like vertebrates (order Selachii) and including the large plankton-feeding basking sharks and predacious sharks.

shoal—A submerged ridge, bank, or bar consisting of or covered by unconsolidated sediments (mud, sand, gravel) which is near enough to the water surface to be a hazard to navigation. If composed of coral or rock, it is called a reef.

shoreface—The narrow zone seaward from the low tide shoreline permanently covered by water, over which beach sands and gravels actively oscillate with changing wave conditions.

shoreline—The boundary line between a body of water and the land at high tide (usually mean high water). *See* coast.

sial—*See* granite.

sill—The lower part of the ridge or rise separating ocean basins from one another or from the adjacent sea floor.

**silt*—An unconsolidated sediment whose particles are between clay and sand sizes.

sima—*See* basalt.

sinking—(or downwelling). A downward movement of surface water generally caused by converging currents or when a water mass becomes more dense than the surrounding water. *See* upwelling.

sofar—An acronym derived from the expression "*so*und *f*ixing *a*nd *r*anging." It is applied to a system in which position is determined by measuring, at shore listening stations, the difference in the time of reception of sound signals produced in a sound channel in the sea.

**solution*—The state in which a substance, or solute, is homogeneously mixed with a liquid called the solvent. Thus, pure water is a solvent and seawater is a solution of many substances.

sonar—An acronym derived from the expression "*so*und *n*avigation *a*nd *r*anging." The method or the equipment for determining by underwater sound techniques the presence, location, or nature of objects in the sea.

**sound*—The periodic variation in pressure, particle displacement, or particle velocity in an elastic medium. *See* sound velocity.

sounding—Measurement of the depth of water beneath a ship.

**sound velocity*—The rate at which sound energy moves through a medium, usually expressed in feet per second. The velocity of sound in seawater is a function of temperature, salinity, and the changes in pressure associated with changes in depth. Increasing any of these factors tends to increase the velocity. Sound is propagated at a speed of 4,742 ft/sec at 32°F, one atmospheric pressure, and a salinity of 35 ‰.

South Equatorial Current—Any of several ocean currents driven by the southeast trade winds flowing over the tropical oceans of the southern hemisphere.

**specific gravity*—The ratio of the density of a given substance to that of distilled water usually at 4°C and at a pressure of one atmosphere.

**specific heat*—The heat capacity of a system per unit mass, that is, the ratio of the heat absorbed (or released) by unit mass of the system to the corresponding temperature rise (or fall). The amount of heat required to raise the temperature of 1 g of water by 1°C. The specific heat of water, which for pure water at 17.5°C (63.5°F) is 1 cal/g, decreases with increasing temperature and salinity.

spicule—A minute needle-like calcareous or siliceous body in sponges, radiolarians, primitive chitons, and echinoderms.

spit—A small point of land or narrow shoal projecting into a body of water from the shore.

sponge—One of a phylum (Porifera) of solitary or colonial, sessile animals of simple construction. Sponges are of many sizes and forms and varied in color.

**spring tide*—Tide of increased range, which occurs about every two weeks when the moon is new or full.

squid—One of an order (Decapoda) of cephalopods in which the body is cigar shaped or globose and bears ten arms, eight of which are of equal length with suckers along the entire length and two of which are longer with suckers only on a broad terminal portion; shell, in most, is embedded in the body or absent. The giant squid is the largest known invertebrate and a food of the sperm whale.

standard depth—A depth below the sea surface at which water properties should be measured and reported. The internationally accepted depths (in meters) are: 0, 10, 20, 30, 50, 75, 100, 150, 200, 250, 300, 400, 500, 600, 800, 1000, 1200, 1500, 2000, 2500, 3000, 4000, 5000, 6000, 7000, 8000, 9000, 10000, to which National Oceanographic Data Center had added 125, 700, 900, 1100, 1300, 1400, and 1750.

**standing wave*—A type of wave in which the surface of the water oscillates vertically between fixed points, called nodes, without progression. It may result from two equal progressive wave trains traveling through each other in opposite directions.

starfish—One of a class (Asteroidea) of echinoderms having a flat, usually five-armed body.

**storm surge*—A rise above normal water level on the open coast due only to the action of wind stress on the water surface. A storm surge resulting from a hurricane or other intense storm also includes the rise in level due to atmospheric pressure reduction as well as that due to wind stress. A storm surge is more severe when it occurs in conjunction with a high tide.

strait—A narrow sea channel which separates two landmasses.

subbottom reflection—The return of sound energy from a discontinuity in material below the sea bottom surface.

sublittoral—The benthic region extending from mean low water to a depth of about 200 m, or the edge of a continental shelf.

submarine canyon—A relatively narrow deep undersea depression with steep slopes, the bottom of which grades continuously downward.

substrate—The base on which an organism lives.

subsurface current—A current usually flowing below the thermocline, generally at slower speeds and frequently in a different direction from the currents near the surface.

Subtropical Convergence—The zone of converging currents generally located in midlatitudes. It is fairly well defined in the southern hemisphere where it appears as an earth-girding region within which the surface temperature increases equatorward.

supralittoral—The shore zone immediately above high tide level, commonly the zone kept more or less moist by waves and spray.

**surf*—A collective term for breakers, or the wave activity between the shoreline and the outermost limit of breakers.

**surface current*—The part of a directly observed water movement that, in nearshore areas, does not extend more than 1 to 3 m below the surface; in deep or open ocean areas, surface currents generally are considered to extend from the surface to depths of about 10 m. When surface currents are computed by theoretical methods, the volume of water in the mixed layer (above the thermocline) from the surface to depths of about 50 to 150 m, generally is referred to as surface current.

surface duct—A zone immediately below the sea surface where sound rays are refracted toward the surface and then reflected. They are refracted because the sound velocity at some depth near the surface is greater than at the surface. The rays alternately are refracted and reflected along the duct to considerable distances from the sound source.

surface tension—A phenomenon peculiar to the surface of liquids, caused by a strong attraction toward the interior of the liquid acting on the liquid molecules in or near the surface in such a way to reduce the surface area. An actual tension results and is usually expressed in dynes per centimeter or ergs per square centimeter.

**surface wave*—A progressive gravity wave in which the disturbance (that is, the particle movement in the fluid mass as well as the surface movement) is confined to the upper limits of a body of water. Strictly speaking this term applies to progressive gravity waves whose speed depends only on wavelength.

sverdrup—A unit of volume transport equal to 1 million m^3/sec.

**swell*—Ocean waves that have traveled out of their generating area. Swell characteristically exhibits a more regular and longer period and has flatter crests than waves within their fetch.

symbiosis—A relationship between two species in which one or both species are benefited and neither is harmed. *See* commensalism, mutualism, protocooperation.

T

tabular iceberg—A flat-topped iceberg, usually calved from an iceshelf formation.

tectonics—The study of origin and development of the broad structural features of the earth.

**temperature-salinity (T-S) diagram*—The plot of temperature versus salinity data of a water column.

tenting—The vertical displacement upward of ice under pressure to form a flatsided arch with a cavity beneath. *See* rafting.

terrace—A bench-like structure bordering an undersea feature.

test—The shell or supporting structure of many invertebrates.

thermistor chain—An instrument-carrying chain (up to 400 m long) generally towed astern to get continuous temperature recordings from upper water layers at sea.

**thermocline*—A vertical negative temperature gradient in some layer of a body of water, which is appreciably greater than the gradients above and below it; also a layer in which such a gradient occurs. The principal thermoclines in the ocean are either seasonal, due to heating of the surface water in summer, or permanent.

thermohaline—Pertaining to both temperature and salinity acting together; for example, thermohaline circulation.

**thermohaline circulation*—Vertical circulation induced by surface cooling, which causes convective overturning and consequent mixing.

**thermohaline convection*—Vertical movement of water observed when seawater, because of its decreasing temperature or increasing salinity, becomes heavier than the water underneath it and a disturbed vertical equilibrium results.

tidal bore—A high breaking wave of water, advancing rapidly up an estuary. Bores can occur at the mouths of shallow rivers if the tide range at the mouth is large. They can also be generated in a river when tsunamis enter shallow coastal water and propogate up the river.

**tidal current*—The alternating horizontal movement of water associated with the rise (flood) and fall (ebb) of the tide caused by the astronomical tide-producing forces.

**tidal prism*—The difference between the mean high water volume and the mean low water volume of an estuary.

tidal wave—In popular usage, any unusually high (and therefore destructive) water level along a shore. It usually refers incorrectly to either a storm surge or a tsunami.

**tide*—The periodic rising and falling of the earth's oceans and atmosphere. It results from the tide-producing forces of the moon and sun acting on the rotating earth. Sometimes, the periodic horizontal movements of the water along coast lines is also called "tide" but is more correctly called tidal current.

tide range—The difference in height between consecutive high and low waters. Where the tide is diurnal the mean range is the same as the diurnal range.

tintinnoid ciliate—Any of a suborder (Tintinnoinea) of microscopic planktonic Protozoa which possess a tubular or vase-shaped outer shell. Several species are luminescent.

**titration*—A chemical method for determining the concentration of a substance in solution. This concentration is established in terms of the smallest amount of the substance required to bring about a given effect in reaction with another known solution or substance. The most common titration is that for chlorinity.

Tongue of the Ocean—A steep-sided, deepwater embayment approximately 100 nautical miles long, 20 nautical miles wide, and one nautical mile deep, connected to the Atlantic Ocean by Northeast Providence Channel and Northwest Providence Channel and trends southeast into the Great Bahama Bank, terminating in a circular cul-de-sac.

toothed whale—A member of the cetacean suborder Odontoceti, which comprises the dolphins, porpoises, killer whales, beaked whales, and sperm whales.

**topography*—The configuration of a surface, including its relief. In oceanology the term is applied to a surface such as the sea bottom or a surface of given characteristics within the water mass.

tracer—A foreign substance mixed with (or attached to) a given substance used to determine distribution or location.

trade winds—The wind system, occupying most of the tropics, which blows from the subtropical highs toward the equatorial trough; a major component of the general circulation of the atmosphere. The winds are northeasterly in the northern hemisphere and southeasterly in the southern hemisphere.

trawl—A bag- or funnel-shaped net to catch bottom fish by dragging along the bottom. Also applied to research nets designed on the same principles.

**trench*—A long, narrow, and deep depression of the sea floor, with relatively steep sides.

**trough*—1. A long depression of the sea floor normally wider and shallower than a trench. 2. *See* wave trough.

**tsunami*—(or tidal wave, seismic sea wave). A long-period sea wave produced by a submarine earthquake or volcanic eruption. It may travel unnoticed across the ocean for thousands of miles from its point of origin and builds up to great heights over shoal water.

tubeworm—*See* serpulid worm.

tunicate—One of a subphylum (Tunicata) of globular or cylindrical, often sac-like animals, many of which are covered by a tough flexible tunic. Many are sessile (such as sea squirts); others are pelagic (such as salps).

**turbidity current*—A highly turbid, relatively dense current carrying large quantities of clay, silt, and sand in suspension, which flows down a submarine slope through less dense seawater.

typhoon—(also called typhon). A severe tropical cyclone in the western Pacific.

U

ultrasonics—The technology of sound at frequencies above the audio range (that is, above 20,000 cycles per second).

ultraviolet radiation—Electromagnetic radiation of shorter wavelength than visible radiation but longer than x-rays; roughly, radiation in the wavelength interval from 1 to 400 millimicrons.

**upwelling*—The process by which water rises from a lower to a higher depth, usually as a result of divergence and offshore currents. *See* sinking. It influences climate by bringing colder water to the surface. The upwelled water, besides being cooler, is richer in plant nutrients, so that regions of upwelling are generally also regions of rich fisheries.

V

vertebrate—A member of the subphylum Vertebrata (phylum Chordata). It contains the fish, amphibians, reptiles, birds, and mammals. Vertebrates have a skull which surrounds a well-developed brain and a skeleton of cartilege or bone.

viscosity—The molecular property of a fluid that enables it to support tangential stresses for a finite time and thus to resist deformation.

W

wake—The region of turbulence immediately behind a boat or other solid object in motion on the water.

**water mass*—A body of water usually identified by its T-S curve or chemical content, and normally consisting of a mixture of two or more water types. (*See* temperature-salinity diagram.)

waterspout—Usually, a tornado occurring over water; rarely, a lesser whirlwind over water, comparable in intensity to a dust devil over land.

**water type*—Seawater of a specific temperature and salinity, and hence defined by a single point on a temperature-salinity diagram.

**wave*—1. A disturbance that moves through or over the surface of the medium (here, the ocean), with speed dependent on the properties of the medium. 2. A ridge, deformation, or undulation of the surface of a liquid.

wave crest—The highest part of a wave.

**wave height*—The vertical distance between a wave crest and the preceding wave trough.

**wavelength*—The distance between corresponding points of two successive periodic waves in the direction of propagation, for which the oscillation has the same phase. Unit of measurement is meters.

wave train—A series of waves moving in the same direction.

wave trough—The lowest part of a wave form between successive wave crests. Also that part of a wave below still water level.

West Greenland Current—The current flowing northward along the west coast of Greenland into Davis Strait.

West Wind Drift—(sometimes called Antarctic Circumpolar Current). The ocean current with the largest volume transport (approximately 110×10^6 cm^3/sec); it flows from west to east around Antarctica and is formed partly by the strong westerly wind in this region and partly by density differences.

whirlpool—Water moving rapidly in a circular path; an eddy or vortex of water.

windward—The direction from which the wind is blowing.

Z

**zooplankton*—The animal forms of plankton. They include various crustaceans, such as copepods and euphausiids, jellyfishes, certain protozoans, worms, mollusks, and the eggs and larvae of benthic and nektonic animals.

Index

0°
15°
30°
45°
60°
75°
90°
105°
120°
135°
150°
165°
180°
165°
150°
75°
70°
60°
45°
30°
15°
0°
15°
30°
45°
60°
70°
GREENLAND SEA
BARENTS SEA
Novaya Zemlya
ARCTIC OCEAN
CHUKCHI SEA
NORWEGIAN SEA
Arctic Circle
BERING SEA
NORTH SEA
BALTIC SEA
Aleutian Basin
SEA OF OKHOTSK
ASIA
EUROPE
Aleutian Trench
Aleutian Abyssal Plain
Kuril Trench
Emperor Seamounts
Chinook Fracture Zone
BLACK SEA
Caspian Sea
SEA OF JAPAN
Japan Trench
YELLOW SEA
MEDITERRANEAN SEA
Bonin Trench
NORTH PACIFIC OCEAN
EAST CHINA SEA
PHILIPPINE SEA
Hawaiian Deep
RED SEA
ARABIAN SEA
Mid-Pacific Mountains
Mariana Trench
CHINA SEA
Philippine Basin
Philippine Trench
Hawaiian Is.
BAY OF BENGAL
AFRICA
Arabian Basin
GULF OF SIAM
Mariana Basin
Marshall Is.
Carlsberg Ridge
Yap Trench
Challenger Deep
Ceylon
Maldive Is.
Somali Basin
Borneo
Solomon Rise
Sumatra
JAVA SEA
New Guinea
Seychelles
Ninety Degree Ridge
Vitiaz Trench
Manihiki Plateau
Angola Abyssal Plain
INDIAN
ARAFURA SEA
Java Trench
Coral Sea Basin
North Fiji Plateau
Samoa
Malagasy
MID INDIAN RIDGE
Great Barrier Reef
CORAL SEA
Fiji I.
Tonga Trench
OCEAN
Mascarene Basin
AUSTRALIA
Cape of Good Hope
Norfolk Ridge
South Fiji Basin
Mozambique Rise
Tasman Basin
Lord Howe Rise
Kermadec Trench
Amsterdam–Naturliste Ridge
Cape Abyssal Plain
Crozet Basin
TASMAN SEA
SOUTH
South Australia Basin
Agulhas Basin
Cape Rise
Tasmania Rise
New Zealand
Chatham Rise
Southwest Pacific Basin
Indian–Antarctic Ridge
Campbell Plateau
Atlantic–Indian Ridge
Kerguelen–Gaussberg Ridge
South Indian Basin
Atlantic–Indian Basin
PACIFIC–ANTARCTIC RIDGE
DAVIS SEA
Balleny Basin
Antarctic Circle
ANTARCTICA

150°
135°
120°
105°
90°
75°
60°
45°
30°
15°
0°
15°
30°
45°
60°
75°
70°
60°
45°
30°
15°
0°
15°
30°
45°
60°
70°
GREENLAND SEA
BARENTS SEA
Novaya Zemlya
BAFFIN BAY
GREENLAND
Norwegian Basin
NORWEGIAN SEA
Iceland
Surtsey I.
Faeroe Is.
HUDSON BAY
LABRADOR SEA
Reykjanes Ridge
Rockall Bank
NORTH SEA
BALTIC SEA
Labrador Basin
Ireland
England
NORTH
NORTH ATLANTIC OCEAN
EUROPE
AMERICA
Hudson Canyon
Grand Banks
Flemish Cap
Biscay Abyssal Plain
BLACK SEA
Caspian Sea
Mendocino Fracture Zone
Monterey Canyon
Blake Plateau
Corner Seamounts
Azores
Azores Rise
Murray Fracture Zone
Baja California Seamount Province
Tongue of the Ocean
Hatteras Abyssal Plain
Bermuda Rise
Atlantis F. Z.
MID-ATLANTIC RIDGE
Madeira Abyssal Plain
MEDITERRANEAN SEA
Molokai Fracture Zone
Scripps-LaJolla canyons
GULF OF MEXICO
Puerto Rico Trench
Cape Verde Abyssal Plain
RED SEA
Clarion Fracture Zone
Mid America Trench
CARIBBEAN SEA
Demerara
Ceara Abyssal Plains
AFRICA
Carlsberg Ridge
Clipperton Fracture Zone
Cocos Ridge
Lesser Antilles
Sierra Leone Rise
Guinea Abyssal Plain
Galapagos Fracture Zone
Galapagos Is.
Romanche F. Z.
Chain Fracture Zone
Congo Canyon
Seychelles
Marquesas Fracture Zone
Peru Trench
Saint Paul's Rocks
SOUTH
Pernambuco Abyssal Plain
Angola Abyssal Plain
INDIAN
Tuamoto Archipelago
Galapagos Rise
Nasca Ridge
Malagasy
EAST PACIFIC RISE
AMERICA
Trindade Seamounts
OCEAN
Chile Trench
Walvis Ridge
Cape of Good Hope
Easter Fracture Zone
Rio Grande Rise
Mozambique Rise
PACIFIC
SOUTH
Tristan de Cunha
Cape Abyssal Plain
Crozet Basin
Chile Rise
Argentine Basin
ATLANTIC
Cape Rise
Agulhas Basin
OCEAN
South Georgia Rise
Falkland Rise
OCEAN
Atlantic-Indian Ridge
Eltanin Fracture Zone
Cape Horn
West Scotia Basin
South Sandwich Trench
East Scotia Basin
SCOTIA SEA
Atlantic-Indian Basin
Southeast Pacific Basin
Antarctic Circle
DNH 73
ANTARCTICA

The sea is calm to-night.
The tide is full, the moon lies fair
Upon the straits;—on the French coast the light
Gleams and is gone; the cliffs of England stand,
Glimmering and vast, out in the tranquil bay.
Come to the window, sweet is the night air!
Only, from the long line of spray
Where the sea meets the moon-blanched land,
Listen! you hear the grating roar
Of pebbles which the waves draw back, and fling,
At their return, up the high strand,
Begin, and cease, and then again begin,
With tremulous cadence slow, and bring
The eternal note of sadness in.

Matthew Arnold